In Focus – Special Book Series

Innovations, Advances and Future Challenges for the Hydrological Community

Editors

Elena Volpi, Elena Cristiano, Marco Peli, Martina Siena and Giulia Zuecco

IWA PUBLISHING

Published by **IWA Publishing**
Unit 104-105, Export Building
1 Clove Crescent
London E14 2BA, UK
Telephone: +44 (0)20 7654 5500
Fax: +44 (0)20 7654 5555
Email: publications@iwap.co.uk
Web: www.www.iwaponline.com

First published 2024

British Library Cataloguing in Publication Data
A CIP catalogue record for this book is available from the British Library

ISBN: 9781789064896

Innovations, Advances and Future Challenges for the Hydrological Community

Contents

© 2023 The Authors
doi: 10.2166/nh.2023.003

Editorial: Innovations, advances, and future challenges for the hydrological community

The 10th edition of the 'Hydrology Days' of the Italian Hydrological Society (SII-IHS) was held in Bologna (Italy) from 16 to 19 September 2019. During this event, hydrologists and stakeholders in the field of water resources management shared the results of their research and discussed the most relevant challenges that the hydrological community must face nowadays. With the celebration of the 10th anniversary, the SII-IHS promoted a Special Issue of *Hydrology Research* open to the participants of the 'Hydrology Days' as well as to other contributions related to the role of the hydrologist in a changing environment and the consequent challenges posed to the hydrological community.

This book includes 10 peer-reviewed papers reflecting inter- and multidisciplinary aims and topics discussed during the 'Hydrology Days' of the SII-IHS. The contributions are organized into the following three main topics: (1) *Climate change and geo-hydrological risk*, (2) *Tools and techniques for improving the comprehension of catchment hydrology and river hydraulics*, and (3) *Sustainable use of water resources*.

The first group of papers focuses on climate change and its implications on hydrological regimes, water availability, and on geo-hydrological risk. Biondi *et al.* (2023) presented an operational tool for the assessment of geo-hydrological risk associated with the combined effects of multiple forcing elements. The analysis of a wide database of damaging events that occurred in Italy within the time frame 2004–2011 allowed them to classify the most recurring triggering factors, the related representative phenomena, and the consequently damaged assets and causes of fatalities. The proposed methodology can be applied for a detailed characterization of past events, as well as for the prediction of potentially damaging effects under prescribed scenarios. Boyko *et al.* (2022) analyzed climatic projections of precipitation from 19 models of the Climate Model Intercomparison Project 5, predicting densities of precipitation for selected prognostic time windows for the Po River valley, in Italy. Their results showed that the annual precipitation volume over the study area will not change notably in the future, while its temporal distribution is predicted to be affected. These potential changes are expected to generate a considerable strain on the area's natural storage capacity and on the water flows that are needed to satisfy irrigation demand as well as hydro-electric and thermal energy production.

The second group of papers investigates different tools and techniques aimed at improving the comprehension of both catchment hydrology and river hydraulics. Bahmanpouri *et al.* (2023) analyzed two hydraulic complexity metrics, M1 and M2, using laboratory data and field data collected along the Tiber River (Italy). The authors showed that the investigation of the hydraulic complexity parameters could be useful to explain the heterogeneity of the habitats, as well as to evaluate ecological and biological patterns in rivers with noteworthy flood rates, such as the Tiber River. Datta *et al.* (2022) investigated how Digital Elevation Models (DEMs), DEM resolutions, and area threshold values influence the delineation of three watersheds in Bangladesh (Halda, Sangu, and Chengi). Results showed that DEM source influences the slope, and DEM resolution affects the basin perimeter of a watershed, while size and shape are independent of the area threshold values. Longobardi & Villani (2023) investigated the Base Flow Index (BFI), one of the most common indices of low flows, and the climatic and catchment properties influencing this metric. The authors observed a prominent role of geological catchment characteristics on the BFI, with poorly and well-drained catchments behaving differently in response to similar climate-forcing variability. Furthermore, the authors showed that the correlation between the BFI and precipitation, at the annual scale, is affected by both climate variability and catchment properties, and that the BFI variability is strongly conditioned by climate intra- and inter-annual variability. Petroselli *et al.* (2022) investigated the performance of STORAGE (STOchastic RAinfall GEnerator), a recently developed version of the Neyman–Scott Rectangular Pulse model for rainfall continuous-time simulation at the sub-daily scale. Making use of the Vistula catchment (Poland) case study, the authors showed how the model correctly reproduced the statistical behavior of rainfall extremes as a function of duration. STORAGE seems to be a robust synthetic rainfall generator for design purposes, able to produce continuous rainfall data that can be used as input for a continuous rainfall–runoff approach.

The last group of papers investigates different strategies to ensure sustainable management of water resources, both in urbanized and rural environments. Concerning urban areas, Cristiano *et al.* (2022) presented an innovative nature-based solution, which combines the retention capacity of traditional green roofs with the retention capacity of a rainwater harvesting system, and investigates the potential benefit of this tool in an urban environment, in terms of water management and thermal properties. The potential of these nature-based solutions is also explored by Mobilia *et al.* (2023), who investigated the effectiveness of green roof retrofitting as a measure to mitigate the effects of damaging hydrological events. The authors took advantage of remote sensing (SAR) images over the Sarno River basin, an area characterized by high population density and high exposure to floods. This contribution identified a clear correlation between the temporal evolution of urbanization in the region (which resulted in a severe increase in the extent of the impervious surface) and the frequency of harmful hydrological events over the last two decades. Different scenarios of green roof retrofitting proved to be effective in reducing stormwater production. As far as rural areas are concerned, Forzini *et al.* (2022) proposed a methodology to select suitable locations for *pokharis*, traditional Nepalese water ponds, based on a geographic information system (GIS), multi-criteria decision-making, and participatory research. In the context of climate changes, the restoration of existing *pokharis* or the creation of new ones could improve the population's economic and living conditions, increasing water availability. Finally, Peli *et al.* (2023) proposed a novel, GIS-based application of Benfratello's method to estimate the climatic water deficit and the irrigation deficit at the field and catchment scale. By using a limited number of parameters, the authors estimated the irrigation demand in the Bonifica della Capitanata consortium (Italy), proving the effectiveness of the tool for the management of water resources in semiarid agricultural areas, in the context of climatic and land use changes.

The Editors would like to acknowledge all reviewers for their valuable contributions in the peer-review process and Emma Gulseven, Emma Buckingham at IWA Publishing, and the Subject Editor, Professor Elena Volpi, for their helpful guidance and support.

Guest Editors
Elena Cristiano
University of Cagliari, Cagliari, Italy

Marco Peli
University of Brescia, Brescia, Italy

Martina Siena
Politecnico di Milano, Milan, Italy

Giulia Zuecco
University of Padova, Legnaro, Italy

REFERENCES

Bahmanpouri, F., Termini, D., Barbetta, S., Dionigi, M. & Moramarco, T. 2023 Significance of hydraulic complexity parameters M_1 and M_2 based on the laboratory and field data. *Hydrology Research* **54** (3), 303–312. https://doi.org/10.2166/nh.2023.089.

Biondi, D. & Scarcella, G. E. 2023 Pasquale Versace; CERCA (Cascading effects in risk consequences assessment): an operational tool for geo-hydrological scenario risk assessment and cascading effects evaluation. *Hydrology Research* **54** (2), 189–207. https://doi.org/10.2166/nh.2023.088.

Boyko, O., Reggiani, P. & Todini, E. 2022 Post-processing climate projections of precipitation for the Po river basin: will Italy's North become water-constrained? *Hydrology Research* **53** (11), 1414–1427. https://doi.org/10.2166/nh.2022.063.

Cristiano, E., Annis, A., Apollonio, C., Pumo, D., Urru, S., Viola, F., Deidda, R., Pelorosso, R., Petroselli, A., Tauro, F., Grimaldi, S., Francipane, A., Alongi, F., Noto, L. V., Hoes, O., Klapwijk, F., Schmitt, B. & Nardi, F. 2022 Multilayer blue-green roofs as nature-based solutions for water and thermal insulation management. *Hydrology Research* **53** (9), 1129–1149. https://doi.org/10.2166/nh.2022.201.

Datta, S., Karmakar, S., Mezbahuddin, S., Hossain, M. M., Chaudhary, B. S., Hoque, E., Abdullah Al Mamun, M. M. & Baul, T. K. 2022 The limits of watershed delineation: implications of different DEMs, DEM resolutions, and area threshold values. *Hydrology Research* **53** (8), 1047–1062. https://doi.org/10.2166/nh.2022.126.

Forzini, E., Piemontese, L., Bresci, E., Barthod, B., Bielser, F., Sylvestre, M., Adhikari, N., Pun, S. & Castelli, G. 2022 Identification of suitable sites for traditional *pokhari* water harvesting in mountain rural communities of the Himalaya. *Hydrology Research* **53** (11), 1340–1356. https://doi.org/10.2166/nh.2022.027.

Longobardi, A. & Villani, P. 2023 Baseflow index characterization in typical temperate to dry climates: conceptual analysis and simulation experiment to assess the relative role of climate forcing features and catchment geological settings. *Hydrology Research* **54** (2), 136–148. https://doi.org/10.2166/nh.2023.026.

Mobilia, M., Longobardi, A., Amitrano, D. & Ruello, G. 2023 Land use and damaging hydrological events temporal changes in the Sarno River basin: potential for green technologies mitigation by remote sensing analysis. *Hydrology Research* **54** (3), 277–302. https://doi.org/10.2166/nh.2023.036.

Peli, M., Rapuzzi, C., Barontini, S. & Ranzi, R. 2023 Application of Benfratello's method to estimate the spatio-temporal variability of the irrigation deficit in a Mediterranean semiarid climate. *Hydrology Research* **54** (4), 451–474. https://doi.org/10.2166/nh.2023.081.

Petroselli, A., De Luca, D. L., Młyński, D. & Wałęga, A. 2022 Modelling annual maximum daily rainfall with the STORAGE (STOchastic RAinfall GEnerator) model. *Hydrology Research* **53** (4), 547–561. https://doi.org/10.2166/nh.2022.100.

doi: 10.2166/nh.2023.089

Significance of hydraulic complexity parameters M_1 and M_2 based on the laboratory and field data

Farhad Bahmanpouri[a,*], Donatella Termini[b], Silvia Barbetta[a], Marco Dionigi[a] and Tommaso Moramarco[a]
[a] Research Institute for Geo-Hydrological Protection, National Research Council (CNR), Perugia 06128, Italy
[b] Dipartimento di Ingegneria, University of Palermo, Viale delle Scienze, Palermo 90128, Italy
*Corresponding author. E-mail: farhad.bahmanpouri@irpi.cnr.it

FB, 0000-0001-7037-3548

ABSTRACT

Hydraulic complexity metrics referred to as M_1 and M_2 play an important role when it comes to the analysis of habitat metrics. In the present paper, the significance of these parameters is analysed by using laboratory data as well as field observations along the Tiber River in Italy. Based on the laboratory data, the estimated parameters allow us to characterise the high/low-velocity areas. Based on field observations, larger magnitudes of M_1 are linked to the zones with large changes in cross-sectional flow velocity. Larger magnitudes of M_2 are observed at the left bank of the channel for all flow conditions, suggesting locations with larger kinetic energy consumption for aquatic organisms. Overall, the findings of the present research would be of particular interest in quantifying biologically important flow patterns occurring at different spatial scales within different streams and flow conditions.

Key words: ecohydraulics, hydraulic complexity, river hydrodynamics, Tiber River

HIGHLIGHTS

- Investigating the hydraulic complexity parameters M_1 and M_2.
- Evaluating the hydraulic complexity parameters based on the laboratory data.
- Evaluating the hydraulic complexity parameters based on the field data of the Tiber River in Italy.

1. INTRODUCTION

In-stream habitat is defined as the physical habitat or 'living' space of in-stream biota that includes the channel's physical structure and the spatial and temporal dynamics of the flow regime (Maddock 1999). Local variables associated with topography and in-stream environments, such as bed slope, bed deformation, and near-bed velocity, affect habitat structure. The similarity among streams within river systems reflects the variations in hydrologic regimes at the regional level (Pedersen *et al.* 2004). In-stream habitat heterogeneity may be linked to both substrate and flow characteristics such as water depth or flow velocity. The spatial heterogeneity or variability of flow, associated with the hydraulic complexity within a stream, is extremely important for the aquatic habitat (Clark *et al.* 2008). Blettler *et al.* (2016) addressed the benthic macroinvertebrate distributions of the Bermejo River confluence in Argentina and the Paraguay River and showed that the macroinvertebrate diversity increased slightly downstream of the confluence. Gualtieri *et al.* (2017, 2020) investigated the hydraulic complexity parameters through the confluence of the Negro and Solimões rivers on the Amazon River in Brazil and their results highlighted that the strength of flow rotation was related to the presence of helical cells forming at the confluence itself.

The present work aims to uncover the significance of metrics M_1 and M_2 introduced by Crowder & Diplas (2002) for quantifying in-stream hydraulic complexity and describing the relationship between hydraulic complexity and fish habitat preferences. While the first parameter, M_1, represents the spatial change in a flow kinetic energy and is a measure of the amount of power expended in moving from one location to another the second parameter, M_2, is a measure of how much more energy an organism must expend if it moves from the lower velocity to the higher ones. Thus, the estimation of metrics M_1 and M_2 would give us the opportunity to evaluate the hydraulic conditions favouring the aquatic habitat without the detailed knowledge of flow turbulence characteristics. We refer the reader to Lacey *et al.* (2012) to find out more about the swimming mechanics and the behaviour of fish in turbulent flows.

To this end, laboratory and field data, collected in the Tiber River (Italy) for different flow conditions, are considered. The paper is organised as follows: first, the laboratory setup and field sites are introduced in Section 2, then the theoretical description of the hydraulic complexity parameters is addressed in Section 3. The results, based both on laboratory observations and field data measures, are presented in Section 4. The paper ends with the conclusion in Section 5.

2. EXPERIMENTAL AND FIELD CASE STUDY AND DATASET

2.1. Laboratory data

The laboratory experiments were carried out in a 7-m long and 0.40-m wide recirculating straight flume, constructed at the hydraulic laboratory of the University of Palermo (Italy). The data were collected during previous research works aiming at analysing the flow velocity field and the evolution of turbulent flow structures along the channel (Termini & Sammartano 2008; Termini 2015). Figure 1 shows the plane view of the laboratory flume.

The channel banks are of Plexiglas strips and the bed is rigid and of quartz sand (D_{50} = median sediment diameter = 0.65 mm). The velocity data were collected during an experimental run (i.e. the Smooth Banks (SB) run described in Termini & Sammartano (2008) and in Termini (2015) with flow discharge of 0.013 m^3/s and water depth $h = 5.15$ cm. During the run, the instantaneous streamwise velocity component was measured by using a 2D side-looking Acoustic Doppler Velocimeter (ADV) by SonTek Inc. For each cross-section, five verticals with a distance of 7 cm were selected, and along each vertical, 14 points with a vertical space of 3 mm were chosen.

2.2. Field data: upper Tiber River (central Italy)

The velocity dataset at Ponte Nuovo gauged site is used for the analysis. Ponte Nuovo is the outlet of the Upper Tiber basin and velocity data consist of velocity points whose number along the sampled verticals is sufficient (more than eight for some verticals) to shape the vertical velocity profile, even in the presence of secondary currents. A current-meter operated by a cableway system was used to record the streamwise flow velocity section (Tarpanelli *et al.* 2013). Specifically, a mechanical current-meter was used and the flow velocity was inferred from the rotation of a bucket-wheel with a diameter equal to 13 cm and a horizontal-axis impeller. The used current-meter was able to measure flow velocity in the range of 0.06–8 m/s (Moramarco *et al.* 2019). Three flow measurements carried out during different hydraulic conditions were considered for the analysis; they refer to low, medium, and high flow conditions and are summarised in Table 1. Each measurement consists of the number of verticals, either 6 or 8, and a number of velocity points, 56–79, which guarantee a good representation of the velocity profile.

3. SPATIAL METRICS FOR RIVER FLOW COMPLEXITY: THEORETICAL BACKGROUND

The velocity gradient is a fundamental component of a spatially varying flow pattern. Velocity gradients are present near banks, boulders, and other obstructions and are potentially useful measures for quantifying and distinguishing between flows having similar depth and velocity values, but surrounded by different spatially varying flow patterns. Two parameters

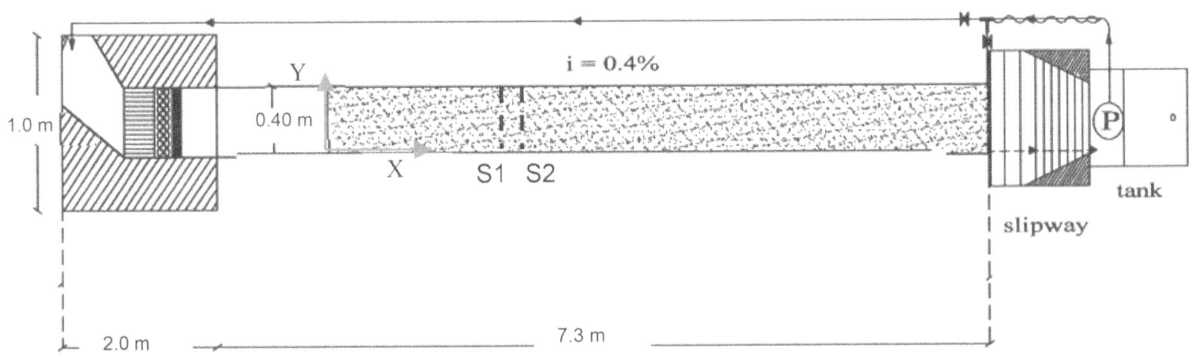

Figure 1 | Plane view of the laboratory channel.

Table 1 | Main characteristics of the flow velocity measurements at the Ponte Nuovo site (Upper Tiber River, central Italy)

Date (flow condition)	Number of verticals	Number of velocity points	Maximum water depth (m)	Discharge (m^3/s)
May 2003 (Low flow)	8	56	1.56	22
March 2000 (Medium flow)	8	79	4.01	274
November 1996 (High flow)	6	68	6.44	463

that incorporate local velocity gradients are evaluated as potential habitat metrics and they are inferred as follows (Crowder & Diplas 2000, 2002; Gualtieri *et al.* 2017):

$$M_1 = \frac{\partial}{\partial y}\frac{V^2}{2} = V\frac{\partial V}{\partial y} \cong V_{ave}\frac{V_2 - V_1}{\Delta_y} \tag{1}$$

$$M_2 = \frac{(\partial/\partial y)(V^2/2)}{(V^2/2)} \cong \frac{\left(\frac{1}{2}\right)(V_2^2 - V_1^2/\Delta_y)}{\left(\frac{V_{min}^2}{2}\right)} = \frac{2V_{ave}((V_2 - V_1)/\Delta_y)}{V_{min}^2} = \frac{2\,M_1}{V_{min}^2} \tag{2}$$

where V is the velocity and V_1 and V_2 are magnitudes of streamwise velocity between the two consecutive velocity points 1 and 2, whose distance can be defined as Δ_y (for the present data $\Delta_y = 0.08$ and 1.0 m for the laboratory and field data, respectively) in the transverse direction; V_{ave} is the average of V_1 and V_2; V_{min} is the minimum value of V_1 and V_2. The first parameter M_1 represents the spatial change in a flow's specific kinetic energy (i.e. per unit mass and unit length (J/kgm)). The second parameter M_2 scales the spatial change in the kinetic energy of the flow at the point having the lower velocity magnitude. Thus, M_2 represents the normalised spatial change in a flow's specific kinetic energy. In other words, according to Crowder & Diplas (2000), the first parameter (M_1) is proportional to the drag force on an organism and it is a measure of the amount of power expended in moving from one location to another. Based on fluid dynamics, drag is the force acting against the relative movement of an object moving relative to a surrounding environment. This functionality is complicated and depends upon the shape of the object, its size, its velocity, and the fluid it is in. The second parameter (M_2) is a measure of how much more energy an organism must expend if it moves from a lower velocity to a higher velocity location. Figure 2 depicts the concept of hydraulic complexity parameters considering the aquatic organisms.

The primary reason for choosing the first parameter is that kinetic energy per unit mass ($V_2/2$) multiplied by a drag coefficient and a frontal area provides the drag force acting on a specific object or organism (Munson *et al.* 1990).

Hayes & Jowett (1994) and Fausch & White (1981) provided data that can be used to estimate the minimum and maximum possible values of $2V_{ave} \times |(V_2 - V_1)/\Delta y|/V_{min}^2$ that existed at the fish locations observed in their studies. Specifically, both studies provided a range of focal point velocity values (V_{min}), 'velocity shears', and 'velocity differences' that existed at fish locations.

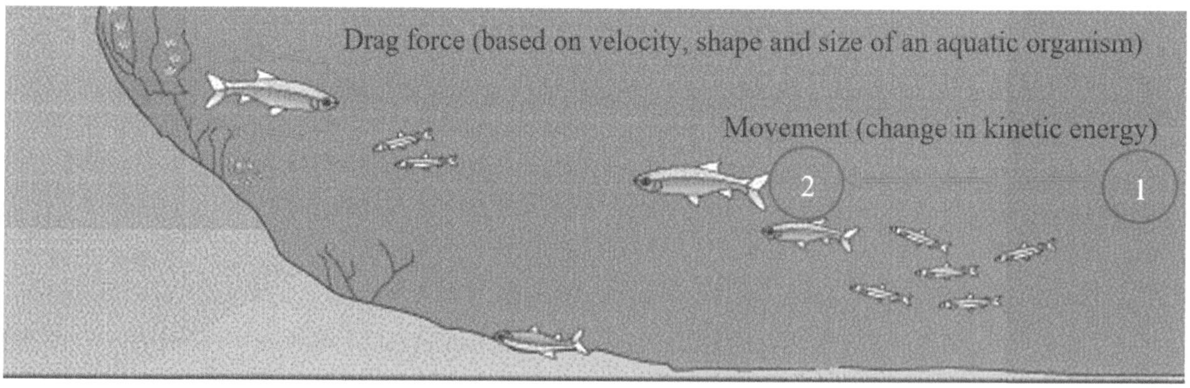

Figure 2 | Description of the hydraulic complexity parameters based on conceptual definitions.

The selected parameters M_1 and M_2 allow the evaluation of the velocity gradient and the variation in a flow's scaled specific kinetic energy, respectively, and this is important for distinguishing between flows characterised by similar values of water depth and velocity but different surrounding velocity gradients. Shields & Rigby (2005) based on field data from Crowder & Diplas (2000, 2002), reported an approximate range of observed magnitudes $0.0 < M_1 < 0.2$ J/kgm and $0.9 < M_2 < 1.3$ 1/m, respectively, for some specific flow conditions.

4. RESULTS AND DISCUSSION

4.1. Results based on the laboratory data

Figure 3 reports the distribution of flow velocity in the considered sections S1 and S2. This figure shows the regions of high/low-flow velocity: in section S1, the highest value of velocity occurs close to the free surface, at a distance $Y = 12.5$ cm from the right bank, and this region of high velocity tends to extend through the flow depth moving towards the bed; a similar behaviour can be observed in section S2 at a distance $Y = 12.5$ cm from each bank although close to the left bank (i.e. at $Y = 27.5$ cm) the high-velocity region extends more towards the bed. This behaviour is consistent with results obtained by Termini & Sammartano (2008) which highlighted the formation of alternating low/high-speed flow regions determining the evolution of alternating turbulent bursting events along the channel.

Figures 4 and 5 report the distributions of the parameters M_1 and M_2 estimated in sections S1 and S2 by using Equations (1) and (2). Figure 4 indicates that in section S1 high spatial velocity gradient, M_1, occurs in the right side of the cross-section and a lower peak value is found close to the free surface at a distance of around 16 cm from the left bank.

In section S2, the peak value of M_1 is found close to the bed on the right side of the cross-section and especially close to the free surface on the left side of the cross-section. In these locations, the organism must spend more energy to move.

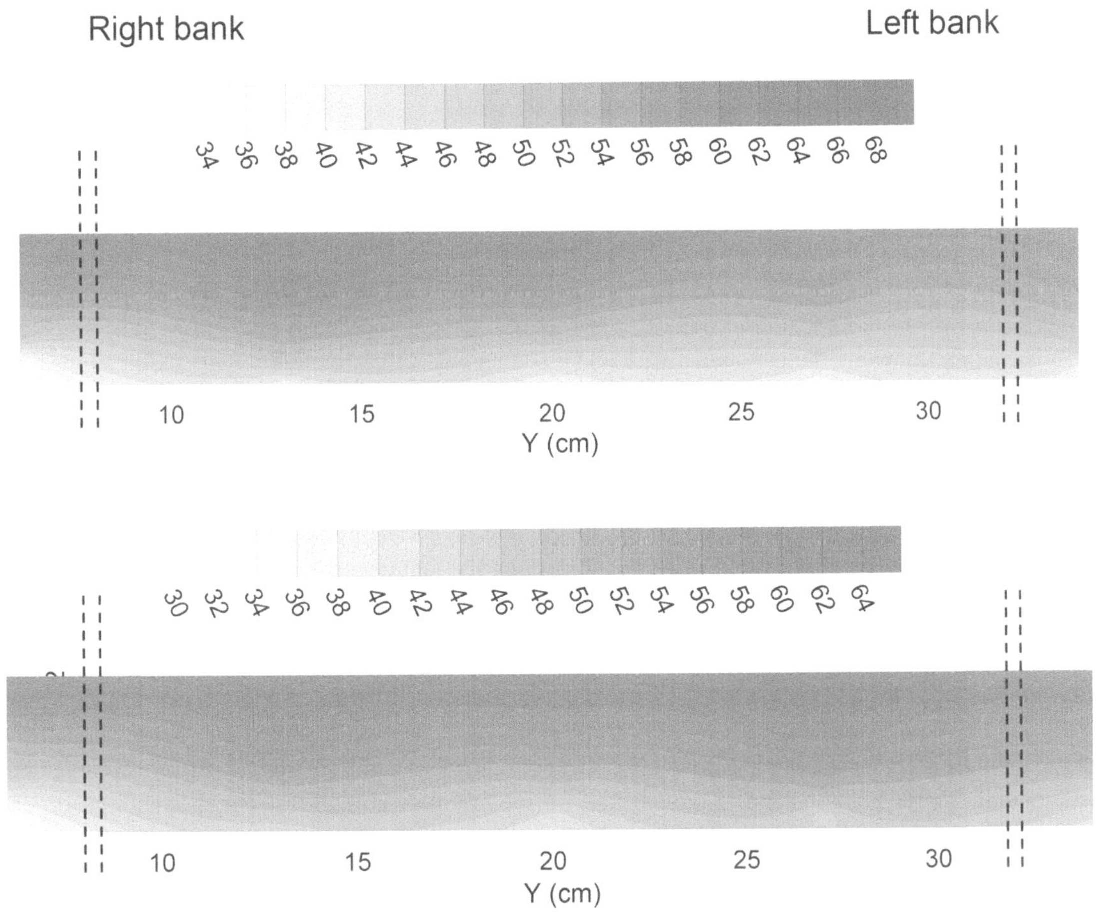

Figure 3 | Distribution of the flow velocity (cm/s) in analysed sections.

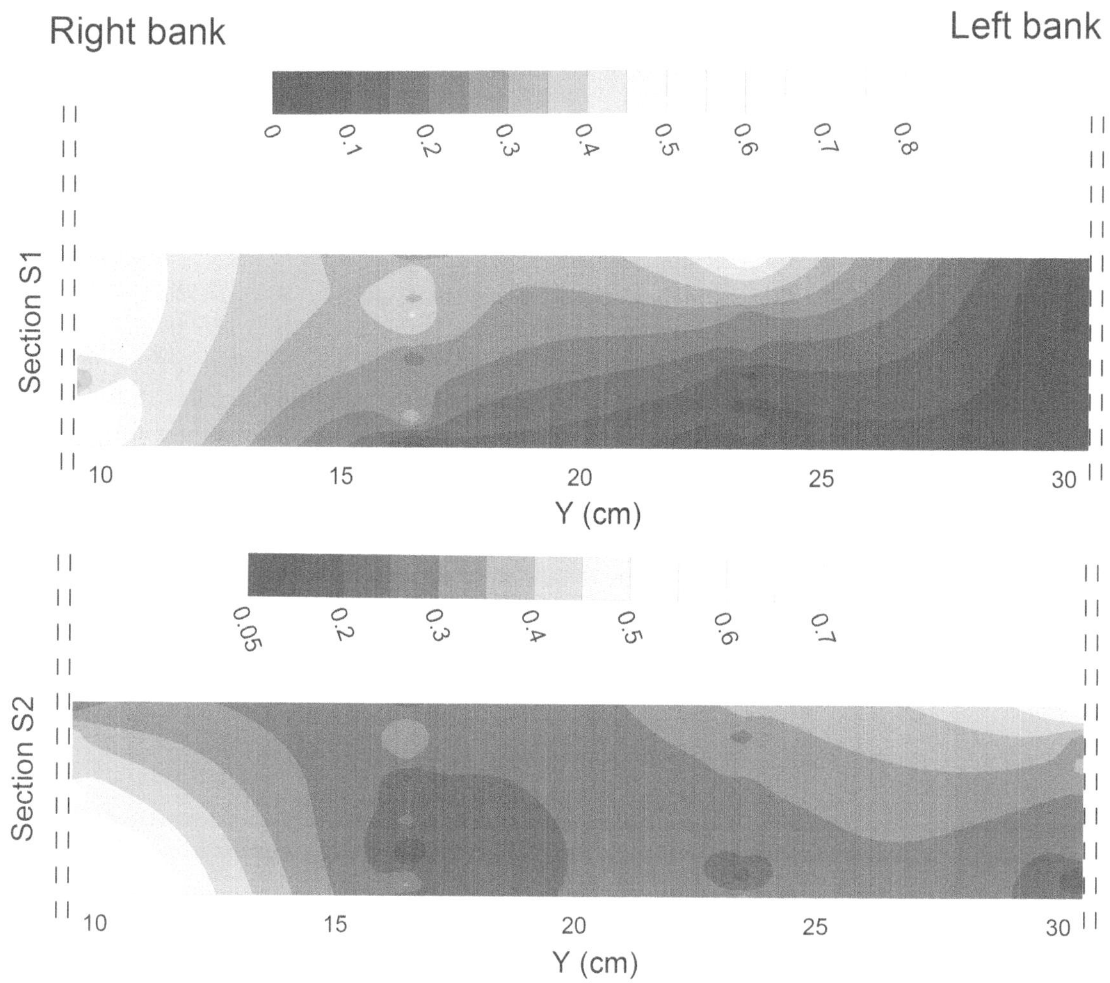

Figure 4 | Distribution of the parameter M_1 (J/kgm) in the analysed sections.

According to Equation (2), while the parameter M_1 could assume equal values with different combinations of V_1, V_2, and V_{ave}, the values of parameter M_2 may differ from those of M_1 depending on the value of V_{min}. Figure 5 shows the distribution of M_2 in sections S1 and S2. As it can be seen, the higher values of M_2 are found on the right side of both sections, but on the left side of the section, S2 values of M_2 greater than those observed in section S1 are obtained.

Thus, Figures 4 and 5 indicate that M_1 and M_2 allow us to identify the accelerating/decelerating areas and the spatial distribution of the kinetic energy, which is defined by the alternating flow velocity pattern in the examined cross-sections. Therefore, they suggest conditions either favouring or not favouring the maintenance of the aquatic habitat.

4.2. Results based on field observations

Then, the significance of the selected parameters M_1 and M_2 is investigated by using also the data collected for the Ponte Nuovo section on the Tiber River. Figure 6 presents the cross-sectional distribution of the velocity for the Tiber River for high, moderate, and low flow conditions, carried out by the current-meter. The black dash-dot line highlights the bathymetry for each cross-section. As seen, for the Ponte Nuovo cross-section the larger magnitudes of velocity are observed at the centre for high and moderate flow conditions and at the left half for the low flow conditions. The discharge in general and the summer discharge, in particular, are of significant importance for maintaining high physical habitat quality (Pedersen *et al.* 2004).

Changing the flow conditions may affect the lives of organisms, for example, as velocity (or discharge) increases, macroinvertebrate diversity, and quality increase (Pedersen *et al.* 2004). The important point is that by increasing the discharge,

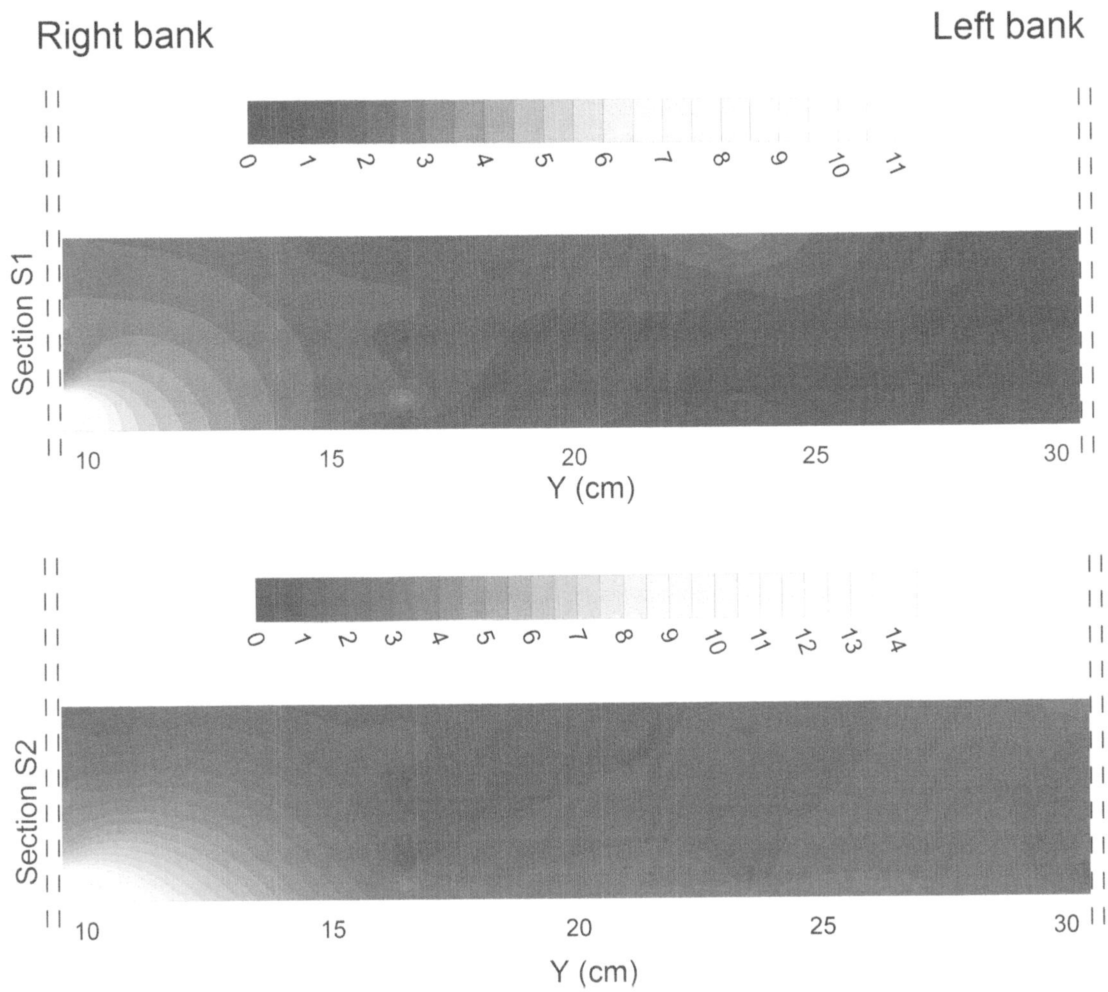

Figure 5 | Distribution of the parameter M_2 (1/m) in the analysed sections.

the velocity dip (the maximum velocity below the water surface) is formed which is more remarkable for the high-flow condition. It is believed that for the present cross-section, the dominant parameter for the formation of velocity dip is the aspect ratio (Moramarco *et al.* 2017; Kundu & Ghoshal 2019; Termini & Moramarco 2020; Bahmanpouri *et al.* 2022a, 2022b).

The knowledge of the velocity distribution allows the investigation of the two selected spatial metrics for river flow complexity. Figure 7 depicts the cross-sectional distribution of the parameter M_1 at Ponte Nuovo.

At the Ponte Nuovo site, the scale of parameter M_1 is 0.0–0.25 and 0.0–0.20 J/kgm for high and moderate flow conditions, respectively, while for low flow conditions, the scale is 0.0–0.02 J/kgm. The larger magnitudes of parameter M_1 are linked to the zones with a larger change in the velocity magnitude. That is, by changing the flow condition from low level to high level, larger magnitudes of drag force are applied to aquatic organisms. Gualtieri *et al.* (2017) reported an average magnitude of $0.01 < M_1$ (J/kgm) < 0.07 between Rio Negro and Rio Solimões in the Amazon River with $0.5 < \Delta_y$ (m) < 2.0. The kinetic energy in these flow structures can be used by swimming fish (Videler *et al.* 1999; Enders *et al.* 2003).

Figure 8 depicts the cross-sectional distribution of parameter M_2 for three different flow conditions. Based on parameter M_2, the kinetic energy at the point with the lower velocity magnitude represents the amount of energy an organism must expend if it moves from the lower velocity location to the higher velocity location. Larger magnitudes of M_2 are observed near the left bank, corresponding to lower velocity zones. A similar result was reported by Gualtieri *et al.* (2017) for the confluence of Rio Negro and Rio Solimões on the Amazon River and Shields & Rigby (2005) for the Tallahatchie River, Mississippi.

The minimum, average, and maximum magnitudes of M_2 are 0.01, 0.12, 1.13 (1/m) for high flow conditions, 0.01, 0.12, 0.7 (1/m) for moderate flow conditions, and 0.01, 0.09, 0.7 (1/m) for low flow conditions, respectively, at Ponte Nuovo section.

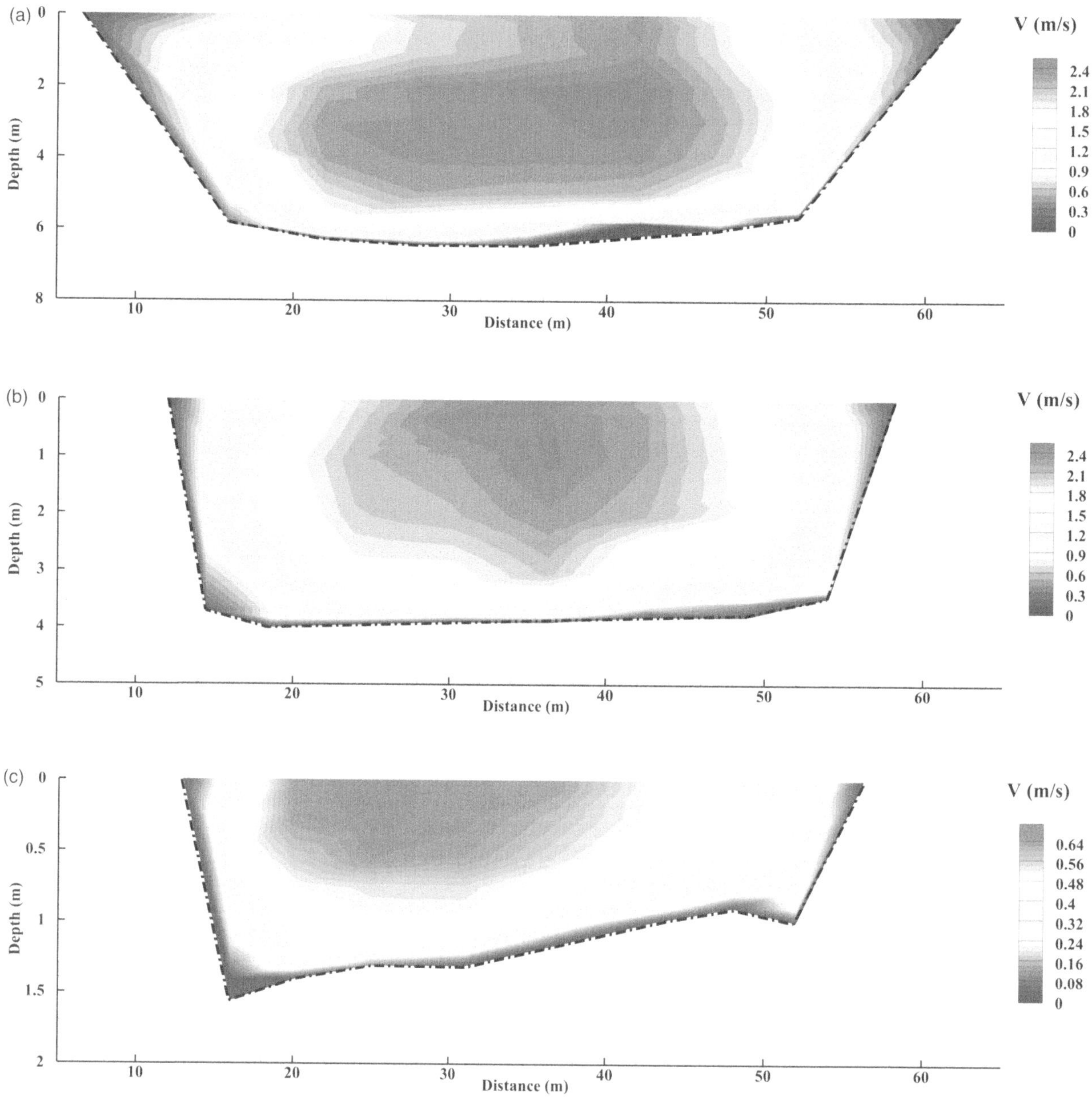

Figure 6 | Ponte Nuovo (Tiber River): cross-sectional distribution of the velocity: (a) High-flow conditions (Nov. 1996), (b) Moderate-flow conditions (Mar. 2000), (c) Low-flow conditions (May 2003).

Gualtieri *et al.* (2017) reported an average magnitude of $0.09 < M_2$ (1/m) < 0.37 at the confluence between Rio Negro and Rio Solimoes in the Amazon River. Other researchers calculated the minimum and maximum magnitudes of M_2 as 0.08 and 0.34 1/m (Fausch & White 1981), 3.1 and 40.7 1/m (Hayes & Jowett 1994), and 0.002 and 97.3 1/m (Crowder & Diplas 2000). The likeness of the identified ranges suggests the same flow patterns that may provide suitable habitats for the organisms such as fish. The difference between the magnitudes for different studies would be first because of different directions and distances between the velocity measurements and second because of the different flow kinematic conditions that make a complete comparison difficult. The results of a field study conducted by Kozarek *et al.* (2010) on Staunton River, Canada, showed that for brook trout, although velocity or depth characteristics are insufficient to explain fish habitat preferences in the complex flows created by boulders and other flow obstructions, fish density generally increases with increasing the flow complexity.

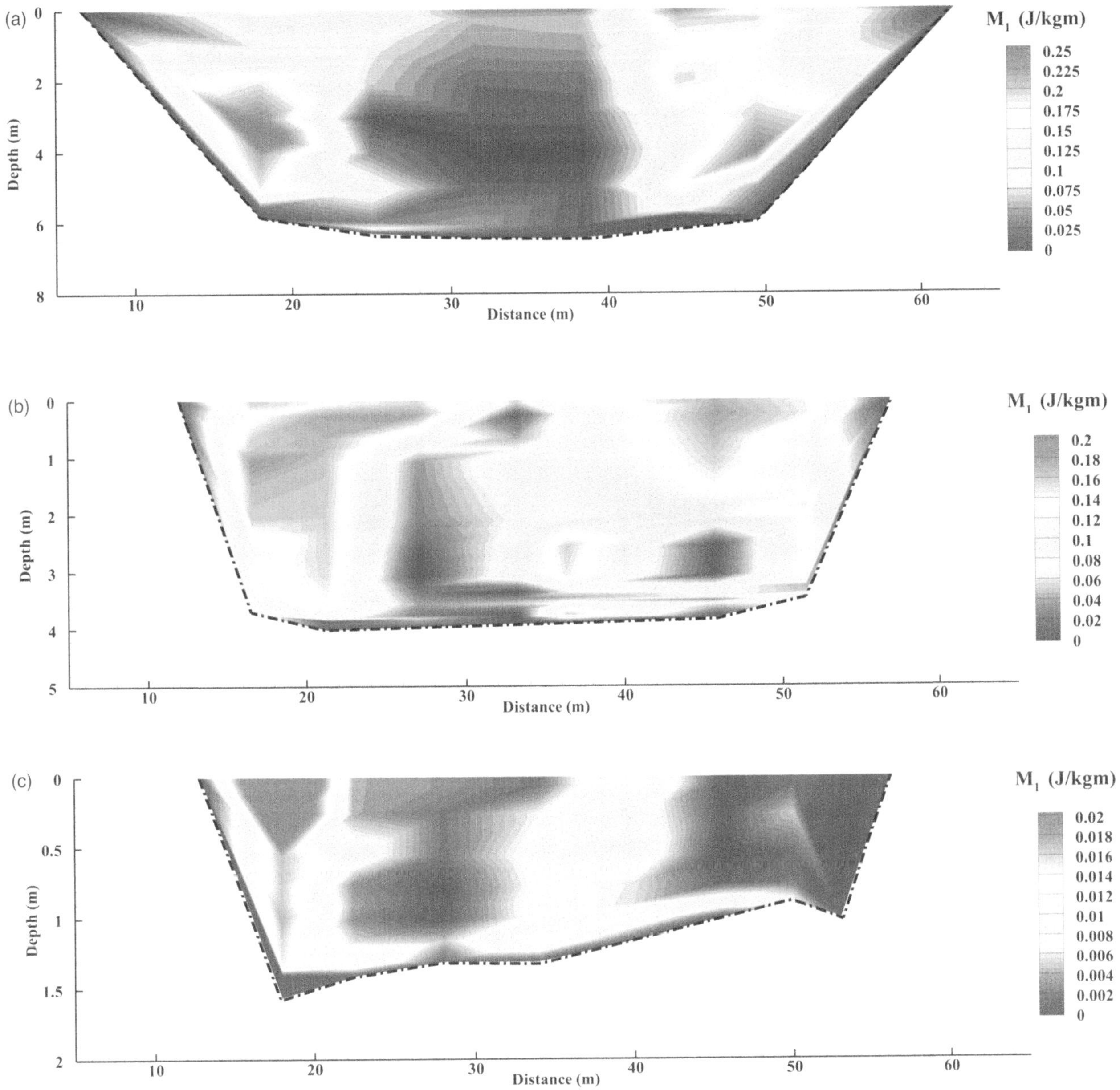

Figure 7 | Ponte Nuovo (Tiber River): cross-sectional distribution of parameter M_1: (a) High-flow conditions, (b) Moderate-flow conditions, (c) Low-flow conditions.

5. CONCLUSIONS

The hydraulic complexity metrics M_1 and M_2 are investigated for velocity datasets collected in a laboratory experiment as well as at gauged sites through the Tiber River in Italy. M_1 and M_2 enable us to calculate the kinetic energy used by aquatic organisms to move in the flow. Based on the laboratory data, the estimated parameters M_1 and M_2 adequately indicate the formation of high/low-velocity areas, along with the accelerated–decelerated zones, within each examined section. According to field observations, larger magnitudes of M_1 are linked to the zones with a larger change in velocity rates, which, for high and moderate flow conditions, it occurs at the centre. Furthermore, larger magnitudes of M_2 are observed at the left bank of the channel suggesting locations with larger kinetic energy consumption for aquatic organisms.

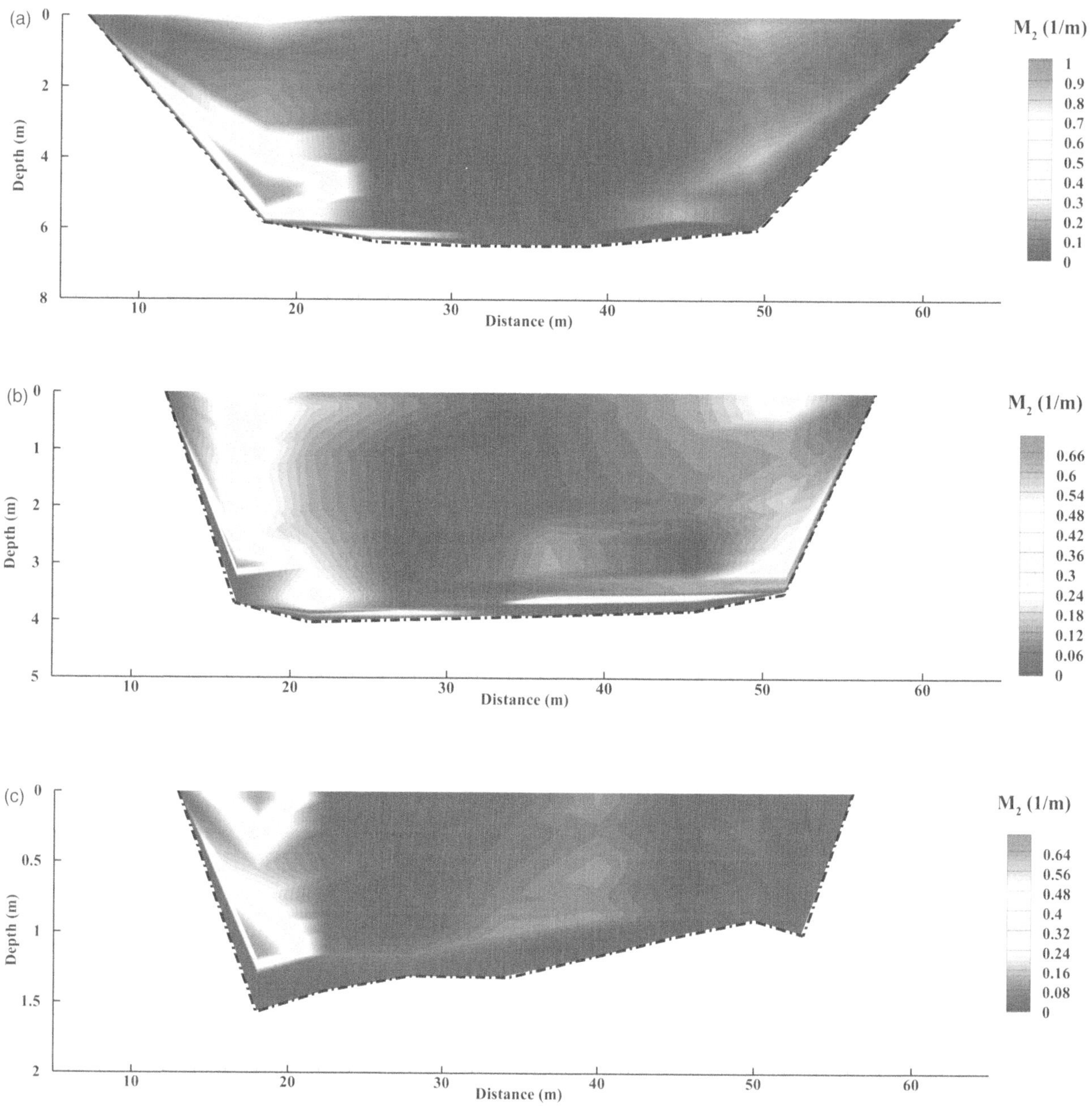

Figure 8 | Ponte Nuovo (Tiber River): cross-sectional distribution of parameter M_2: (a) High-flow conditions, (b) Moderate-flow conditions, (c) Low-flow conditions.

Overall, the findings suggest that the hydraulic complexity parameters may add some insights to explain the habitat heterogeneity as well as evaluate ecological and biological patterns in rivers with remarkable flood rates such as the Tiber River. As the next step of the present research, hydraulic complexity parameters will be calculated based on the application of Entropy theory considering either all measured surface velocities or only a single measured surface velocity.

ACKNOWLEDGEMENTS

This work has been supported by the Italian National Research Programme PRIN 2017, with the project '*IntEractions between hydrodyNamics flows and bioTic communities in fluvial Ecosystems: advancement in dischaRge monitoring and understanding of Processes Relevant for ecosystem sustaInability by the development of novel technologies with fIeld observatioNs and laboratory testinG* (ENTERPRISING)'.

DATA AVAILABILITY STATEMENT

All relevant data are included in the paper or its Supplementary Information.

CONFLICT OF INTEREST

The authors declare there is no conflict.

REFERENCES

Bahmanpouri, F., Barbetta, S., Gualtieri, C., Ianniruberto, M., Filizola, N., Termini, D. & Moramarco, T. 2022a Prediction of river discharges at confluences based on entropy theory and surface-velocity measurements. *J. Hydrol.*, 127404. https://doi.org/10.1016/j.jhydrol.2021.127404.

Bahmanpouri, F., Eltner, A., Barbetta, S., Bertalan, L. & Moramarco, T. 2022b Estimating the average river cross section velocity by observing only one surface velocity value and calibrating the entropic parameter. *Water Resour. Res.* https://doi.org/10.1029/2021WR031821.

Blettler, M. C. M., Amsler, M. L., Ezcurra De Drago, I., Drago, E., Paira, A., Espinola, L. A., Eberle, E. G. & Szupiany, R. 2016 Fine sediment input and benthic fauna interactions at the confluence of two large rivers. *Int. J. Environ. Res.* **10** (1), 65–76.

Clark, J. S., Rizzo, D. M., Watzin, M. C. & Hession, W. C. 2008 Spatial distribution and geomorphic condition of fish habitat in streams: an analysis using hydraulic modelling and geostatistics. *River Res. Appl.* **24** (7), 885–899.

Crowder, D. W. & Diplas, P. 2000 Evaluating spatially explicit metrics of stream energy gradients using hydrodynamic model simulations. *Can. J. Fish. Aquat. Sci.* **57** (7), 1497–1507.

Crowder, D. W. & Diplas, P. 2002 Vorticity and circulation: spatial metrics for evaluating flow complexity in stream habitats. *J. Fish. Aquat. Sci.* **59** (4), 633–645.

Enders, E. C., Boisclair, D. & Roy, A. G. 2003 The effect of turbulence on the cost of swimming for juvenile Atlantic salmon (Salmosalar). *J. Fish. Aquat. Sci.* **60** (9), 1149–1160.

Fausch, K. D. & White, R. J. 1981 Competition between brook trout (Salvelinus fontinalis) and brown trout (Salmo trutta) for position in a Michigan stream. *Can. J. Fish. Aquat. Sci.* **38**, 1220–1227.

Gualtieri, C., Ianniruberto, M., Filizola, N., Santos, R. & Endreny, T. 2017 Hydraulic complexity at a large river confluence in the Amazon basin. *Ecohydrology* **10** (7), e1863.

Gualtieri, C., Abdi, R., Ianniruberto, M., Filizola, N. & Endreny, T. A. 2020 A 3D analysis of spatial habitat metrics about the confluence of Negro and Solimões rivers, Brazil. *Ecohydrology* **13** (1), e2166.

Hayes, J. W. & Jowett, I. G. 1994 Microhabitat models of large drift-feeding brown trout in three New Zealand rivers. *N. Am. J. Fish. Manage.* **14** (4), 710–725.

Kozarek, J., Hession, W., Dolloff, C. & Diplas, P. 2010 Hydraulic complexity metrics for evaluating in-stream brook trout habitat. *J. Hydraul. Eng.* **136** (12), 1067–1076.

Kundu, S. & Ghoshal, K. 2019 An entropy based model for velocity-dip-position. *J. Environ. Inform.* **33** (2), 113–128.

Lacey, R. J., Neary, V. S., Liao, J. C., Enders, E. C. & Tritico, H. M. 2012 The IPOS framework: linking fish swimming performance in altered flows from laboratory experiments to rivers. *River Res. Appl.* **28** (4), 429–443.

Maddock, I. 1999 The importance of physical habitat assessment for evaluating river health. *Freshwater Biol.* **41** (2), 373–391.

Moramarco, T., Barbetta, S. & Tarpanelli, A. 2017 From surface flow velocity measurements to discharge assessment by the entropy theory. *Water* **9** (2), 120.

Moramarco, T., Barbetta, S., Bjerklie, D. M., Fulton, J. W. & Tarpanelli, A. 2019 River bathymetry estimate and discharge assessment from remote sensing. *Water Resour. Res.* **55** (8), 6692–6711.

Munson, B. R., Young, D. F. & Okiishi, T. H. 1990 *Fundamentals of Fluid Mechanics*, 2nd edn. John Wiley and Sons, Inc., New York.

Pedersen, M. L., Friberg, N. & Larsen, S. E. 2004 Physical habitat structure in Danish lowland streams. *River Res. Appl.* **20** (6), 653–669.

Shields, F. D. & Rigby, J. R. 2005 River habitat quality from river velocities measured using acoustic Doppler current profiler. *Environ Manage. (N.Y.)* **36** (4), 565–575.

Tarpanelli, A., Barbetta, S., Brocca, L. & Moramarco, T. 2013 River discharge estimation by using altimetry data and simplified flood routing modeling. *Remote Sens.* **5** (9), 4145–4162. https://doi.org/10.3390/rs5094145.

Termini, D. 2015 Experimental analysis of horizontal turbulence of flow over flat and deformed beds. *Arch. Hydroengineering Environ. Mech.* **63** (3–4), 77–99. doi:10.1515/heem-2015-0021.

Termini, D. & Moramarco, T. 2020 Entropic model application to identify cross-sectional flow effect on velocity distribution in a large amplitude meandering channel. *Adv. Water Resour.* **143**. https://doi.org/10.1016/j.advwatres.2020.103678.

Termini, D. & Sammartano, V. 2008 Experimental analysis of relation between coherent turbulent structures and formation of bed-forms. *Arch. Hydroengineering Environ. Mech.* **55** (3–4), 125–143.

Videler, J. J., Muller, U. K. & Stamhuis, E. J. 1999 Aquatic verte-brate locomotion: wakes from body waves. *J. Exp. Biol.* **202** (23), 3423–3430.

First received 15 August 2022; accepted in revised form 6 February 2023. Available online 27 February 2023

doi: 10.2166/nh.2023.088

CERCA (Cascading Effects in Risk Consequences Assessment): An operational tool for geo-hydrological scenario risk assessment and cascading effects evaluation

Daniela Biondi *, Graziella Emanuela Scarcella and Pasquale Versace

Department of Informatics, Modelling, Electronics and System Engineering, University of Calabria, 87036 Rende (CS), Italy
*Corresponding author. E-mail: daniela.biondi@unical.it

DB, 0000-0002-3570-9605

ABSTRACT

Building multiple, complex risk scenarios is a priority for the improvement of the effectiveness of early warning systems and technical countermeasure designs to detect phenomena associated with severe weather events, such as floods and landslides. This study presents CERCA (Cascading Effects in Risk Consequences Assessment), a methodology for the characterisation of event scenarios that is consistent with the current Italian Civil Protection Guidelines on the national warning system for weather-related geo-hydrological and hydraulic risks. The aim is to propose a simple, effective, multi-scale operational tool that can be adapted to multiple purposes. CERCA is structured as a tool for a typical 'scenario analysis' in a multi-hazard context through the qualitative assessment of cascading effects and consequences for different categories of elements at risk, particularly in terms of human losses. The framework is assessed on a case study concerning a local event in Rossano (Calabria, Italy) and on a number of damaging events that occurred in Italy during the period 2004–2021. The proposed approach can be effective in processing post-disaster information, monitoring the real-time evolution of critical situations, creating priority lists for decision-making, and providing general dependency matrices to be used for 'ex-ante' definitions of scenarios.

Key words: cascading effects, damaging events catalogue, emergency management and planning, multi-risk scenarios

HIGHLIGHTS

- CERCA is consistent with Italian Civil Protection Guidelines.
- Cascading effects and consequences for elements at risk have been qualitatively assessed.
- Datasets of 297 events for weather-related geo-hydrological and hydraulic risks are presented.

GRAPHICAL ABSTRACT

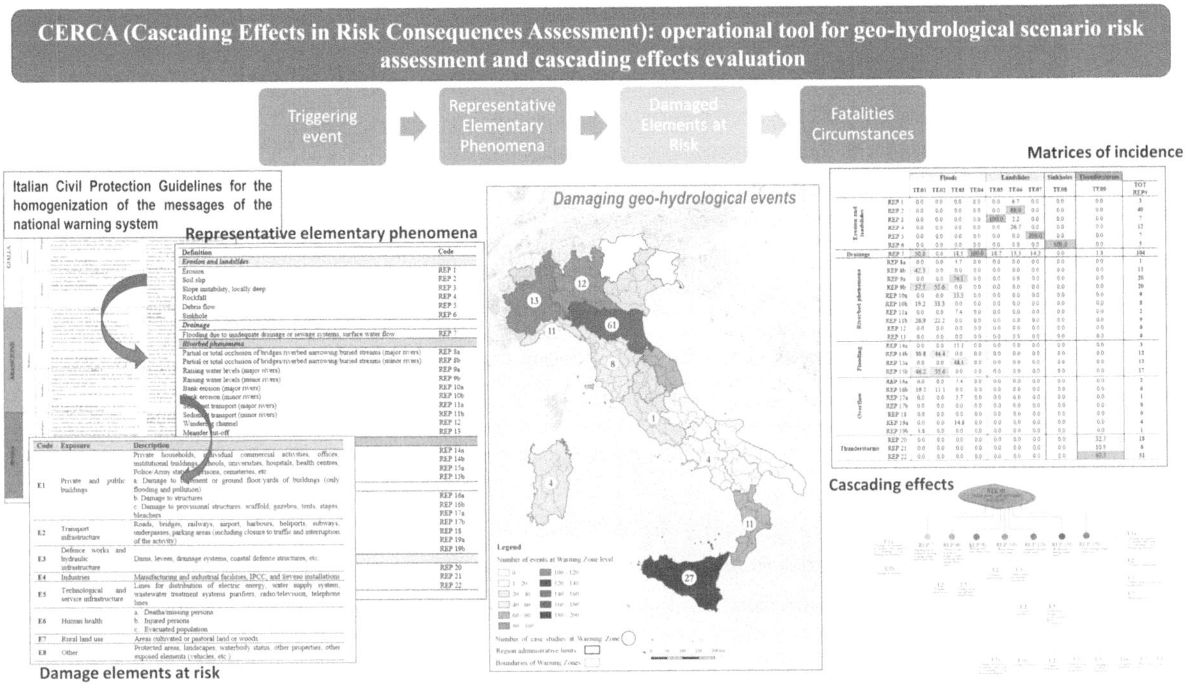

1. INTRODUCTION

The risks associated with severe weather events, such as floods and landslides, are major natural hazards that affect Italian territory and cause causalities, severe damage to human property and infrastructure, and economic losses. In recent years, there has been a growing awareness that systematic and comprehensive investigations into the development of damaging events are crucial to gain experience and knowledge for efficient forecasting and technical countermeasure designs. This involves an advancement towards a more holistic understanding of the sources, processes, and consequences of exposed elements in order to implement effective strategies and options for risk reduction.

Regarding floods in the USA, surveys have been undertaken to report flash flood events, including details about their magnitude and spatio-temporal extent as well as their disastrous consequences, mainly reported in two datasets: the Severe Hazards Analysis and Verification Experiment (SHAVE) and the US National Weather Service (NWS) Storm Data, which consist of trained spotter reports (Gourley *et al.* 2010; Calianno *et al.* 2013). In the European context, the standardisation of procedures for the survey and analysis of events has developed notably due to the Floods Directive (FD) of the European Parliament of 2007 (Floods Directive 2007), which specifically addresses the need for national catalogues of past flood events and the initiatives that are supported by European projects such as FLOODsite (FLOODsite Project Deliverable D9.1 2007) and Hydrometeorological Data Resources and Technologies for Effective flash flood forecasting (HYDRATE) (Gaume *et al.* 2009). Nevertheless, it is noteworthy that there are some inconsistencies between definitions and taxonomy of the types of hazards, as well as criteria and classification of loss indicators across different available reporting schemes (e.g., FD Reporting Schema; Guidance for Reporting under the Floods Directive – Guidance Document No. 29; the INSPIRE Natural Hazard Classification, https://inspire.ec.europa.eu/id/document/tg/nz; the JRC-DLD Guidance – Joint Research Centre, Guidance for Recording and Sharing Disaster Damage and Loss Data, https://publications.jrc.ec.europa.eu/repository/handle/JRC95505).

In Italy, the Civil Protection Department makes available to Regions and River Basin Districts Authorities a web-GIS platform, named FloodCat, which aims to function as a catalogue of flood events to address the mentioned requirements of the FD. Moreover, regarding the post-flood surveys accommodating both quantitative and qualitative analyses, several research groups have evaluated the social and economic consequences of flood events (e.g., Ballio *et al.* 2015; Carrera *et al.* 2015; Menoni *et al.* 2016; Amadio *et al.* 2019; Molinari *et al.* 2020) or, more generally, the damaging geo-hydrological events resulting from rain-triggered floods and landslides. The research mostly focused on a single event or a limited area (Aceto *et al.* 2016; Caloiero *et al.* 2017) or is no longer updated (Guzzetti *et al.* 1994).

Nevertheless, the collection of information relating to events and their consequences are not standardised and varies between subjects who collect data for different purposes (e.g., performance evaluation of the warning system, damage recognition to infrastructure, recovery plans, and insurance refunds). Building this knowledge base at the national level can be time-consuming and demanding in terms of human resources. It requires accessibility to data, which is often insufficient and uncertain, and suffers from a lack of homogeneity on a large scale, which is necessary for the creation of multi-event databases with respect to the various needs imposed by integrated risk management. The National Warning System for real-time forecasting of this type of natural hazard requires harmonic development of all its different components: forecast/monitoring of the forcing, simulation/monitoring of phenomena, warning, and estimation of consequent impacts.

Standardisation of data formats for emergency classification, improvement of the risk scenario and events knowledge, and improvement of the exchange of information between the various stakeholders involved in risk management are some of the requirements that emerged in interviews with different organisms involved in the Civil Protection and emergency management in the Italian context within the RECIPE (Reinforcing Civil Protection capabilities into multi-hazard risk assessment under climate change) project (RECIPE Deliverable 2.2 2020).

Single-hazard approaches in which hazards are treated as isolated and independent phenomena are common. However, their prevalence can distort management priorities or underestimate risks (Gill & Malamud 2016). It is evident that natural hazards intersect with one another and/or interact with other possible hazards, increasing the complexity of the overall impacts of disasters on populations and economies. This calls for a reformulation and paradigm shift from a single-hazard perspective to a multi-hazard and systemic risk perspective, a notion supported by the national commitments to the UNISDR Sendai Framework for Disaster Risk Reduction. Multi-hazard assessments may also recognise the spatio-temporal overlap of natural hazards (Kappes *et al.* 2010; Gill & Malamud 2014), dependence between individual hazards and the interactions with anthropogenic processes and built environment, which often amplify impacts (e.g. Arrighi *et al.* 2021). Many methodologies available in the literature allow for the identification and visualisation of potential natural hazard interactions (e.g., Del Monaco *et al.* 2006; FEMA 2011 – HAZUS multi-hazard analysis levels; Marzocchi *et al.* 2009; Schmidt *et al.* 2011; Gill & Malamud 2016; Ming 2019); tools have also been developed within the context of the EU FP7 project MATRIX, including New Multi-Hazard and Multi-Risk Assessment Methods for Europe (Aspinall *et al.* 2014), although they are sometimes complex and require significant amounts of data.

The importance of a holistic approach to natural hazard assessment is also needed to capture the dimensions of interconnectedness and cascading effects on social, economic, and environmental systems, also known as ripple effects or domino effects (Zuccaro *et al.* 2018). The growing concern over the domino effect and the forensic disaster analysis approach is not new in the incident and industrial risk assessment (highlighted in Europe by the Seveso Directives). This approach has more recently been applied to natural disasters (Mendoza 2019) to understand the fundamental causes of natural disasters and their impacts on society, building resilience, and reducing risk through learning from past experiences (Integrated Research on Disaster Risk IRDR 2011).

In this study, we examined the cascading impacts that can arise from geo-hydrological hazards and provided a methodology for an integrated scenario assessment for strategic risk management in Italy named CERCA (Cascading Effects in Risk Consequences Assessment). The criteria followed were to use and optimise existing data and instruments and to be consistent with the national Civil Protection Guidelines, in order to produce a simple and effective operational tool. The main aim was to define a working framework for the classification of events and related damage, which can be useful for building event scenarios and sufficiently easy to implement in a real-time operational context. This study addresses, from a methodological and experimental point of view, the prediction/assessment of the effects of extreme rain events in complex areas, particularly in densely inhabited and highly anthropic areas. A dense network of physical, cultural, social, and economic interrelations makes these areas particularly vulnerable.

The framework is structured to be used as a tool for typical 'scenario analysis' in a multi-hazard context through the qualitative assessment of cascading effects and consequences on different categories of elements at risk, particularly in terms of human losses, caused by a forcing event. All available information can be further systematised and analysed in a multi-use view of the information itself, purposed to help decision-makers, enhance critical strategic thinking, and improve citizens' preparedness.

The analysis of a significant variety of case studies of past events in the national territory was used to test and verify the proposed methodology at different scales. Information from the *ad hoc* collation of data or available from existing databases was organised with homogeneous criteria to build scenarios (triggering processes and impacts) that are significant for the Italian situation. The selected events ensured a satisfactory coverage of different climatic, geo-lithological, topographical, and land-use conditions of different parts of the national territory, as well as different types of forcing events responsible for the damage. The spatial unit at which events were analysed was determined from the Warning Zones (WZ) of the national warning system, which relies on a mosaic over the entire national territory (average size of approximately 2,000 km^2) that represents markedly homogeneous areas by the type and severity of expected events and related effects.

In this study, the CERCA approach was assessed to determine if it can be considered an effective and consistent framework:

- to process post-disaster information at the local level; '*ex-post*' surveys help to identify site-specific dependencies based on local hazard proneness and exposure and vulnerability conditions as well as to identify effective countermeasures;
- for the analysis and efficient active surveillance of the real-time evolution of critical situations, helping operative structures of civil protection to continuously update the picture of phenomena occurring in the WZ;
- to provide general dependency matrices to be used in the definition of '*ex-ante*' scenarios and recurring cascading event trees through analysis of numerous past events.

In summary, the study seeks to:

- define a method for collecting and organising information relating causal relationships to the effects associated with extreme rain events;
- enhance the knowledge, documentation, and understanding of past events;
- propose examples of the use of the method according to different objectives in relation to the scale of analysis and the needs of stakeholders.

In the following sections, we first outline the concept of the proposed framework and its components. This is followed by a description of applications at various levels, local and national, to a dataset of past events with significant adverse impacts that occurred in Italy during the period 2004–2021.

2. METHODS

2.1. Italian Civil Protection Operational Guidelines

The 'Operational Guidelines of February 10, 2016' (OG) were issued by the Italian Civil Protection Department (https://www.protezionecivile.gov.it/it/normativa/indicazioni-operative-per-l-omogeneizzazione-dei-messaggi-di allertamento-e-delle-relative-fasi-operative-per-rischio-meteo-idro) in order to promote standard languages, timelines, and operating procedures for the warning system on the national territory and the mosaic of its 20 regions, thus facilitating an effective exchange of information within the territorial levels of civil protection and communication with citizens.

We will focus on Annex 1 'Guidelines for the homogenization of the messages of the national warning system: levels of criticality and warning and related event scenarios', which contains the 'Table of weather related geo-hydrological and hydraulic warning and criticality'. The table describes the expected impacts and damage with a criticality-level alert for different risks.

Criticality levels are distinguished as ordinary, moderate, and high, in addition to a reference level characterised by the absence of relevant phenomena in relation to geo-hydrological (GH) or hydraulic (H) risk. The criticality of H risk arises from floods that affect large rivers, for which it is possible to predict the evolution of events based on monitoring of water levels. The criticality of GH risk arises from local phenomena such as certain types of landslides that can be triggered and mobilised by water, outflows in urban areas, and floods affecting small river networks. Furthermore, the OG introduce the criticality for 'geo-hydrological risk due to thunderstorms' (GHT), which is inherently characterised by a high forecast uncertainty in terms of location, timeline, and intensity and therefore cannot be subject to a reliable quantitative forecast. Criticality levels are uniquely associated with warning colour codes: the absence of relevant forecasting phenomena matches a green alert; ordinary, moderate, and elevated criticality correspond to yellow, orange, and red alerts, respectively.

The 'Table of weather related geo-hydrological and hydraulic warning and criticality' in Annex 1, for each criticality alert level, identifies two fields corresponding to expected phenomena and effects and damage. At each level, expected phenomena are characterised in terms of intensity and extent; effects and damage identify possible consequences for assets (i.e., infrastructure, residential buildings, productive activities, and people) and rely on the extent of damaged areas and severity of expected impacts. Regarding impacts on the population, yellow alerts concern 'potential risk for the safety of people with possible loss of life due to accidental causes', orange alerts correspond to 'circumstances that could endanger the safety of people with possible loss of life', and red alerts correspond to 'high risk for the safety of people with possible loss of life'.

2.2. CERCA overview

The CERCA methodology frames the problem as a typical scenario analysis through the assessment of possible cascading effects and consequences characterised by a cause/effect relationship produced by a triggering event (TE). In our scheme, we considered weather-related geo-hydrological and hydraulic events to be TEs. From this perspective, we considered the chain to start from the type of event generated by intense or prolonged rain and only implicitly assumed precipitation to be the main cause of the cascade of effects.

A 'cascade scenario' corresponds to the chain, or event-tree, starting from the TE, possible different phenomena interacting or independently generated by the same originating event, and the damage caused to different categories of elements at risk. The proposed conceptual framework is divided into four stages, represented in Figure 1, referring to the assessment of:

- Triggering Events (TE),
- cascading effects in terms of Representative Elementary Phenomena (REP),
- cascading effects in terms of Damaged Elements at Risk (DEAR),
- Fatalities Circumstances (FC).

A TE can induce a single event-tree, consisting of a single chain of elements or more branches in parallel in the case of originating events inducing different phenomena or damage. For example, the cascading effects of a primary event such as a flood, characterised by different observed parallel phenomena (e.g., levee failure, flooding, bank erosion, etc.) affecting one or more classes of exposures, are described by different independent paths originating from the same TE.

Nevertheless, in a multi-hazard approach perspective, more than one TE can occur at the same time: the simplest case is to have independent non-interacting TEs and treat them as 'parallel hazards' (Liu *et al.* 2016) each having a separate event-tree; a second multi-hazard relationship between different event types consists of those that occur in chains when 'an adverse event triggers one or more sequential events' (e.g., landslides impacting on a river network and generating a flood); these are also called 'domino effects', 'concatenated', or 'cascading hazards' (Marzocchi *et al.* 2009).

The analysis of a relevant number of past events and related cascading effects trees can be also used to provide general dependency matrices between different elements of the chain for the assessment of *ex-ante* scenarios. As will be illustrated in Sections 4.2.1 and 4.2.2, matrices of cascading dependencies can be obtained representing the relative frequencies or conditional probabilities of occurrences of an element at a certain stage of the chain, given the occurrence of an element at the previous stage. This approach allows to visualise the interactions for all the possible combination of elements at different stages of the chain and to identify recurring cascading trees.

2.2.1. Triggering event

The first step of cascade scenario assessment is to identify the type of TE.

Nine categories of TE (Table 1) were identified. The main distinction was made with respect to the type of event (i.e., floods and landslides), but further distinctions were made based on the extent, evolution time, and impact on specific elements.

Figure 1 | Overview of the steps involved in CERCA for the assessment of the cascading effects.

Table 1 | Triggering events

Triggering event	Code	Description
Flash flood and other rapid onset flood (time of concentration $\leq$6 h)	TE01	Flood, caused by extreme rainfall events in a short period of time over a relatively small area, that rises and falls quite rapidly with little or no advance warning
Medium onset flood (time of concentration <12 h)	TE02	A river flood occurring in basins with time of concentration less than 12 h
Slow onset flood (time of concentration $\geq$12 h)	TE03	A river flood occurring in basins with time of concentration larger than 12 h
Pluvial flood	TE04	Extreme rainfall event creating a flood independent of an overflowing water body includes urban storm water, rural overland flow, or excess water
Deep-seated slope failure	TE05	Rainfall induced deep-seated translational slide; rotational slide; wedge failure, or compound failure
Shallow landslides	TE06	Downhill surface slope movements of soil, rock, and organic materials triggered by rainfall; soil erosion
Debris flow	TE07	Fast-moving landslides that mainly affect slope but can also enter a stream channel or hidden channel
Sinkhole	TE08	Depression in the ground caused by some form of collapse of the surface layer (when underlying bedrock is naturally dissolved by circulating groundwater)
Thunderstorm	TE09	Severe short-lived weather disturbance associated with lightning, thunder, wind, heavy rain, or hail

Fluvial or river floods that occur when the water level in a river or stream rises and overflows onto the surrounding banks, shores, and neighbouring land have been differentiated into three types based on the time of concentration of the contributing watershed (TE01, TE02, and TE03).

In addition, the 'Pluvial flood' events (TE04) that occur when heavy rainfall creates a flood independent of an overflowing water body were considered and described as surface water flooding that occurs when the drainage or sewer system cannot cope with the speed of rainfall and floods into roads and nearby structures. Landslides were differentiated into 'Deep-seated slope failure', 'Shallow landslide', 'Debris flow', and 'Sinkhole' (TE05, T06, TE07, and TE08). Moreover, 'Thunderstorm' (TE09) was included to account for heavy localised rain events, which are often accompanied by strong winds, hail, and lightning.

2.2.2. Representative Elementary Phenomena

The characterisation of the consequences associated with a TE was accomplished through the process of identification and classification of the representative phenomena described in the OG. Phenomena support event descriptions in terms of mechanisms, characteristics, or localisation. An event scenario can exhibit more than one associated phenomenon.

Twenty-two REP were identified (Table 2). These phenomena could be related to a specific risk (GH, GHT, or H). Some REPs that pertained to minor rivers were differentiated from those that were observed in major rivers to separate REPs related to GH risk from those related to H risk.

2.2.3. Impacts on human health and damage to property

The next step in the chain pertains to associating potential effects on human health and damage to properties and infrastructure. Damage assessments of natural hazards provide crucial information for decision support and policy development in the fields of natural hazard management and adaptation planning in response to climate change (Merz *et al.* 2010). Smith & Ward (1998), following the damage loss schematisation by Schuster & Fleming (1986) with reference to floods, classified damage into direct and indirect losses. Direct losses are caused by the physical contact of the floodwater with humans or property and indirectly affect networks and social activities, causing extra costs for actions taken to prevent other losses (e.g., loss of production for companies). Furthermore, it is possible to distinguish between immediate- or long-term consequences (e.g., interruption of communication networks and critical infrastructure) and tangible or intangible consequences

Table 2 | Representative elementary phenomena

Definition	Code
Erosion and landslides	
Erosion	REP1
Soil slip	REP2
Slope instability, locally deep	REP3
Rockfall	REP4
Debris flow	REP5
Sinkhole	REP6
Drainage	
Flooding due to inadequate drainage or sewage systems, surface water flow	REP7
Riverbed phenomena	
Partial or total occlusion of bridges/riverbed narrowing/buried streams in major (minor) rivers	REP8a (REP8b)
Raising water levels in major (minor) rivers	REP9a (REP9b)
Bank erosion in major (minor) rivers	REP10a (REP10b)
Sediment transport in major (minor) rivers	REP11a (REP11b)
Wandering channel	REP12
Meander cut-off	REP13
Flooding	
Flooding of areas close to streams for major (minor) rivers	REP14a (REP 4b)
Widespread flooding for major (minor) rivers	REP15a (REP15b)
Overflow	
Overflow of river banks in major (minor) rivers	REP16a (REP16b)
Bridges and infrastructures overtopping in major (minor) rivers	REP17a (REP17b)
Levee piping	REP18
Levee damage/failure in major (minor) rivers	REP19a (REP19b)
Thunderstorms	
Hailstorm	REP20
Lightning	REP21
Wind	REP22

(e.g., psychological effects of loss of life, displacement, and property damage). Similar concepts apply to landslide consequences (e.g., Petrucci & Gullà 2010) and can be valid for other natural hazards.

Methods for assessing these impacts range from quantitative to qualitative or descriptive. Here, we referred to a descriptive approach to possible direct impacts.

The major types of impacts included in the OG were clustered into eight main categories of exposed elements (Table 3). Particular attention was paid to the harmonisation with existing classification (e.g., FloodCat) and definitions in the sharing of damage and loss data among EU countries and institutions (like the Guidance for Reporting under the Floods Directive 2013).

The last additional level in the chain, when fatalities occurred, was represented by an analysis of the circumstances of death. A classification of the most common fatal accident occurrences was proposed to understand the impact of disastrous events on the population and to design recommendations for self-protective actions (see Petrucci (2022) for a comprehensive review of the classifications in literature; Papagiannaki *et al.* 2022). In the presented scheme, this step constituted an additional level of information that detailed exposure, E6a, referring to death or missing people.

In Table 4, various categories of circumstances that adequately explain the place and dynamics of the majority of deaths were proposed.

Table 3 | Elements at risk

Code	Exposure	Description
E1	Private and public buildings	Private households, individual commercial activities, offices, institutional buildings, schools, universities, hospitals, health centres, police/army stations, prisons, cemeteries, etc. (a) Damage to basement or ground floor/yards of buildings (only flooding and pollution) (b) Damage to structures (c) Damage to provisional structures: scaffold, gazebos, tents, stages, bleachers
E2	Transport infrastructure	Roads, bridges, railways, airport, harbours, heliports, subways, underpasses, and parking areas (including closure to traffic and interruption of the activity)
E3	Defence works and hydraulic infrastructure	Dams, levees, drainage systems, coastal defence structures, etc.
E4	Industries	Manufacturing and industrial facilities, IPCC, and Seveso installations
E5	Technological and service infrastructure	Lines for the distribution of electric energy, water supply system, wastewater treatment systems purifiers, radio/television, telephone lines
E6	Human health	(a) Deaths/missing persons (b) Injured persons (c) Evacuated population
E7	Rural land use	Areas cultivated or pastoral land or woods
E8	Other	Protected areas, landscapes, waterbody status, other exposed elements (vehicles, etc.)

Table 4 | Common circumstances of fatalities caused by geo-hydrological disasters

Typical circumstances
R01: People inside buildings in flooded rooms at or below the street level
R02: Vehicles or motor-vehicles in flooded roadways
R03: Pedestrians dragged by water or people who have fallen/have been swept into river
R04: Vehicles or motor-vehicles in underpass or tunnel
R05: Weather-induced fatal vehicles incidents
R06: People involved in partial or total collapse of buildings caused by landslides or dragged/buried by mud or debris flow
R07: Vehicles hit by landslides or dragged/buried by mud or debris flow
R08 People outdoors hit by landslides or dragged/buried by mud or debris flow
R09: People outdoors involved in collapse of embankments, bridges, etc.
R10: Side effects (heart attacks, infectious diseases, wind-related deaths)

3. STUDY AREA AND DATA

A catalogue of past events with significant adverse impacts that occurred in Italy during the period 2004–2021 was built. Only events for which certified information was available (accredited source) regarding the type and extent of the event and the damage that occurred were considered.

The major sources of information were:

- event reports drawn up by the Functional Centre of Civil Protection, concerning the meteorological forcing of the event, its hydro-pluviometric features, and the impact on the territory;
- declarations of state of emergency;
- the POLARIS website (https://polaris.irpi.cnr.it/) that publishes accurate and detailed information on geo-hydrological risks (Salvati *et al.* 2016);
- the Emergency Events Database (EM-DAT; https://www.emdat.be/) and the International Disaster Database by the Centre for Research on the Epidemiology of Disasters (CRED);

- the European Severe Weather Database (https://eswd.eu/) that contains important weather events that can endanger people or cause damage, and
- products from the Emergency Management Service (EMS) of the Copernicus Program (list of EMS Rapid Mapping activations).

The entire catalogue contains 297 events that occurred in 18 regions in the period 2004–2021. The events involved 953 WZ (considering multiple occurrences of different events over the same zone) interested by consequences and damage (Table 5). The number of catalogued events varied from region to region according to their significance and the availability of necessary information. Overall, the catalogued events caused 246 fatalities, with 16 unknown circumstances of death (Figure 2).

From this dataset, only past events with relevant information describing cascading paths from the main TE to adverse consequences were selected to identify hazard/consequence dependencies. The methodology was thus applied to 49 case studies at the regional level, which involved 152 WZ. Figure 2 highlights the distribution of the selected events on the national territory, grouped at the regional level, after a refinement process.

4. RESULTS

4.1. Identifying local compatibility and dependencies. Case study: Rossano, 12 August 2015 historical event

We assessed the suitability of the approach for the event that occurred on 12 August 2015 in the north-east of Calabria (Figure 3); the application also exemplifies the work that we have done for each damaging event in Section 4.2. The Corigliano–Rossano municipality (established in 2018 by the merger of Corigliano and Rossano administrations), which covers 346 km^2 and has a population of 74,850 inhabitants (2021 Italian National Statistical Institute census database), was hit by a short-lived and localised rainstorm with more than 230 mm of rain in less than 24 h, causing damage exceeding 100 million euros (Rago *et al.* 2021). The affected area fell within WZ Cala-5 of the region.

The investigated area is frequently affected by high-intensity rainfall, commonly characterised by short duration and small extent (Ferrari & Terranova 2004), which historically caused flash floods and widespread triggering of landslides with loss of

Table 5 | Main characteristics of catalogued events

Region	No. of events	No. of involved WZ	No. of fatalities
Abruzzo	8	18	2
Basilicata	4	8	1
Calabria	23	66 (11)	25 (6)
Campania	6	11 (4)	6 (2)
Emilia-Romagna	36	146 (61)	6 (3)
Lazio	9	29 (1)	4 (1)
Liguria	22	53 (11)	25 (20)
Lombardy	15	89 (12)	9 (4)
Marche	23	63	9
Molise	4	6	0
Piedmont	25	101(13)	10 (1)
Puglia	8	19	7
Sardinia	9	27 (4)	32(3)
Sicily	63	187 (27)	69 (0)
Tuscany	13	50 (8)	26(12)
Umbria	7	27	1
Aosta Valley	11	30	3
Veneto	11	23	11
Total	**297 (49)**	**953 (152)**	**246 (52)**

Numbers in brackets refer to events considered in the application of the methodology.

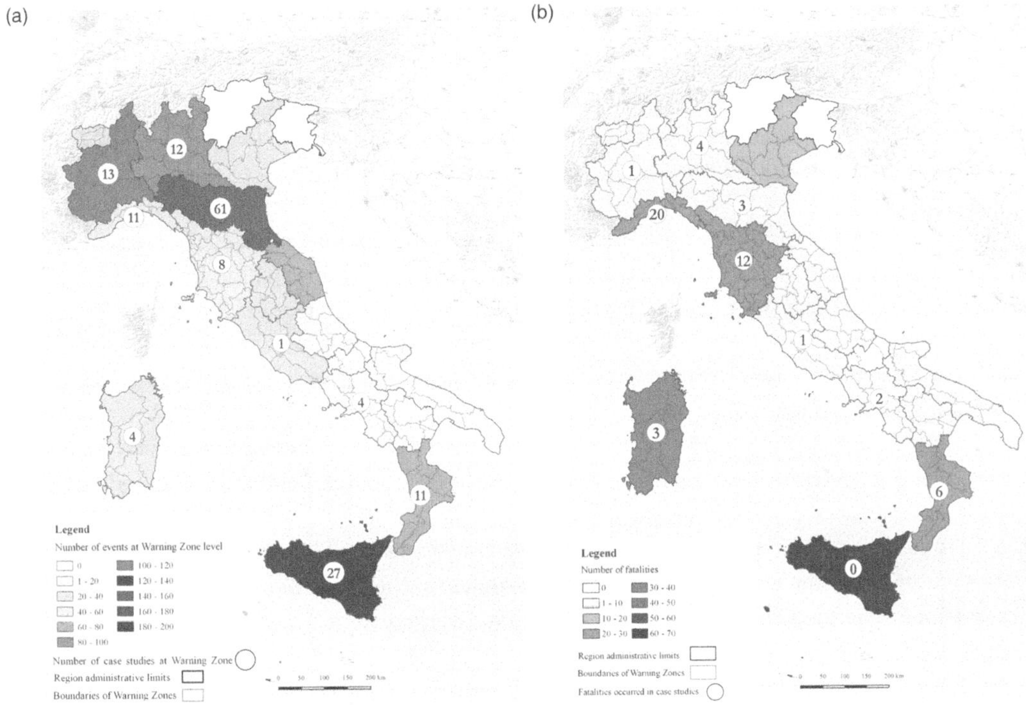

Figure 2 | (a) Distribution on the national territory of the catalogued events at the WZ level, grouped per region. Numbers in the map refer to events that have been investigated with the proposed approach. (b) Regional distribution of fatalities on the national territory. The investigated period was 2004–2021. Numbers in the map refer to fatalities that occurred in the investigated events.

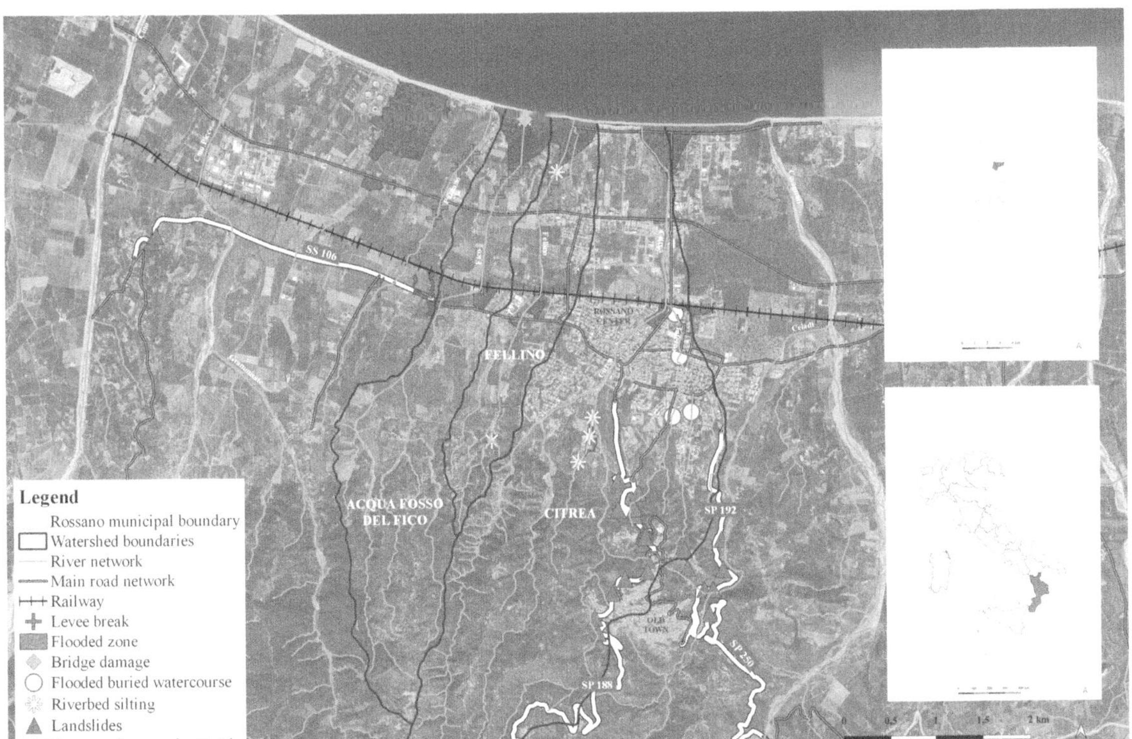

Figure 3 | Study area in Rossano municipality: localisation of the main ground effects of the event that occurred on 12 August 2015.

lives and severe effects on infrastructure and agricultural activities (the most significant events occurred in 1925, 1946, 1975, 1980, and 2000; Petrucci & Versace 2005).

The orography is peculiar; elevation ranges from sea level to 1,475 m with relatively high gradients; watersheds are small, extending from a minimum of 2.5 km^2 (Fellino creek) to a maximum of 66 km^2 (Colognati creek). The coastal zones are crossed by the Ionian Railway 'Taranto-Reggio Calabria', the State Road E90 (SS106 and SS106bis), and a number of provincial roads connected to the main urbanised areas and the inner mountainous zone.

The event of 12 August 2015 was one of the most severe events in the recent history of the study area.

Heavy rainfall caused flooding of various streams, rivers, and widespread shallow landslides. The rain data from the raingauge network of the Regional Functional Centre (Arcuri *et al.* 2015) showed rain maxima for durations of 3, 6, and 12 h, which can be considered as a 100-year return period event: the gauge in Corigliano recorded a cumulative rainfall of approximately 230 mm (from 00:00 am to 11:00 pm), with 155 mm occurring in only 5 h.

The area of the Citrea Stream basin (approximately 11 km^2) in the former municipality of Rossano was greatly affected by rainstorm-induced ground effects (Figure 3).

In the early morning of 12 August at around 8:00 am, a levee breach occurred in correspondence with a directional change of the Citrea Stream, along a stretch of the stream confined to artificial (concrete) embankments before it flowed into the Ionian Sea.

Flat areas in the surroundings of the failure were hit by massive floods and mud; hotels and bathing establishments located in the seafront area were severely damaged (mostly at Lido Sant'Angelo at the outlet of the Citrea Stream). As a consequence of the heavy rains, a tributary of the Citrea, the San Paolo Stream, overflowed in the centre of Rossano, particularly in the urbanised areas of Rossano-Stazione. The San Paolo Stream, which mainly flowed artificially under streets and houses, damaged numerous commercial activities and houses, dragged cars, disrupted circulation, and endangered human lives.

The breaking of embankments and silting also affected Fellino, Acqua Fosso del Fico, San Mauro, and Gennarito creeks at several locations causing overflows in the surrounding floodplain areas.

According to the event report (AdB 2015), the village of Momena was severely hit as a consequence of the flooding of the Fellino (area of approximately 2.62 km^2) and Acqua Fosso del Fico (area of approximately 5.27 km^2) creeks due to the sudden collapse of embankments. The most severe damage to homes and commercial activities occurred in Lussemburgo and Stati Uniti Street, where the water mixed with mud and reached a height of one and a half metres. Immediately after the event, a large accumulation of sediment in the riverbed was observed, with an estimated volume of approximately 20,000 m^3.

Furthermore, widespread slope erosion and shallow landslides occurred, approximately 5,300 (Rago *et al.* 2021), mainly soil slip and debris flows had been mapped, particularly along road SP250, which lead to the old town of Rossano.

Fortunately, no fatalities occurred, but more than 650 people had to be evacuated and housed in emergency facilities. Additionally, there was serious damage to tourist services, agricultural infrastructure, several interruptions to roads and the Jonian Railway, as well as severe problems with water and sewerage networks.

TEs identified for the case study were: TE01 '*Flash flood and other rapid onset flood*' and TE06 '*Shallow landslides*'.

REPs and damage were associated with each TE.

Concerning TE01, REPs are:

- ***REP7***: Flooding due to inadequate drainage or sewage systems, surface water flow;
- ***REP8b***: Partial or total occlusion of bridges/riverbed narrowing/buried streams (minor rivers);
- ***REP9b***: Raising water levels (minor rivers);
- ***REP10b***: Bank erosion (minor rivers);
- ***REP11b***: Sediment transport (minor rivers)
- ***REP15b***: Widespread flooding (minor rivers);
- ***REP19b***: Levee damage/break (minor rivers).

Damage reported pertaining to different assets:

- ***E1***: Public and private buildings;
- ***E2***: Transport infrastructure;
- ***E3***: Defence works and hydraulic infrastructure;
- ***E5***: Technological and service infrastructure;

- ***E6***: Human health;
- ***E7***: Rural land use;
- ***E8***: Other.

Figure 4 shows the chains originating from scenarios associated with TE01.

With regard to TE06 'Shallow landslides', the cascade of elementary phenomena and damage to exposed elements is illustrated in Figure 5. Examples for REPs and damaged assets of the Rossano event are represented in Figures 6 and 7, respectively.

4.2. Characterising scenarios for TEs at the national scale

The frequencies of the categorised TEs resulting from the analysis performed on 152 WZ are displayed in Table 6. In 78 of the 152 cases (over 50% of cases), more than one TE was observed; most of the time, a flood event was accompanied by a landslide type of event.

Among the total of 257 TEs, 54% were fluvial or pluvial floods, 23% were soil erosion or landslides, and 21% were thunderstorms. Only 2% were sinkholes.

TE04 'Pluvial floods' was the most frequent TE, representing 30% of all occurrences, followed by TE09, 'Thunderstorms' (21%) and TE06, 'Shallow landslides' (17%). The incidence of the other TEs was less than 10%.

4.2.1. Elementary phenomena

Table 7 shows the dependencies between the TEs and REPs resulting from the back-analysis performed on the selected damaging events in the form of the matrix of interactions. According to the findings regarding the occurrence of identified REPs (last column, TOT REP#, Table 7), REP7 'Flooding due to inadequate drainage or sewage system, surface water flow' was the most observed phenomenon; REP22, REP2, REP9a, REP9b followed with 20 or more occurrences; REP12, REP13, and REP18 were never observed among the collective events.

The occurrence percentage of each REP for each TE was evaluated from the ratio between the number of REP occurrences observed for a specific TE and the total number of TE occurrences. The occurrence percentage of each REP for each TE was evaluated from the ratio between the number of REP occurrences observed for a specific TE and the total number of the TE

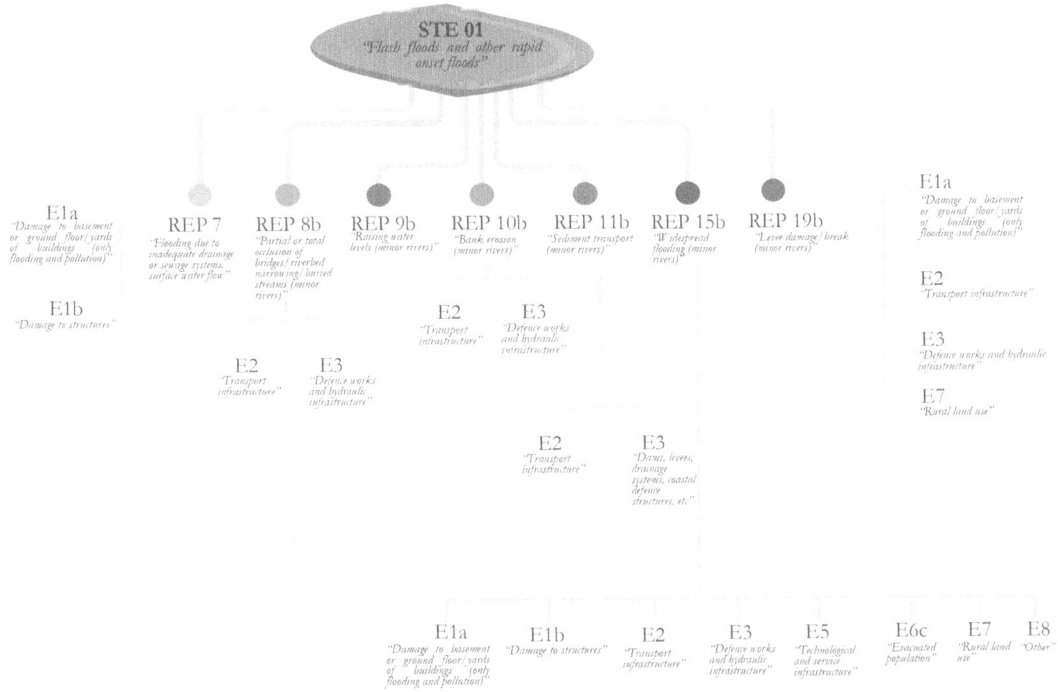

Figure 4 | The cascade scenario for TE01 from the case study of Rossano (CS), 12 August 2015.

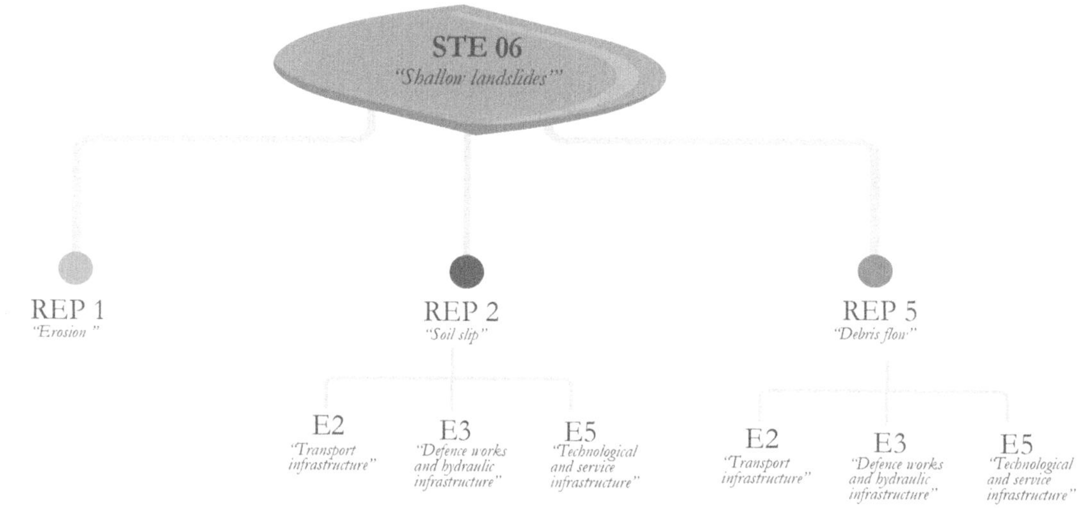

Figure 5 | The cascade scenario for TE06 from the case study of Rossano (CS), 12 August 2015.

occurrences. Specifically, the occurrence percentage of having the *i*th REP given the occurrence of the *j*th TE can be evaluated considering the relative frequency $f_{ij}[\mathrm{REP}_i, \mathrm{TE}_j]/f_j[\mathrm{TE}_j]$, where f_{ij} is the number of occurrences of *i*th REP and *j*th TE; f_j is the total number of *j*th TE occurrences. For example, the value that can be read in Table 7 at the crossing of REP2 (soil slip) and TE06 (shallow landslides) means that nearly 89% of the cases where TE06 was considered as the main TE (namely 45,

Figure 6 | Examples of REPs from the Rossano event.

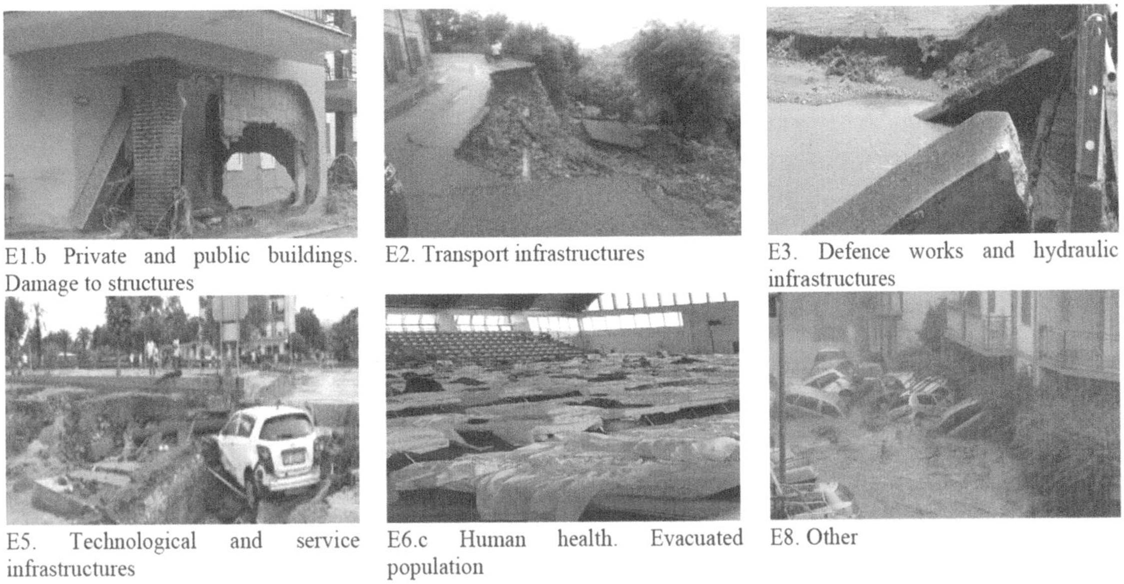

E1.b Private and public buildings. Damage to structures

E2. Transport infrastructures

E3. Defence works and hydraulic infrastructures

E5. Technological and service infrastructures

E6.c Human health. Evacuated population

E8. Other

Figure 7 | Examples of damaged assets from the Rossano event.

Table 6 | Number and percentages of catalogued TEs

Triggering events (TE)									
Floods				Landslides			Sinkholes	Thunderstorms	
TE01	TE02	TE03	TE04	TE05	TE06	TE07	TE08	TE09	Total #
26	9	27	77	6	45	7	5	55	257
10.1%	3.5%	10.5%	30.0%	2.3%	17.5%	2.7%	2.0%	21.4%	

from Table 6), soil slips were observed. In some cases, there is a complete alignment between a REP and a TE because of an evident correspondence (e.g., REP6 and TE08 both refer to sinkholes). Overall, TEs generally implied more REPs: nine REPs were associated with TE01, six of them with an occurrence of more than 20%.

The matrix of incidence helps to evaluate the level of dependencies identified and to develop the first general qualitative evaluation of expected REPs based on the frequency of occurrence (low, medium, and high) for each TE. The assessment performed on specific territorial contexts through specialisations and customisations that consider events and specific local vulnerability conditions can provide reliable information about the chain of ground effects that are most likely to occur as a result of a given TE.

4.2.2. Damage to exposures

Table 8 shows the damaged assets for each REP. The percentage of occurrences of each type of exposed element related to each REP was calculated, as for Table 7, by the ratio between the number of damage occurrences for a specific REP and the number of the REP occurrences.

Overall, the most frequent damaged exposures were transport infrastructure (E2) and private and public buildings, particularly damage to the basement or ground floor/yards of buildings (only flooding, E1a). Impacts on transport infrastructure included either physical damage to the exposed elements or closure to traffic and operational interruption owing to imminent risk to stability (e.g., bridge pile scour) or to stakeholders (e.g., pedestrians and drivers). Only in 6% of cases REP occurred without consequent damage.

The percentage of damage to transport infrastructure was consistently above 50%, excluding REP1, 11b, 12, 13, 18, 20, and 21. Damage to the basement or ground floor/yards of buildings was mostly affected by REPs related to flooding and overflow (i.e., 15a, 15b, 16a, 16b, 19a, and 19b with percentages higher than 50%).

Table 7 | Percentage (%) of occurrence of REPs for each TE and total number (#) of REPs

		Floods				Landslides			Sinkholes	Thunderst.	
		TE01	TE02	TE03	TE04	TE05	TE06	TE07	TE08	TE09	TOT REP#
Erosion and landslides	REP 1	0.0	0.0	0.0	0.0	0.0	6.7	0.0	0.0	0.0	3
	REP 2	0.0	0.0	0.0	0.0	0.0	88.9	0.0	0.0	0.0	40
	REP 3	0.0	0.0	0.0	0.0	100.0	2.2	0.0	0.0	0.0	7
	REP 4	0.0	0.0	0.0	0.0	0.0	26.7	0.0	0.0	0.0	12
	REP 5	0.0	0.0	0.0	0.0	0.0	0.0	100.0	0.0	0.0	7
	REP 6	0.0	0.0	0.0	0.0	0.0	0.0	0.0	100.0	0.0	5
Drainage	REP 7	50.0	0.0	18.5	100.0	16.7	13.3	14.3	0.0	1.8	104
Riverbed phenomena	REP 8a	0.0	0.0	3.7	0.0	0.0	0.0	0.0	0.0	0.0	1
	REP 8b	42.3	0.0	0.0	0.0	0.0	0.0	0.0	0.0	0.0	11
	REP 9a	0.0	0.0	74.1	0.0	0.0	0.0	0.0	0.0	0.0	20
	REP 9b	57.7	55.6	0.0	0.0	0.0	0.0	0.0	0.0	0.0	20
	REP 10a	0.0	0.0	33.3	0.0	0.0	0.0	0.0	0.0	0.0	9
	REP 10b	19.2	33.3	0.0	0.0	0.0	0.0	0.0	0.0	0.0	8
	REP 11a	0.0	0.0	7.4	0.0	0.0	0.0	0.0	0.0	0.0	2
	REP 11b	26.9	22.2	0.0	0.0	0.0	0.0	0.0	0.0	0.0	9
	REP 12	0.0	0.0	0.0	0.0	0.0	0.0	0.0	0.0	0.0	0
	REP 13	0.0	0.0	0.0	0.0	0.0	0.0	0.0	0.0	0.0	0
Flooding	REP 14a	0.0	0.0	11.1	0.0	0.0	0.0	0.0	0.0	0.0	3
	REP 14b	30.8	44.4	0.0	0.0	0.0	0.0	0.0	0.0	0.0	12
	REP 15a	0.0	0.0	48.1	0.0	0.0	0.0	0.0	0.0	0.0	13
	REP 15b	46.2	55.6	0.0	0.0	0.0	0.0	0.0	0.0	0.0	17
Overflow	REP 16a	0.0	0.0	7.4	0.0	0.0	0.0	0.0	0.0	0.0	2
	REP 16b	19.2	11.1	0.0	0.0	0.0	0.0	0.0	0.0	0.0	6
	REP 17a	0.0	0.0	3.7	0.0	0.0	0.0	0.0	0.0	0.0	1
	REP 17b	0.0	0.0	0.0	0.0	0.0	0.0	0.0	0.0	0.0	0
	REP 18	0.0	0.0	0.0	0.0	0.0	0.0	0.0	0.0	0.0	0
	REP 19a	0.0	0.0	14.8	0.0	0.0	0.0	0.0	0.0	0.0	4
	REP 19b	3.8	0.0	0.0	0.0	0.0	0.0	0.0	0.0	0.0	1
Thunderst.	REP 20	0.0	0.0	0.0	0.0	0.0	0.0	0.0	0.0	32.7	18
	REP 21	0.0	0.0	0.0	0.0	0.0	0.0	0.0	0.0	10.9	6
	REP 22	0.0	0.0	0.0	0.0	0.0	0.0	0.0	0.0	92.7	51

In addition, damage to technological and service infrastructure (E5) and evacuated population (E6c) were frequently observed (more than 20% of occurrences for eight REPs).

On 23 occasions, there were fatalities with a total of 52 deaths encompassed in category E6a. Referring to REPs, casualties were likely to occur with widespread flooding (REP15a and REP15b), debris flow phenomena (REP5), or partial or total occlusion of bridges/riverbed narrowing/buried streams (REP8b). A more detailed characterisation of the circumstances of death that further specifies the chain for category E6a is presented in Table 9. We expect that the results of this work will be helpful in designing recommendations for self-protective actions and proactive policies that can contribute to reducing the human death toll. The majority were flood fatalities (82%) occurring both indoors and outdoors and landslide-related fatalities (11%) occurring outdoors. Numerous people were affected unexpectedly along roads (35%) and travelling in vehicles (37%).

5. DISCUSSION AND CONCLUSIONS

This study introduces CERCA, a framework for event scenario assessment that is coherent with the current Italian Civil Protection OG for the national warning system. The methodology was developed for weather-related geo-hydrological and hydraulic hazards and categories of elements at risk. The framework allows for flexibility in multi-purpose and multi-scale applications and proposes a common set of criteria and standards that can be valuable in the definition of *ex-ante* and *ex-post* risk mitigation strategies.

Table 8 | Percentage (%) of damage occurrence on exposed elements for each REP

	FER 2.0	E1.a	E1.b	E1.c	E2	E3	E4	E5	E6.a	E6.b	E6.c	E7	E8
Erosion and landslides	REP 1	0	0	0	33	0	0	0	0	0	0	33	0
	REP 2	5	3	0	100	3	0	13	5	3	3	0	8
	REP 3	14	14	14	71	14	14	29	0	0	57	0	0
	REP 4	0	8	0	83	0	0	17	0	0	0	0	0
	REP 5	38	13	0	100	0	0	25	25	25	25	0	13
	REP 6	0	20	0	80	0	0	20	0	20	40	0	20
Drainage	REP 7	49	1	1	95	3	3	8	1	0	4	5	5
Riverbed phenomena	REP 8a	0	0	0	100	0	100	0	0	0	0	0	0
	REP 8b	33	8	0	83	8	0	8	17	0	0	0	8
	REP 9a	20	5	0	60	10	5	10	0	0	20	0	0
	REP 9b	30	20	0	50	5	0	10	5	0	20	0	10
	REP 10a	0	0	0	56	11	0	22	0	0	0	0	0
	REP 10b	25	25	0	63	25	0	13	0	0	13	0	0
	REP 11a	0	0	0	50	0	0	0	0	0	0	0	0
	REP 11b	0	22	0	44	22	0	0	0	0	0	0	0
	REP 12	0	0	0	0	0	0	0	0	0	0	0	0
	REP 13	0	0	0	0	0	0	0	0	0	0	0	0
Flooding	REP 14a	33	0	0	67	0	0	0	0	0	33	0	33
	REP 14b	33	8	0	83	0	8	8	8	0	17	25	8
	REP 15a	92	23	15	100	31	46	69	46	8	54	54	46
	REP 15b	82	41	0	100	24	18	53	47	0	35	29	53
Overflow	REP 16a	50	0	0	100	0	0	0	0	0	0	0	0
	REP 16b	57	0	0	86	0	0	0	0	0	0	14	14
	REP 17a	0	0	0	100	0	100	0	0	0	0	0	0
	REP 17b	0	0	0	100	0	0	0	0	0	0	0	0
	REP 18	0	0	0	0	0	0	0	0	0	0	0	0
	REP 19a	75	0	0	75	50	25	0	0	0	0	75	0
	REP 19b	100	0	0	100	100	0	0	0	0	0	100	0
Thunderstorms	REP 20	6	0	17	39	0	0	0	0	6	0	44	11
	REP 21	0	0	0	0	0	0	67	0	0	0	0	17
	REP 22	0	0	47	73	0	6	22	0	6	4	12	20

Table 9 | Characterisation of circumstances of fatal incidents

	# Fatalities
R01: People inside buildings in flooded rooms at or below the street level	14
R02: Vehicles or motor-vehicles in flooded roadways	12
R03: Pedestrians dragged by water or people who have fallen/have been swept into a river	16
R04: Vehicles or motor-vehicles in underpasses or tunnels	0
R05: Weather-induced fatal vehicles incidents	1
R06: People involved in partial or total collapse of buildings caused by landslides or dragged/buried by mud or debris flow	0
R07: Vehicles hit by landslides or dragged/buried by mud or debris flow	6
R08 People outdoors hit by landslides or dragged/buried by mud or debris flow	0
R09: People outdoors involved in the collapse of embankments, bridges, etc.	2
R10: Side effects (heart attacks, infectious diseases, wind-related deaths)	1
Total	**52**

The proposed methodology was assessed using a case study concerning a local event in Rossano (Calabria, Italy) and 152 events in the WZ of the Italian territory that occurred during the period 2004–2021.

The first case details a post-disaster survey at a local level and is aimed at illustrating CERCA functionality in describing cascading effects.

The same framework can have a potentially relevant application as a tool for real-time analysis and monitoring of the evolution of phenomena for civil protection purposes. During an event or in the emergency phase, civil protection coordination centres receive and collect reports of impacts and damage from the population and involved stakeholders. CERCA, by using a base of standard classification, can support the work of real-time classification and understanding of the causal sequence of phenomena, particularly when dealing with multi-hazard events. Taking advantage of the organisation of reports and real-time direct observations about what is happening can lead to improvements in monitoring and managing the emergency from different perspectives and particularly by better coordinating all activities, prioritising interventions, taking the decision about emergency procedures (e.g., evacuation, closing of activities, reinforcement of protection works, etc.), and driving field surveillance towards emerging critical situations. An accurate verification and deepening investigation at different times in the post-event should be considered as mandatory to confirm data gathered in real time.

A second achievement concerns the possibility to identify site-specific dependencies based on hazard proneness, exposure, and vulnerability conditions and to highlight risks from cascade effects that are rarely predicted with the traditional 'decomposition' approach based on Hazard-Vulnerability-Exposure multiplication through simple maps overlay (e.g., consequences related to levee break as for the Rossano event). In this regard, customisation of the general theoretical framework to site-specific cases is a necessary step to produce reliable scenarios according to the vulnerability of the affected area and the magnitude of the forcing event, its duration and extension.

Highly detailed descriptions about precise location and the severity of observed phenomena as well as fatal circumstances should be considered in the development of site-specific applications aiming to effectively support decision-makers in their preparedness and disaster mitigation strategies.

At the same time, the chain here illustrated should be linked to characteristics of the forcing rainfall so as to furnish a comprehensive view of possible hazard/impact scenarios that can be used in the forecasting phase.

The national-scale back-analysis, although incomplete, offers an overview of the chains generated by TEs and the information summarised in matrices provides an overall picture of dependencies limited to the number and type of investigated events. The analysis showed that in over 50% of investigated cases, more than one TE was observed, most of the time floods accompanied by landslides, confirming the necessity for multi-risk analysis. TE04 'Pluvial flood', particularly affecting urban areas, was the most frequent TE with 30%, mainly causing damage to the basement or ground floor/yards of public and private buildings (E1a) and to transport infrastructure (E2) (49 and 95% of the times, respectively). This preliminary assessment is valuable to support further analyses that could explore also the dependency from the severity of the event and, as stated above, from the forcing rainfall. In this regard, the proposed approach is part of a long-term cooperation with Civil Protection that aims at creating a comprehensive and easy-to-use framework for data storage and analysis based on common criteria and standards. Ongoing research based on the CERCA framework concerns the attribution of a severity class to REP and DEAR in order to develop a procedure for an ex-post evaluation of an event severity colour code based on ground impacts coherent with the national Civil Protection OG. We believe that this assessment is crucial to adequately define cascading effects scenarios as well as to properly assess the performance of the forecasting and warning system.

The analysis of death circumstances is instrumental in addressing risk mitigation and by helping in conveying clear and effective messages about self-protective behaviours to enhance population preparedness for specific TEs. In several cases, accident responsibility can be attributed to a victim's hazardous behaviour: indoor victims were hit by flood while trying to save belongings or seeking shelter in garages, basement or underground levels rather than on roofs or upper floors. In this perspective, well-detailed datasets offer the opportunity to implement specifically designed measures and recommendations for self-protective actions, e.g., awareness campaign, risk education, road network management, risk signage, and impact-based warnings represent possible risk mitigation policies and initiatives regarding vehicle-related fatalities (37% in the investigated dataset). Further studies about extending the dataset and better detailing the fatal circumstances (e.g., location; victim activity, age, and health conditions; vehicle type, etc.) are underway. A non-negligible limitation of the approach to large-scale application, that is common to most of the existing tools and databases, lies in the availability of detailed and reliable data regarding the event evolution, cascading effects, and circumstances surrounding fatal incidents. As a consequence, results may suffer from epistemic uncertainty that refers to uncertainty associated with incomplete or imperfect knowledge about the processes involved. Nevertheless, a systematic process of collection, accurate verification, and organisation of data about weather-related adverse events can help to create a virtuous circle, minimise biases, and improve the quality of the outcomes.

ACKNOWLEDGEMENTS

This work was partially carried out within the cooperation agreement between Italian National Civil Protection and Camilab-DIMES Unical on 'Criteri di allertamento, valutazione delle conseguenze e preannuncio di eventi idrogeologici estremi relativi al rischio da frana e da inondazione'.

DATA AVAILABILITY STATEMENT

Data cannot be made publicly available; readers should contact the corresponding author for details.

CONFLICT OF INTEREST

The authors declare there is no conflict.

REFERENCES

Aceto, L., Caloiero, T., Pasqua, A. A. & Petrucci, O. 2016 Analysis of damaging hydrogeological events in a Mediterranean region (Calabria). *J. Hydrol.* **541**, 510–522.

AdB, Autorità di Bacino 2015 *Rapporto di evento precipitazioni del 30 ottobre–2 novembre 2015.* UOA Protezione Civile and Autorità di Bacino, Regione Calabria.

Amadio, M., Scorzini, A. R., Carisi, F., Essenfelder, A. H., Domeneghetti, A., Mysiak, J. & Castellarin, A. 2019 Testing empirical and synthetic flood damage models: the case of Italy. *Nat. Hazards Earth Syst. Sci.* **19** (3), 661–678.

Arcuri, S., Fusto, F., Marsico, L. & Rotundo, R. 2015 *Evento meteopluviometrico del 12 agosto 2015.* Rapporto speditivo di evento. ARPACAL, Centro Funzionale Multirischi della Calabria, Regione Calabria, p. 17.

Arrighi, C., Pregnolato, M. & Castelli, F. 2021 Indirect flood impacts and cascade risk across interdependent linear infrastructures. *Nat. Hazards Earth Syst. Sci.* **21**, 1955–1969. https://doi.org/10.5194/nhess-21-1955-2021.

Aspinall, W., Bengoubou-Valerius, M., Desramaut, N., Di Ruocco, A., Fleming, K., Garcia-Aristizabal, A., Gasparini, P., Gehl, P., Khazai, B., Komendantova, N., Laskowski, S., Liu, Z., Marti, J., Marzocchi, W., Mignan, A., Monfort-Climent, D., Mrzyglocki, R., Nadim, F., Parolai, S., Patt, A., Réveillère, A., Scolobig, A., Tyagunov, S., van Gelder, P., Vidar Vangelsten, B., Vinchon, C., Vorogushyn, S., Wang, J., Wenzel, F., Woo, G. & Zschau, J. 2014 *MATRIX Reference Reports (Scientific Technical Report; 14/13) Potsdam: Deutsches GeoForschungsZentrum GFZ*, p. 104. https://doi.org/10.2312/GFZ.b103-14137.

Ballio, F., Molinari, D., Minucci, G., Mazuran, M., Arias Munoz, C., Menoni, S., Atun, F., Ardagna, D., Berni, N. & Pandolfo, C. 2015 The RISPOSTA procedure for the collection, storage and analysis of high quality, consistent and reliable damage data in the aftermath of floods. *J. Flood Risk Manage.* **11**, 5604–5615.

Calianno, M., Ruin, I. & Gourley, J. 2013 Supplementing flash flood reports with impact classifications. *J. Hydrol.* **477**, 1–16.

Caloiero, T., Pasqua, A. A., Perrotta, P., Petrucci, O., Russo, L. & Tansi, C. 2017 Civil protection and damaging hydrogeological events: comparative analysis of the 2000 and 2015 events in Calabria (southern Italy). *Adv. Geosci.* **44**, 101–113.

Carrera, L., Standardi, G., Bosello, F. & Mysiak, J. 2015 Assessing direct and indirect economic impacts of a flood event through the integration of spatial and computable general equilibrium modelling. *Environ. Modell. Softw.* **63**, 109–122.

Delmonaco, G., Margottini, C. & Spizzichino, D. 2006 *ARMONIA Methodology for Multi-Risk Assessment and the Harmonisation of Different Natural Risk Maps.* Deliverable 3.1.1, ARMONIA.

FEMA (Federal Emergency Management Agency) 2011 *Multi-Hazard Loss Estimation Methodology: Flood Model.* Technical Manual, Department of Homeland Security, Federal Emergency Management Agency, Mitigation Division, Washington, D.C.

Ferrari, E. & Terranova, O., 2004 Non-parametric detection of trends and change point years in monthly and annual rainfalls. In: *Proceedings of 1st Italian-Russian Workshop New Trends in Hydrology* (Gaudio, R., ed.). CNR Publications 2823, Rende, 24–26 September 2004, pp. 177–188.

Floods Directive 2007/60/EC of the European Parliament and of the Council of 23 October 2007 on the Assessment and Management of Flood Risks.

FLOODsite 2007 *T09-06-01 Evaluating Flood Damages: Guidance and Recommendations on Principles and Methods, Project Deliverable D9.1.*

Gaume, E., Bain, V., Bernardara, P., Newinger, O., Barbuc, M., Bateman, A., Blaškovičová, L., Blöschl, G., Borga, M., Dumitrescu, A., Daliakopoulos, I., Garcia, J., Irimescu, A., Kohnova, S., Koutroulis, A., Marchi, L., Matreata, S., Medina, V., Preciso, E., Sempere-Torres, D., Stancalie, G., Szolgay, J., Tsanis, I., Velasco, D. & Viglione, A. 2009 A compilation of data on European flash floods. *J. Hydrol.* **367** (1–2), 70–78.

Gill, J. C. & Malamud, B. D. 2014 Reviewing and visualizing the interactions of natural hazards. *Rev. Geophys.* **52**, 680–722.

Gill, J. C. & Malamud, B. D. 2016 Hazard interactions and interaction networks (cascades) within multi-hazard methodologies. *Earth Syst. Dyn.* **7** (3), 659–679.

Gourley, J., Erlingis, J. M., Smith, T. M., Ortega, K. L. & Hong, Y. 2010 Remote collection and analysis of witness reports on flash floods. *J. Hydrol.* **394** (1–2), 53–62.

Guidance for Reporting under the Floods Directive 2000/60/EC 2013 *Guidance Document N° 29*. Technical Report. A Compilation of Reporting Sheets adopted by Water Directors Common Implementation Strategy for the Water Framework Directive 2000/60/EC.

Guzzetti, F., Cardinali, M. & Reichenbach, P. 1994 The AVI project: a bibliographical and archive inventory of landslides and floods in Italy. *Environ. Manage.* **18**, 623–633.

Integrated Research on Disaster Risk. Forensic Investigations of Disasters 2011 *The FORIN Project (IRDR FORIN Publication No. 1)*. Integrated Research on Disaster Risk, Beijing.

Kappes, M., Keiler, M. & Glade, T. 2010 From single- to multi-hazard risk analyses: a concept addressing emerging challenges. In: *Proc. of the 'Mountain Risks' International Conference: Bridging Science to Society*, 24–26 November, Firenze, Italy.

Liu, B., Siu, Y. L. & Mitchell, G. 2016 Hazard interaction analysis for multi-hazard risk assessment: a systematic classification based on hazard-forming environment. *Nat. Hazards Earth Syst. Sci.* **16**, 629–642.

Marzocchi, W., Mastellone, M. L., Di Ruocco, A., Novelli, P., Romeo, E. & Gasparini, P. 2009 *Principles of Multi-Risk Assessment. Interaction Amongst Natural and Man-Induced Risks*. European Commission, Directorate-General for Research, Environment Directorate, Luxembourg, p. 72.

Mendoza, M. T. 2019 *Forensic Investigation of Post-Flood Damage Data to Support Spatial Planning*. PhD Thesis, Politecnico di Milano.

Menoni, S., Molinari, D., Ballio, F., Minucci, G., Mejri, O., Atun, F., Berni, N. & Pandolfo, C. 2016 Flood damage: a model for consistent, complete and multipurpose scenarios. *Nat. Hazards Earth Syst. Sci.* **16**, 2783–2797.

Merz, B., Kreibich, H., Schwarze, R. & Thieken, A. 2010 Assessment of economic flood damage. *Nat. Hazards Earth Syst. Sci.* **10**, 1697–1724.

Ming, X. 2019 *An Integrated Assessment Framework for Quantifying and Forecasting Water-Related Multi-Hazard Risk. PhD Thesis*, Newcastle University.

Molinari, D., Scorzini, A. R., Arrighi, C., Carisi, F., Castelli, F., Domeneghetti, A., Gallazzi, A., Galliani, M., Grelot, F., Kellermann, P., Kreibich, H., Mohor, G. S., Mosimann, M., Natho, S., Richert, C., Schroeter, K., Thieken, A. H., Zischg, A. P. & Ballio, F. 2020 Are flood damage models converging to 'reality'? Lessons learnt from a blind test. *Nat. Hazards Earth Syst. Sci.* **20**, 2997–3017.

Papagiannaki, K., Petrucci, O., Diakakis, M., Kotroni, M., Aceto, L., Bianchi, C., Brázdil, R., Gelabert, M. G., Inbar, M., Kahraman, A., Kılıç, O., Krahn, A., Kreibich, H., Llasat, M. C., Llasat-Botija, M., Macdonald, N., de Brito, M., Mercuri, M., Pereira, S. & Řeh, J. 2022 Developing a large-scale dataset of food fatalities for territories in the Euro-Mediterranean region, FFEM-DB. *Scientific Data* **9**, 166. https://doi.org/10.1038/s41597-022-01273-x.

Petrucci, O. 2022 Factors leading to the occurrence of flood fatalities: a systematic review of research papers published between 2010 and 2020. *Nat. Hazards Earth Syst. Sci.* **22**, 71–83.

Petrucci, O. & Gullà, G. 2010 A simplified method for landslide damage scenario assessment based on historical data. *Nat. Hazards* **52**, 539–560.

Petrucci, O. & Versace, P. 2005 *Frane e Alluvioni in Provincia di Cosenza agli Inizi del '900: Ricerche Storiche nella Documentazione del Genio Civile*. I. Quaderni dell'Osservatorio di Documentazione Ambientale, Bios, Rende (CS), Italy.

Rago, V., Lupiano, V., Chiaravallo, F., Chiodo, G., Gabriele, S., Pellegrino, A. D., Terranova, O. G. & Iovine, G. 2021 Geomorphic effects caused by heavy rainfall in the Corigliano-Rossano area (NE Calabria, Italy) on 12 August 2015. *J. Maps* **17** (2), 280–289.

RECIPE Deliverable 2.2. 2020 *Report on Civil Protection and Emergency Management Requirements to Face Natural Hazards RECIPE – 874402*. UCPM-2019-PP-AG.

Salvati, P., Pernice, U., Bianchi, C., Marchesini, I., Fiorucci, F. & Guzzetti, F. 2016 Communication strategies to address geohydrological risks: the POLARIS web initiative in Italy. *Nat. Hazards Earth Syst. Sci.* **16**, 1487–1497.

Schmidt, J., Matcham, I., Reese, S., King, A., Bell, R., Henderson, R., Smart, G., Cousins, J., Smith, W. & Heron, D. 2011 Quantitative multi-risk analysis for natural hazards: A framework for multi-risk modelling. *Nat. Hazards* **58** (3), 1169–1192.

Schuster, R. L. & Fleming, R. W. 1986 Economic losses and fatalities due to landslides. *Bull. Assoc. Geol.* **23**, 11–28.

Smith, K. & Ward, R. 1998 *Floods: Physical Processes and Human Impacts*. Wiley, Chichester, UK.

Zuccaro, G., De Gregorio, D. & Leone, M. F. 2018 Theoretical model for cascading effects analyses. *Int. J. Disaster Risk Reduct.* **30**, 199–215. https://doi.org/10.1016/j.ijdrr.2018.04.019.

First received 9 August 2022; accepted in revised form 15 January 2023. Available online 4 February 2023

doi: 10.2166/nh.2022.063

Post-processing climate projections of precipitation for the Po river basin: will Italy's North become water-constrained?

Oleksiy Boyko [a], Paolo Reggiani [a,*] and Ezio Todini [b]

[a] Research Institute for Water and Environment, Department of Civil Engineering, University of Siegen, Siegen, Germany
[b] Italian Hydrological Society, Bologna, Italy
*Corresponding author. E-mail: paolo.reggiani@uni-siegen.de

OB, 0000-0002-3481-4294; PR, 0000-0001-9989-9885; ET, 0000-0003-0847-1379

ABSTRACT

Surface and groundwater resource availability depends on precipitation patterns. Climatic change may alter not only future annual totals of precipitation but also its temporal distribution. In regions depending strongly on snow accumulation for steady water supply, this can lead to water constraints. We process climatic projections of precipitation from 19 models of the Climate Model Inter-comparison Project 5 for the Po river, Italy. The study area hosts Italy's most important lakes and reservoirs and is inhabited by 16 million people. The river basin is also known for its productive areas of irrigated agriculture. We apply a Bayesian processor of uncertainty, which we calibrate on a comprehensive set of high-resolution gridded observations. The processor outputs predictive densities of precipitation for selected prognostic time windows. These densities can be used in conjunction with an utility function to estimate potential losses and/or evaluate the benefits of mitigating actions. For the study area, annual precipitation will not change notably in the future for both an optimistic and a pessimistic scenario. The temporal distribution of precipitation will become affected. These potential changes result in considerable strain on storage capacity and water flows needed to satisfy irrigation demand as well as hydroelectric and thermal energy production.

Key words: Bayesian analysis, climate projections, CMIP5, precipitation, predictive uncertainty, Po river

HIGHLIGHTS

- We present a 21st-century precipitation and temperature climate analysis for Northern Italy using CMIP5 Scenarios RCP4.5 and RCP8.5.
- The projections show a rather small change in seasonal precipitation totals.
- The temperature increases considerably over the analysis period.
- Snow depth and area extent will most likely decrease on average.
- Italy will need to address future challenges in water resource management in the Po basin.

GRAPHICAL ABSTRACT

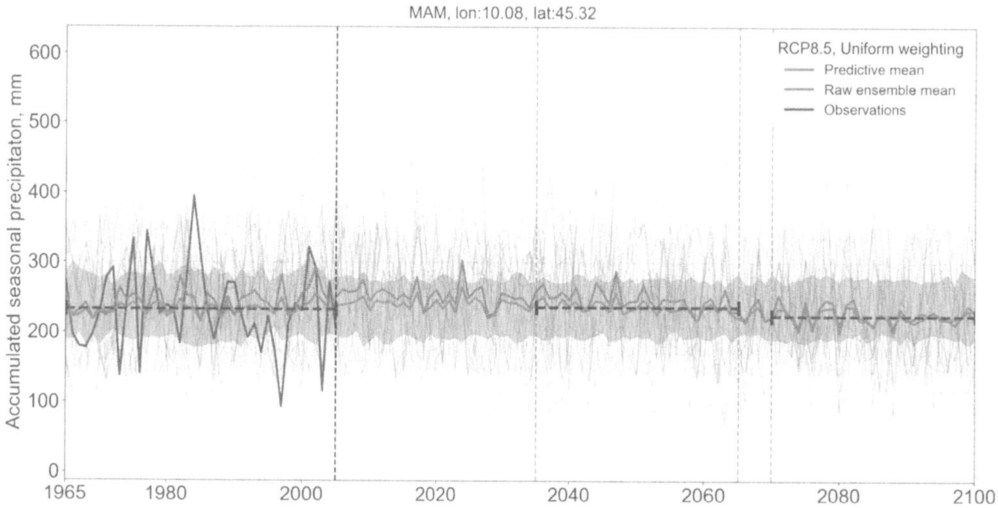

1. INTRODUCTION

Currently, we observe climatic changes that are linked to global warming caused by greenhouse gas emissions and are predicted to cause temperature and precipitation changes across many regions of the globe (IPCC 2021, 2022). While the land surface of continental masses may become warmer over the 21st century, precipitation patterns may also change, with some regions becoming, on average, drier and others wetter. Especially the frequency of extreme hydrometeorological events will change, with intense precipitation events becoming more numerous and drought increasingly extensive. Related natural hazards and risks may most likely increase.

The assessment of climate change-induced risks (Aven & Renn 2015) and the planning of mitigating actions (Strzepek *et al.* 2011) rely on multiple numerical integrations of Earth system models, which simulate, as realistically as possible, the interaction of atmosphere, land, ocean, and sea-ice processes. The physics in these models is described by a system of highly nonlinear governing equations, simulated on a three-dimensional grid with finite spatial resolution (IPCC 2013). Due to the chaotic nature of the Earth system (Lorenz 1963), the integration of the physical equations is affected by initial and boundary conditions, including pre-defined carbon emission scenarios (Palmer 2000). The solutions are, therefore, sensitive to model uncertainties, owing to the fact that real processes are simulated approximately and that the equations are resolved on a finite grid. As the forecast lead time increases, model uncertainties become dominant. Several multi-model ensemble simulations of future climate have been performed as part of the Climate Model Inter-comparison Project (CMIP), which to date include multiple executive phases (Meehl *et al.* 2000, 2007; Eyring *et al.* 2012; Taylor *et al.* 2012). Specialized studies on the Mediterranean region that were executed in the frame of the Coordinated Climate Experiment Project (CORDEX), a joint effort on downscaling CMIP5 (and more recently also CMIP6) projections to regional scales, point towards intensification of hydrometeorological extremes and considerable challenges ahead for the region (Pieri *et al.* 2017; Tramblay & Somot 2018).

The Po river valley is the largest single river basin in Italy and hosts the country's largest freshwater lakes. The densely populated basin area (225 inhabitants/km^2) is also the largest contiguous and prolific agricultural area in Italy. It also hosts important service industries and manufacturing sites, including hydropower plants as well as thermal power stations sited along the river for cooling. Altogether, these activities are characterized by intense water use, whereby 87% of surface water use and 17% of groundwater use are attributable to irrigation alone. Economic activities in the Po basin contribute to 34% of the added value created in Italy (Musolino *et al.* 2018). The hydrology of the basin is dominated by liquid precipitation, which occurs mainly in autumn and spring, and snowpack accumulation in the high Alpine and Apennine ranges, which is ablated during spring, adding meltwater to river flows. Some four large natural lake systems at the foothills of the Alps are fed by runoff from alpine catchments and constitute large natural reservoirs, which have been artificially regulated during the first half of the 20th century. In conjunction with a series of much smaller artificial multi-purpose reservoirs in the Alps and the Apennines, they constitute important storage capacity for the regulation of surface water flows for the Po river and its tributaries, thus ensuring sufficient irrigation water supply during the dry season.

During the summer of 2022, the Po river experienced the largest drought of the last 70 years (Levantesi 2022), caused mainly by a lack of snow accumulation in the alpine part of the basin during the winter of 2021/2022 and the absence of liquid precipitation during several consecutive winter and spring months. At the beginning of August, the flow rate at the basin closing station Pontelagoscuro reached a low point of 100 m^3/s, and far below the long-term minimum of 400 m^3/s usually reached towards the end of the irrigation season. Simultaneously, warm air flows from the African continent across the Mediterranean-enhanced evaporation, putting ecosystems under particular water stress. If such types of extreme meteorological conditions become regular, it will cause severe water scarcity events that affect the entire socioeconomic and productive system in the basin. Besides the lack of water for consumption and irrigation, there are collateral effects like electric energy shortages due to reduced river flows for hydropower generation and cooling of thermal plants and saltwater intrusions from the Adriatic coast into the Po delta area, causing the salinization of fertile land. Mitigating actions would need to be devised, which require a credible prognosis of future precipitation and surface runoff, groundwater recharge, and temporal pattern of snowfall. To this end, we use climate projections from the Climate Model Inter-comparison Project Five (CMIP5), which need to be post-processed for bias correction and uncertainty quantification. While most studies agree on the future increase in temperature, results on local precipitation trends are conflicting, with some studies suggesting a considerable decrease in precipitation (Padulano *et al.* 2020), while others indicate no significant past and/or future trend (Ciccarelli *et al.* 2008; Braca *et al.* 2019).

Climate model projections are supposed to produce the general statistics of underlying climate variables (Knutti *et al.* 2010) but are likely to be biased in terms of mean and variance for the reasons mentioned previously. Hence, systematic post-processing of climate ensemble output is needed. Different techniques have been progressively proposed (Palmer & Räisänen 2002; Giorgi & Mearns 2003; Furrer *et al.* 2007; Rougier 2007; Tebaldi & Sansò 2009; Reggiani *et al.* 2021; Koutsoyiannis & Montanari 2022) and applied among others to variables of particular hydrological relevance, such as temperature and precipitation. However, most of these approaches assume that the spread of the ensemble projection is equal to the uncertainty of future climate.

The main objective of this work is to estimate the predictive density for precipitation in the Po valley in Italy. We perform a separate study on the precipitation of the river Po, in which we employ a large CMIP5 model output ensemble, which we elaborate through a Bayesian processor to correct on hand of past observations for bias of mean and variance and obtain the predictive distributions of future precipitation for the temporal windows 2035–2065 and 2070–2100. The predictive distribution is a conditional density that expresses the probability of future occurrences, given climate model projections that serve as pseudo-observations. The knowledge of the full predictive distribution of precipitation is an essential constituent for any decision-making process in undertaking mitigating actions (Reggiani *et al.* 2022). In this context, one must also address the issue of non-stationarity of the climate process. Non-stationarity can be addressed differently, either by assuming underlying linear or nonlinear trends (Buser *et al.* 2009), or, as done here, by assuming climate as a weakly stationary process, for which stationarity holds firm over the time windows of the preset length.

The proposed Bayesian processor was originally presented by Reggiani *et al.* (2021) and applied to temperature projections for the Po river basin. Here, we change the focus of that work by post-processing seasonal projections of precipitation. We first calibrate the processor by making use of comprehensive gridded datasets of precipitation, which have been generated by interpolating more than 50 years of rain gauge data from a dense observing network by the aggregation to seasonal totals. The calibrated processor yields bias-corrected estimates of seasonal precipitation and variance that are propagated by the climate models to the end of the 21st century. In this study, we choose two future climate scenarios, referred to as Representative Concentration Pathways (RCPs) by the IPCC (IPCC 2013), namely RCP4.5 as the more optimistic warming scenario and RCP8.5 as the pessimistic one.

The manuscript is structured into six sections. Section 2 introduces definitions and explains the methodology; Section 3 describes data and application; Section 4 presents the results; Section 5 discusses the water resource management implications in the Po river basin and Section 6 reports the conclusions. Supplementary Material is provided in Appendix A.

2. DEFINITIONS AND METHODOLOGY

2.1. Predictive uncertainty

The predictive density is defined as the probability distribution of a quantity, conditional on all presently available information (Hamill & Whitaker 2006):

$$f(\mathbf{x}|\hat{\mathbf{x}}) \tag{1}$$

where $\mathbf{x}$ is the vector of climatic observations (past or future) and $\hat{\mathbf{x}}$ is a vector of predictions of $\mathbf{x}$. These become the known quantities, which enhance our knowledge by conditioning the probability distribution of the observations $\mathbf{x}$. Future, not-yet observed, occurrences are indicated by $\mathbf{x}^*$, and their respective predictions by $\hat{\mathbf{x}}^*$.

Climate projections are error-affected estimates of the future Earth system states serving as decisional knowledge support in reducing uncertainty through bias-removal and sharpening (Gneiting *et al.* 2007) of the probability density of $\mathbf{x}^*$ conditional on $\hat{\mathbf{x}}^*$. This conditional density is essential for decision-making in any downstream impact analysis.

2.2. Stochastic properties of climate projections

The principal difference between handling climate projections and classical short- and medium-range forecasts is that presently available projections may be capable of preserving the probability distributions of the simulated Earth system variables, but not their observed auto- and cross-correlation structure (Giorgi & Francisco 2000; Palmer 2000; Vidale *et al.* 2003; Déqué *et al.* 2007), a characteristic named asynchronicity (Stoner *et al.* 2013). For example, projections of precipitation are known to exhibit stronger autocorrelation than observed rainfall (Ines & Hansen 2006).

2.3. Conditioning of variates

Instead of conventional conditioning of a predictor on a predictand, we prefer 'weakly' conditioning the two variables similar to distribution matching (DM) (Wood *et al.* 2002; Drusch *et al.* 2005; Boé *et al.* 2007). This approach is used regularly for bias-correcting or downscaling Earth system model output or remotely sensed data against observations. DM (which we denote with (:) instead of conventional conditioning indicated by |) is a form of conditioning, in which the mean does not depend on correlation, and hence, we call it *weak conditioning*. DM differs from quantile matching (QM), which uses empirical cumulative distribution functions (ECDFs) instead of fitted distribution models and can be advantageous for certain climatic variables (Themeßl *et al.* 2011). Both methods are also used for downscaling climate model output, a topic that is not of relevance here, where we deal with upscaling instead. When weakly conditioning two arbitrarily distributed variables, we apply probability matching between the ECDFs or by fitting appropriate parametric distributions F and G and by inverting one of the two (Hirsch 1982). In this way, one distribution morphs into the other. By setting the cumulative distribution function (CDF) of the observations equal to that of a predictor:

$$G(\mathbf{x}) = F(\hat{\mathbf{x}}) \tag{2}$$

and then inverting G:

$$\hat{\hat{\mathbf{x}}} = G^{-1}(F(\hat{\mathbf{x}})) = T(\hat{\mathbf{x}}). \tag{3}$$

We obtain a transformation T. This is equivalent to introducing a new variable with the same moments of the variate it is calibrated against. If the original variate is Gaussian, denoted with the Greek letter $\boldsymbol{\eta}$, T becomes:

$$\hat{\hat{\boldsymbol{\eta}}} = T(\hat{\boldsymbol{\eta}}) = \mu_\eta + \sigma_\eta \frac{(\hat{\boldsymbol{\eta}} - \mu_{\hat{\eta}})}{\sigma_{\hat{\eta}}} \tag{4}$$

with distribution:

$$f(\hat{\hat{\boldsymbol{\eta}}}) \sim N(\mu_\eta, \sigma_\eta^2) \tag{5}$$

The predictive density of the conditioned variable for the model j at time i is, hence, Gaussian:

$$f_{i,j}(\eta_i{:}\hat{\eta}_{i,j}) \sim N(\mu_{\eta_i:\hat{\eta}_{i,j}}, \sigma^2_{\eta_i:\hat{\eta}_{i,j}}) \tag{6}$$

with parameters:

$$\begin{cases} \mu_{\eta_i:\hat{\eta}_{i,j}} = E(\eta_i{:}\hat{\eta}_{i,j}) = \mu_{\boldsymbol{\eta}} + \sigma_{\boldsymbol{\eta}} \dfrac{(\hat{\eta}_{i,j} - \mu_{\hat{\boldsymbol{\eta}}_j})}{\sigma_{\hat{\boldsymbol{\eta}}_j}} \\ \sigma^2_{\eta_i:\hat{\eta}_{i,j}} = \mathrm{Var}(\eta_i{:}\hat{\eta}_{i,j}) = 2\sigma^2_{\boldsymbol{\eta}}(1 - \rho_{\boldsymbol{\eta}\hat{\boldsymbol{\eta}}_j}) \end{cases} \tag{7}$$

where $E(\cdot{:}\cdot)$ and $\mathrm{Var}(\cdot{:}\cdot)$ are taken over all possible realizations of η_i at time i. The predictive distribution of the projected variable conditional on model output is obtained in total analogy and it yields a series of probability distributions of the observations $f_j(\boldsymbol{\eta}^*{:}\hat{\boldsymbol{\eta}}^*_j)$ that are weakly conditional (:) on each projection $\hat{\boldsymbol{\eta}}^*_j$. For this distribution, the mean $\mu_{\boldsymbol{\eta}}$ and variance $\sigma^2_{\boldsymbol{\eta}}$ derived from observations in Equation (7) are used because the bias with respect to observations and the variance of observations are assumed to hold also well into the future (Collins *et al.* 2006).

Weak conditioning through matching of Gaussian distributions has the advantage that one retains a parametric expression of type (6) over the entire distribution, including the tails, which allows one to better represent the predictive probability distribution for decision-making on adaptation measures (Reggiani *et al.* 2022).

2.4. Retrospective prediction and projections

As explained in Section 2.3., we use weak conditioning for retrospective predictions and projections. A prediction/projection is performed by estimating the density (1) at a time i. For a climate projection ensemble with multiple predictors, one can adopt the concept of mixture distribution, a weighted mean of univariate conditional distributions. For an arbitrarily (i.e. not necessarily Gaussian) distributed variate $\mathbf{x}$, the mixture distribution is $f(\mathbf{x}{:}\hat{\mathbf{x}}_j)\forall j \in 1, \ldots, m$:

$$f(x_i{:}\hat{\mathbf{x}}_i) \sim E_{(j)}[f_j(x_i{:}\hat{x}_{i,j})] = \sum_{j=1}^{m} w_j f_j(x_i{:}\hat{x}_{i,j}) \tag{8}$$

The similarity sign indicates that the predictive distribution is 'approximated' by the mixture, while the hatted bold symbol $\hat{\mathbf{x}}_i$ indicates the m-dimensional vector $[\hat{x}_{i,1}, \ldots, \hat{x}_{i,m}]$ of predictions j at time i. Expressions for the mean and variance of the mixture distribution are given in Supplementary Material, Appendix A. The weights w_j are the probability of the jth ensemble member being a valid prediction. One can adopt two approaches in assessing the weights: (1) Bayesian Model Averaging (BMA) and (2) uniform weighting (UW), as shown in Reggiani *et al.* (2021).

BMA (Raftery *et al.* 2005) assumes non-uniform weights w_j in (8) and maximizes the log-likelihood of observations:

$$\begin{cases} \max\limits_{w_j} \log\mathcal{L}(w_1, \ldots, w_m) = \max\limits_{w_j} \sum\limits_{n}^{i=1} \log \sum\limits_{m}^{j=1} w_j f_j(x_i{:}\hat{x}_{i,j}) \\ \text{sub j. to} \sum\limits_{j=1}^{m} w_j = 1; \; > w_j \geq 0 \end{cases} \tag{9}$$

Maximum likelihood estimation (MLE) of weights is a special case of maximum a posteriori (MAP) estimation that assumes a uniform prior distribution of the weights. An important assumption underlying BMA is the independence of variates. Some degree of dependency is inevitably present (Knutti *et al.* 2017) because the climate models of the CMIP5 ensemble use similar initial conditions. Nevertheless, because of the generally weak dependency between variates, we continue to assume full independence, as in classical BMA. The predictive mean $\mu_{x_i:\hat{\mathbf{x}}_i}$ and variance $\sigma^2_{x_i:\hat{\mathbf{x}}_i}$ of the mixture distribution for the case of independent variates become (Supplementary Material, Appendix A):

$$\begin{cases} \mu_{x_i:\hat{\mathbf{x}}_i} & = \sum_{j=1}^{m} w_j \mu_{x_i:\hat{x}_{i,j}} = \mathbf{w}^T \boldsymbol{\mu}_{x_i:\hat{\mathbf{x}}_i} \\ \sigma^2_{x_i:\hat{\mathbf{x}}_i} & = \sum_{j=1}^{m} w_j \sigma^2_{x_i:\hat{x}_{i,j}} + \sum_{j=1}^{m} w_j (\mu_{x_i:\hat{x}_{i,j}} - \mu_{x_i:\hat{\mathbf{x}}_i})^2 \end{cases} \tag{10}$$

For the present analysis, we simply assume exclusively UW when mixing the m univariate predictive densities $f_j(\mathbf{x}:\hat{\mathbf{x}}_j)$, therefore giving no prior preference to any member of the climate model ensemble. The final expressions for the mixture can be derived from (8) by setting $w_j = 1/m$, with mean and variance given by (10). The described approaches, which require a transformation of variables into the Gaussian space, overcome the classical problem of hydrological statistical projections, which tend to be linear or nonlinear (Beven 2021), but hardly chaotic as required in a climate-changing environment. In the proposed approach, we use an ensemble of Earth system models to describe the chaotic projections into the future and the associated model projection uncertainty but subsequently use these projections as uncertain measures of future outcomes to finally assess the probability distribution of the actual future, albeit unknown, quantities.

3. DATA AND APPLICATION

Post-processing of precipitation projections is applied for the Po river basin as shown in Figure 1. A range of data pre-processing steps are necessary to prepare the datasets prior to climate output post-processing. The primary effort consists of projecting the CMIP5 output ensemble and observations onto a common regular spatial grid because the various CMIP5 models natively use different spatial resolutions and different grid projections.

3.1. Observed precipitation

Our selected study region covers a rectangular area including the entire Po river basin in Northern Italy, as shown in Figure 1. The area extent is from (46.75°N, 6.5°E) in the N–W corner to (44.0°N, 12.5°E) in the S–E. The ground observations of precipitation, which are needed to calibrate the processor, consist of the Central-Northern Italy daily precipitation dataset ArCIS (Climatological Archive for Central-Northern Italy), a comprehensive gridded precipitation dataset ($\sim 5 \times 5$ km regular grid) covering the Italian territory above latitude 42°20′. The product has been derived from 1,762 rain gauges that belong to

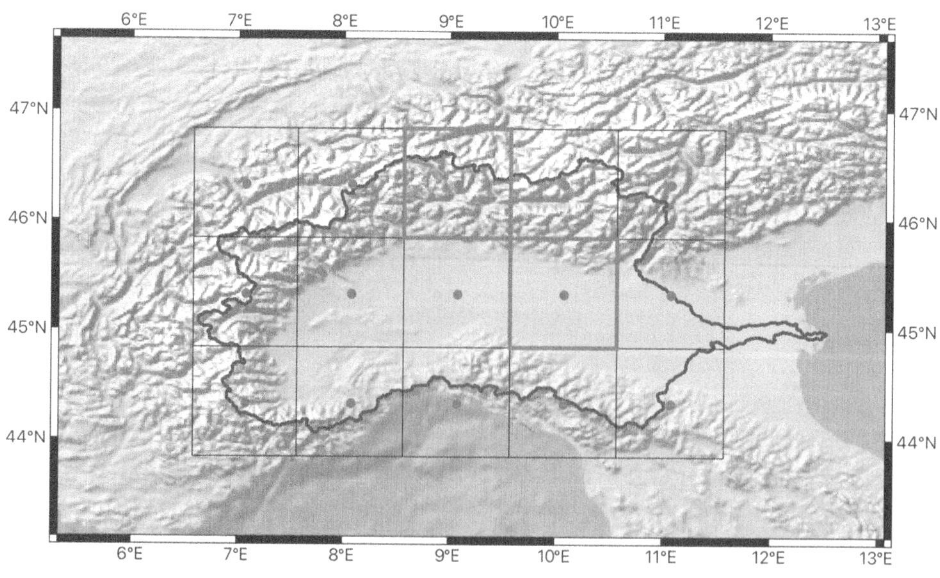

Figure 1 | The study area in the Po river basin in Italy and the common 1.0° × 1.0° climate model grid for projections and observations. The cell numbering increases row-wise starting in the upper left corner, with analysis cells 3 and 9 framed in red and dots indicating cell centres. Please refer to the online version of this paper to see this figure in colour: http://dx.doi.org/10.2166/nh.2022.063.

different networks of 11 Italian administrative regions plus several stations in neighbouring countries. The temporal coverage extends over the 1965–2015 period at daily resolution (Pavan *et al.* 2019; Grazzini *et al.* 2020; ISPRA 2021). To make ArCIS congruent with the spatial resolution of CMIP5 climate output data, the original gridded data need to be upscaled by spatial averaging towards a $1^\circ \times 1^\circ$ spatial grid as shown in Figure 1. The averaging is performed using a 'first-order conservative second-order conservative' method (Jones 1999) included in the Climate Data Operator (CDO) software library (Schulzweida 2021). Finally, we aggregate the daily data series into seasonal totals. In summary, we obtain 15 series of seasonal precipitation totals that are representative of the cells in Figure 1 covering the basin area. The cell centres are indicated by blue dots.

3.2. Projected precipitation

As a next step, we process the outputs of the CMIP5 climate model ensemble. CMIP5 control simulations start on 01 January 1901 and end on 31 December 2005 and are made available as monthly means. The prognostic period reaches from 01 January 2006 through 31 December 2100. We select a $n_b = 41$ years' baseline window (BW), 1965–2005; two $n_p = 31$ years' prognostic windows, 2045–2065 (PW1) and 2070–2100 (PW2), and two RCP scenarios, RCP4.5 and RCP8.5, four cases in total. The monthly precipitation series from the 19 models listed in Table 1 are extracted from the global CMIP5 simulated precipitation output for the study area and aggregated to seasonal totals. The selected CMIP5 models' outputs have different spatial resolutions and need to be aligned by projection onto the selected common $1^\circ \times 1^\circ$ regular grid, in line with what has been done for observations, using once more the 'first-order conservative second-order conservative' mass-preserving interpolation technique (Jones 1999). At the end of processing, we obtain a series of seasonal data that are equally resolved in space and overlap temporally with the observation records for the 1965–2015 period.

3.3. Normalization of distributions

Furthermore, we apply the processor by first comparing the distributions of the upscaled seasonal total observed precipitation for each of the 15 selected cells against those of the simulated seasonal climate for each member of the CMIP5 model ensemble. The predictive distribution (6) is formulated for Gaussian data. However, the distributions of observed and projected

Table 1 | List of 19 global climate models from the CMIP5 experiment used as a climate projection ensemble over two prognostic windows in the 21st century

Model	Source organization
ACCESS1-0	Bureau of Meteorology – CSIRO AU
BCC CSM1 1	Beijing Climate Center, CN
BCC-CSM1-1m	moderate res. 318
BNU-ESM	Bejing National University, CN
CanESM2	Canadian Centre Clim. Modelling and Analysis, CND
CCSM4	Comm. Clim. Sys. Model V4, NCAR-UCAR, USA
CESM1-CAM5	Comm. Earth Sys. ModelV1, Atmos. Sys. ModelV5, USA
CNRM-CM5	Centre National Recherche Météorologique, FR
CSIRO-Mk3-6-0	CSIRO AU
GFDL-ESM2M	Geophysical Fluid Dynamics Lab, USA
GFDL-ESM2G	
INMCM4	Russian Academy of Sciences, RU
IPSL-CM5A-MR	Institut Pierre Simon Laplace, FR
MIROC5	Model for Interdisciplinary Research on Climate, JP
MIROC-ESM	
MIROC-ESM-CHEM	
MPI-ESM-MR	Max Planck Institute, DE
MRI-CGCM3	Meteorological Research Institute, JP
NorESM1-M	Norwegian Climate Center, NO

seasonal precipitation are mostly non-Gaussian, left-limited and need to be fitted with appropriate parametric models H. These are selected for each spatial cell and each temporal season among the log-normal, normal, gamma and Weibull distributions according to statistical hypothesis test outcomes. Figure 2 shows examples of parametric distributions H that were fitted to the 41 data points for selected cells and seasons.

As a next step, the non-Gaussian variates $\mathbf{x}$, whose distributions are modelled as parametric distributions H, need to be mapped into standard Gaussian variates $\boldsymbol{\eta}$ by normal quantile transform (NQT) through DM as in Equations (2) and (3):

$$\boldsymbol{\eta} = Q^{-1}(H(\mathbf{x})) = T(\mathbf{x}) \tag{11}$$

where Q^{-1} denotes the inverse standard normal distribution. Once observations and predictors have been transformed to standard normal variates, they can be conditioned as per Equations (4)–(7).

3.3.1. Post-processing

The **actual** post-processing phase starts by weakly conditioning the standard Gaussian series of the CMIP5 precipitation projection ensemble against those of standard normalized observations. To this end, we evaluate a univariate predictive distribution of type (6) for each of the j models, which is parameterized for each season through separate sets of values for the conditional mean and variance, as per (7). The stochastic relationship between the model and observations, therefore, depends on the mean and variance of observations and the Pearson correlation between standard normal observations and model projection.

Second, the univariate conditional distributions of daily data are weighted linearly as in Equation (8). The predictive Gaussian mean $\mu_{\eta_i:\hat{\boldsymbol{\eta}}_i}$ and variance $\sigma^2_{\eta_i:\hat{\boldsymbol{\eta}}_i}$ for the BW are estimated via Equation (10). Finally, the fully Gaussian predictive distribution needs to be mapped back into the space of origin by inverse NQT:

$$x_i{:}\hat{\mathbf{x}}_i = Q^{-1}(F(\eta_i{:}\hat{\boldsymbol{\eta}}_i)) \tag{12}$$

where F and Q^{-1} are, respectively, normal and the inverse standard normal distribution. We indicate the arbitrary zero-bound predictive distribution of the variate $x_i{:}\hat{\mathbf{x}}_i$ in the space of origin with $\Phi(x_i{:}\hat{\mathbf{x}}_i)$ and the density with $\phi(x_i{:}\hat{\mathbf{x}}_i)$. The latter

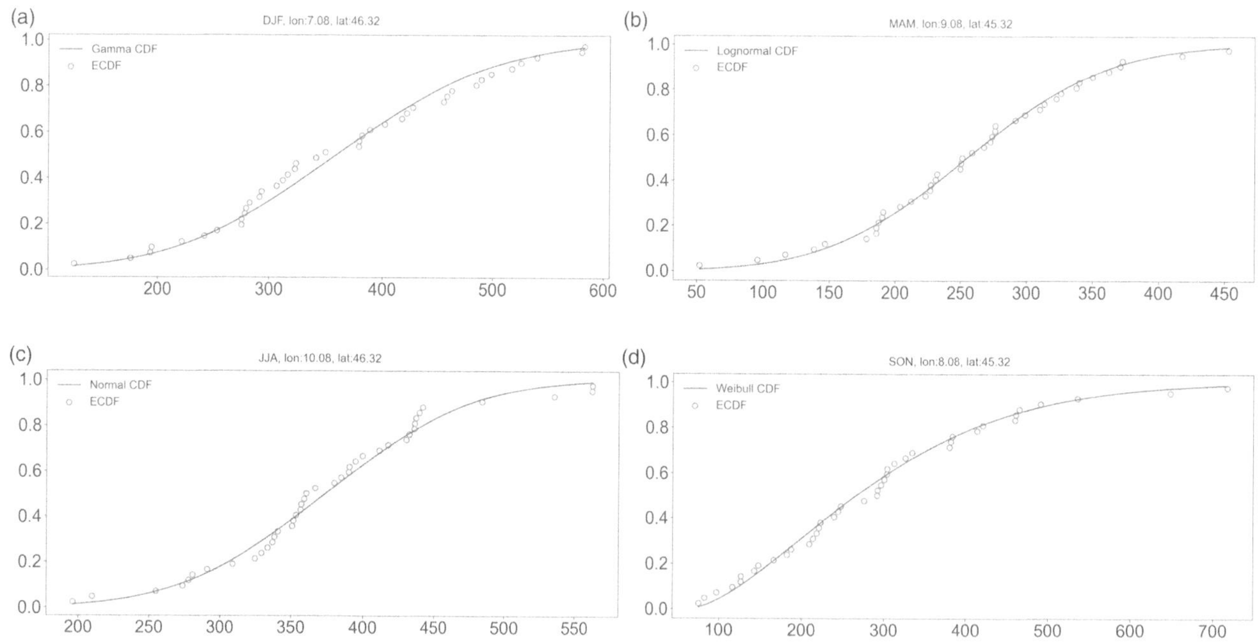

Figure 2 | Parametric distributions of 1965–2005 observed data for selected seasons and cells: (a) gamma; (b) log-normal; (c) normal; and (d) Weibull fits.

corresponds to the predictive density of Equation (1). As is to be expected, it must assign zero probability to below-zero values of precipitation. The distribution Φ has its own, non-necessarily Gaussian, mean $\mu_{x_i:\hat{\mathbf{x}}_i}$ and variance $\sigma^2_{x_i:\hat{\mathbf{x}}_i}$ also given by (10). These PDFs can be traced either as a sample relative frequency histogram or approximated by a pdf estimated through a kernel density estimator, while mean and variance can be easily estimated from respective samples. Figure 3 visualizes the outlined post-processing steps by means of a flow chart.

4. RESULTS

The post-processing results are presented for the four seasons of the year (December, January, February - DJF; March, April, May - MAM; June, July, August - JJA; September, October, November - SON) as the mean distribution for the season over the n years of the BW or the two predictive windows. The window-average distributions have a mean, which is equal to the mean of the predictive mean values $\mu_{x_i:\hat{\mathbf{x}}_i}$, $i = 1, \ldots, n$, making up the window, the so-called mean of means $E[\mu_{x_i:\hat{\mathbf{x}}_i}]$. The window-mean variance is estimated in accordance with the law of total variance (Mahler & Dean 2001):

$$\mathrm{Var}_{(i)}[x_i:\hat{\mathbf{x}}_i] = E_{(i)}\left[\sigma^2_{x_i:\hat{\mathbf{x}}_i}\right] + \mathrm{Var}_{(i)}\left[\mu_{x_i:\hat{\mathbf{x}}_i}\right] \tag{13}$$

where $E_{(i)}$ and $\mathrm{Var}_{(i)}$ are, respectively, the expected value and variance taken over the n years of the window. The total variance (13) equals the mean of the predictive variances for the n individual years of the window plus the dispersion of the means $\mu_{x_i:\hat{\mathbf{x}}_i}$ around the mean of means. Figures 4 and 5 show the post-processed ensemble output for the two red-framed analysis cells 3 and 9 in Figure 1. For cell 3, we selected for cell RCP4.5, winter, and for cell 9 RCP8.5, spring. Both figures show the raw ensemble (pink) and the ensemble mean (purple). The bias-corrected predictive mean is indicated in bright red, whereas observations in the BW are represented in dark blue. The vertical dashed black lines indicate the limitations of the BW (left) and the two predictive windows PW1 (centre) and PW2 (right). The bold black dashed horizontal lines indicate the

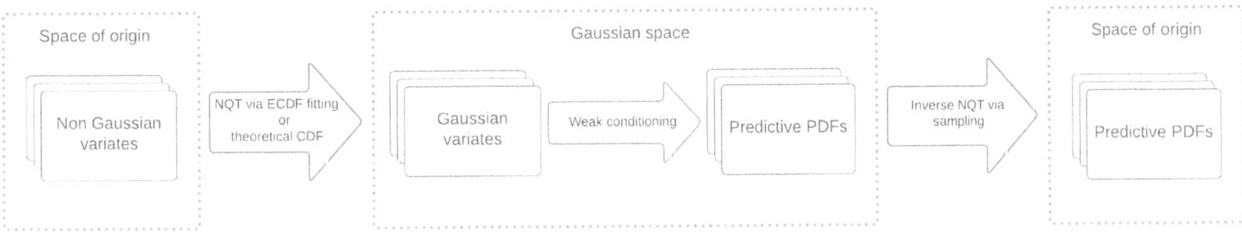

Figure 3 | A scheme of the variable transformation across the post-processing.

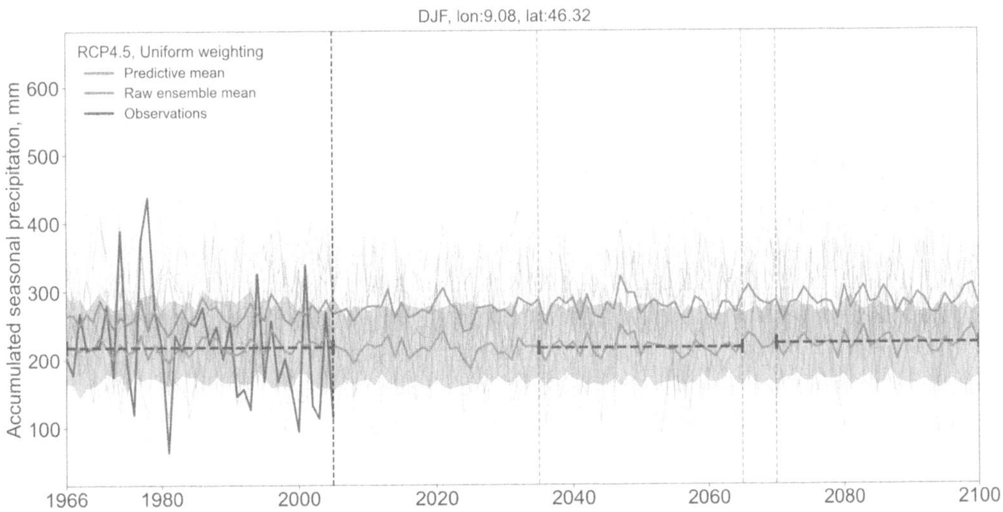

Figure 4 | Cell 3, RCP4.5, winter (DJF) seasonal precipitation totals x_i, unprocessed ensemble output $\hat{x}_{ij}$, raw ensemble mean $E_{(j)}(\hat{x}_{ij})$, predictive mean $\mu_{x_i:\hat{x}_i}$, UW. The shadowed areas indicate the 50–95% credible intervals.

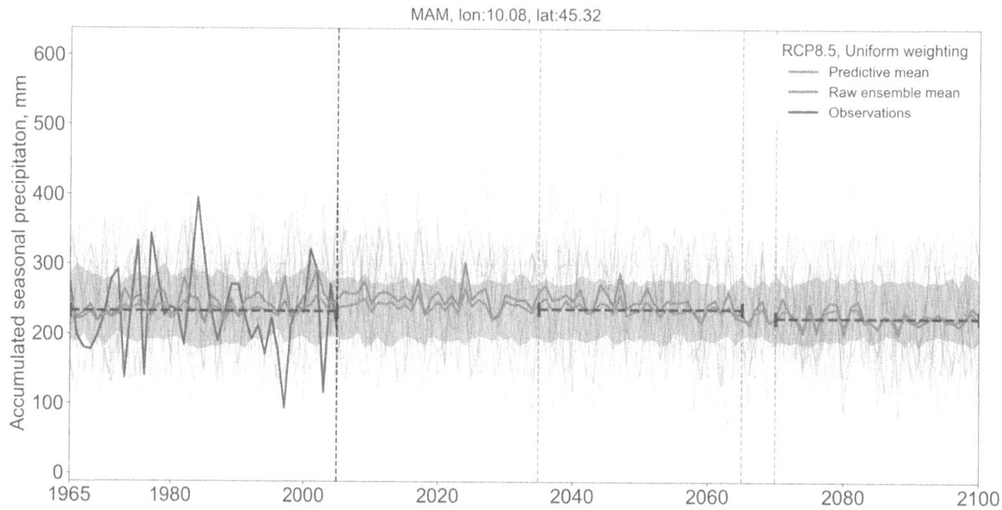

Figure 5 | Cell 9, RCP8.5, spring (MAM) seasonal precipitation totals x_i, unprocessed ensemble output $\hat{x}_{ij}$, raw ensemble mean $E_{(j)}(\hat{x}_{ij})$, predictive mean $\mu_{x_i;\hat{x}_i}$, UW. The shadowed areas indicate the 50–95% confidence intervals.

mean of means for the three windows. The grey shadowed areas show the predictive uncertainty through the 50 and 95% confidence intervals. One notices that post-processing removes the bias of the model ensemble with respect to observations and adjusts the predictive variance.

In the interest of decision-makers, we summarize the results concisely in terms of increments or decrements of total seasonal precipitation, which one can expect across the 21st century. To this end, we calculate the differences between the BW and, respectively, PW1 and PW2. Table 2 reports the percentual and absolute differences of the mean of means between BW and PW1 and between BW and PW2, respectively, for all seasons. Absolute values, expressed in mm, are given between parentheses.

Due to the close geographical proximity of the chosen set of cells, the changes between cells are small. For the more optimistic RCP4.5 emission scenario, one recognizes a mostly neutral pattern in terms of precipitation change until the middle of the century and a slightly noticeable increase during winter and autumn, typically the wettest season, for the latter part of the century. For the more pessimistic RCP8.5 emission scenario, precipitation slightly increases with respect to the reference period until the middle of century across all seasons, while for the latter part of the century, it decreases across all seasons, with the exception of winter. From these results, we conclude that following CMIP5 climate simulations, seasonal total precipitation for both emission scenarios remains essentially stable over the Po valley. This result is congruent with the analysis of Braca *et al.* (2019), who also find no significant changes in total annual precipitation from their analysis of the NCAR Community Climate System Model (CCSM4) output (for reference, see Table 1) over Northern Italy as part of their climate analysis for the entire Italian peninsula and Campania region as a local target.

The near-stationarity of precipitation also justifies our choice of a 30-year analysis window for the application of the weakly stationary process assumption. Contrarily, the temperature for the study area changes more rapidly across the 21st century, a reason why predictive windows with a length of 20 years have to be chosen (Reggiani *et al.* 2021). Nevertheless, the climatic analysis of precipitation in a water resource management context needs to be analysed from a broader perspective, as addressed subsequently.

5. DISCUSSION AND CONCLUSIONS

Despite projected seasonal precipitation totals not seeming to change significantly with respect to the 'current' 1965–2005 period, which we have selected as a control period, there are still implications for water resource management in the Po valley due to projected temperature changes across the 21st century.

Reggiani *et al.* (2021) analysed CMIP5 2-m temperature projections for the Po river basin. These were conditioned and corrected on the 1979–2005 ECMWF ERA5 temperature reanalysis (Hersbach *et al.* 2020) acting as observation proxy and then aggregated over the entire Po river basin area. Temperatures are altitude-corrected from the ERA5 grid cell elevation to a common reference height using the dry adiabatic lapse rate. In this precursory article to the present one, we performed a weighted combination of ensemble projections using UW as well as BMA weighting, with weights changing per month.

Table 2 | Percentual and absolute (mm) (in parenthesis) changes of precipitation between the mean of the baseline (BW) period and the mean of the two predictive windows PW1 and PW2, 4 seasons and 15 cells in Figure 1

		DJF	MAM	JJA	SON	DJF	MAM	JJA	SON
Cell	Coordinates	PW1% (mm)				PW2% (mm)			
RCP4.5									
1	07.08°, 46.32°	−0.56(−2.0)	0.37(1.3)	0.65(1.88)	−0.85(−2.9)	1.12(4.05)	0.17(0.62)	−1.5(−4.31)	1.1(3.75)
2	08.08°, 46.32°	−0.25(−0.56)	0.4(1.69)	0.61(2.2)	−0.71(−3.3)	1.26(2.8)	0.5(2.08)	−1.45(−5.29)	1.28(6.0)
3	09.08°, 46.32°	−0.71(−1.54)	0.03(0.13)	0.98(4.65)	−1.1(−5.37)	1.5(3.26)	0.52(2.193)	−1.28(−6.08)	0.8(4.22)
4	10.08°, 46.32°	−1.1(−1.81)	−0.13(−0.39)	0.54(2.02)	−0.8(−2.86)	2.6(4.36)	0.6(1.71)	−0.6(−2.34)	0.93(3.3)
5	11.08°, 46.32°	−0.8(−1.07)	−0.31(−0.72)	0.53(1.6)	0.54(1.53)	3.07(4.14)	0.67(1.6)	−0.73(−2.19)	1.0(2.9)
6	07.08°, 45.32°	−0.52(−0.94)	−0.29(−0.85)	−0.0(−0.0)	−0.78(−2.18)	1.51(2.74)	−0.1(−0.25)	−0.76(−1.8)	0.3 (0.93)
7	08.08°, 45.32°	−0.7(−1.15)	0.23(0.7)	0.35(0.86)	−0.66(−1.93)	1.5(2.26)	0.3(0.93)	−0.96(−2.4)	0.63(1.84)
8	09.08°, 45.32°	−0.76(−1.37)	0.5(1.27)	0.65(1.53)	−0.13(−0.39)	1.44(2.58)	−0.23(−0.58)	−1.12(−2.65)	0.3(1.09)
9	10.08°, 45.32°	−0.89(−1.5)	0.0(0.0)	−0.04(−0.1)	0.42(1.2)	2.11(3.7)	0.2(0.46)	−0.6(−1.6)	0.4(1.13)
10	11.08°, 45.32°	−0.89(−1.53)	−0.1(−0.24)	0.42(1.04)	0.56(1.49)	2.47(4.26)	−0.19(−0.4)	−1.0(−2.53)	0.3(0.82)
11	07.08°, 44.32°	0.4(0.82)	−0.1(−0.25)	−0.3(−0.55)	−1.41(−3.9)	0.43(0.77)	−0.4(−1.21)	−0.95(−1.93)	0.9(2.57)
12	08.08°, 44.32°	0.02(0.04)	−0.17(−0.42)	0.49(0.71)	−1.14(−3.25)	0.47(0.88)	−0.92(−2.28)	−1.7(−2.5)	1.26(3.58)
13	09.08°, 44.32°	−0.3(−0.97)	0.1(0.3)	−0.53(−1.1)	−0.51(−2.44)	0.21(0.6)	−0.65(−2.0)	−0.7(−1.4)	0.3(1.65)
14	10.08°, 44.32°	−0.62(−2.14)	0.31(0.97)	−0.9(−1.96)	−0.9(−4.47)	0.39(1.34)	−0.8(−2.45)	−0.4(−0.82)	0.8(3.47)
15	11.08°, 44.32°	−0.99(−2.45)	−0.04(−0.11)	−0.5(−0.9)	−0.77(−2.5)	1.07(2.64)	−0.53(−1.3)	−0.79(−1.44)	0.35(1.15)
RCP8.5									
1	07.08°, 46.32°	0.71(2.6)	0.91(3.26)	1.1(3.2)	0.94(3.2)	1.9(7.2)	−3.4(−12.0)	−5.7(−16.7)	−2.88(−9.8)
2	08.08°, 46.32°	0.67(1.5)	1.82(7.73)	0.9(3.4)	1.9(8.8)	3.3(7.3)	−4.7(−20.3)	−6.4(−23.41)	−4.6(−21.7)
3	09.08°, 46.32°	1.2(2.6)	1.7(7.15)	1.1(5.1)	1.37(6.75)	2.3(5.0)	−4.3(−18.15)	−6.4 (−30.3)	−2.8(−13.8)
4	10.08°, 46.32°	1.6(2.6)	1.1(3.27)	0.97(3.6)	1.7(6.02)	2.0(3.3)	−4.1(−12.06)	−5.0(−18.6)	−2.86(−10.15)
5	11.08°, 46.32°	1.2(1.6)	1.17(2.8)	0.5(1.42)	1.6(4.5)	2.1(2.81)	−3.7(−8.65)	−3.8(−11.5)	−2.7(−7.51)
6	07.08°, 45.32°	1.6(2.9)	1.06(3.17)	1.25(2.95)	0.7(1.99)	1.72(3.12)	−4.7(−14.05)	−6.62(−15.5)	−4.5(−12.5)
7	08.08°, 45.32°	3.0(4.7)	1.72(5.27)	0.63(1.56)	0.9(2.63)	0.9(1.45)	−5.3(−16.3)	−6.52(−16.15)	−5.2(−15.2)
8	09.08°, 45.32°	2.3(4.14)	1.8(4.7)	1.37(3.2)	0.6(1.9)	0.8(1.4)	−5.1(−13.0)	−6.99(−16.53)	−3.3(−9.9)
9	10.08°, 45.32°	2.1(3.6)	1.18(2.75)	0.45(1.1)	1.4(4.11)	1.28(2.2)	−3.47(−8.1)	−6.4(−15.7)	−3.23(−9.2)
10	11.08°, 45.32°	2.9(5.0)	1.46(3.32)	0.45(1.1)	0.97(2.6)	1.16(2.0)	−3.83(−8.7)	−6.3(−15.5)	−2.9(−7.7)
11	07.08°, 44.32°	3.3(5.9)	2.12(6.16)	0.96(1.95)	1.0(2.9)	−2.1(−3.73)	−5.9(−17.14)	−9.9(−20.2)	−5.9(−16.4)
12	08.08°, 44.32°	3.7(6.95)	2.1(5.1)	0.3(0.4)	1.1(3.2)	−2.15(−3.99)	−6.75(−16.8)	−9.1(−13.27)	−6.6(−18.7)1
13	09.08°, 44.32°	2.1(6.7)	2.0(6.3)	0.9(1.84)	2.2(10.6)	−1.58(−5.05)	−5.1(−15.85)	−8.4(−17.4)	−4.9(−23.5)
14	10.08°, 44.32°	1.63(5.65)	1.41(4.44)	0.96(1.98)	2.0(9.1)	−1.44(−4.99)	−4.61(−14.5)	−7.97(−16.5)	−4.21(18.9)
15	11.08°, 44.32°	1.43(3.54)	1.05(2.6)	0.98(1.8)	1.4(4.5)	−1.18(−2.91)	−3.8(−9.4)	−7.87(−14.4)	−3.04(−9.87)

Table 3 | Percentual and absolute (°C) (in parenthesis) temperature changes between the mean of the baseline (BW) period and the mean of the 2 predictive windows PW1 and PW2, 4 seasons, Bayesian model averaging and uniform weighting

		RCP4.5		RCP8.5	
Season	**BW** *T* **(°C)**	**PW1** % (Δ*T* (°C))	**PW2** % (Δ*T* (°C))	**PW1** % (Δ*T* (°C))	**PW2** % (Δ*T* (°C))
Bayesian model averaging					
DJF	0.24	458(1.1)	704 (1.7)	608 (1.46)	1,375(3.3)
MAM	8.79	18(1.6)	16.4(2.21)	16.3(2.1)	34(4.4)
JJA	18.73	13.3(2.50)	16.4(3.1)	16.4(3.1)	34.1(6.4)
SON	10.01	10.5(1.05)	24.3(2.4)	22.6(2.2)	46.2(4.6)
Uniform weighting					
DJF	0.24	530(1.2)	745(1.8)	720(1.73)	1,420(3.41)
MAM	8.79	17.3(1.5)	22(1.9)	21.7(1.91)	45(3.9)
JJA	18.73	11.1(2.1)	14.6(2.7)	15(2.80)	28.5(5.3)
SON	10.01	10(1.0)	22.1(2.2)	20.7(2.07)	44(4.39)

Source: Reproduced from Reggiani *et al.* (2021).

Due to the evident non-stationarity of temperature, we had to shrink the predictive window length for PW1 and PW2 to 20 years, to reasonably apply the assumption of a weakly stationary process, instead of using 30-year windows as for precipitation. The analysis in Reggiani *et al.* (2021) clearly shows that spatial average temperature increments between the BW and the 2040–2060 (PW1) and 2080–2100 (PW2) predictive windows are considerable and, as to be expected, higher for the pessimistic RCP8.5 scenario. Table 3 summarizes the basin-average temperature changes as percentages and absolute values for all seasons, as reported in the paper.

From the analysis of future precipitation performed here, in conjunction with the earlier study on future temperature, we can draw the following conclusions concerning climatic impacts on water resources in the Po river basin:

First of all, higher evapotranspiration, caused by higher temperatures, highly reduces the availability of renewable water resources for rain-fed agriculture in the Po valley, which in turn increases the need for artificial irrigation, if production is to be kept up. Moreover, the reduction of winter precipitation associated with higher temperatures, especially during winter and spring, pushes the climatic snow line to higher altitudes with lower amounts of precipitation stored as snowpack. As a result, the natural storage capacity of the snowpack is lost, with several consequences, such as the lack of slow recharge of water tables, which constitute a fundamental renewable water resource in the Po river basin, as well as the loss of water resource availability as surface runoff during the spring melting season. Finally, other observed aspects, which are not part of this study, are also worrisome, such as the reduction of the number of rainy days and the increase in precipitation intensity during storms, which enhances the fast transit of precipitated water towards the sea. Additional small and distributed surface storage capacity would be needed to increase the water availability, given the environmental implications and the lack of suitable low-risk sites. Therefore, even though the annual total precipitation may not change significantly, the availability of water resources will considerably drop, requiring water resource management strategies that are substantially different from those adopted at present. Adaptation strategies will also be required for the abatement of non-revenue water in potable water distribution systems by reducing distributed losses, which, in the Po valley, easily range from 20 to 40%. At the same time, crop types and irrigation practices in agriculture will have to be revisited in the light of preserving and optimizing the use of available water resources.

ACKNOWLEDGEMENTS

We would like to thank the Italian National Institute for Environmental Protection and Research (ISPRA) for making the gridded precipitation data for the Po river basin available. We also acknowledge the CMIP5 community for providing access to the climate output ensembles. This research was partly funded by Deutsche Forschungsgemeinschaft (DFG), Grant Number RE3848/4 awarded to the second author. We would like to thank three anonymous reviewers for their efforts, which have helped to improve the quality of the manuscript.

DATA AVAILABILITY STATEMENT

CMIP5 data can be downloaded from https://pcmdi.llnl.gov/mips/cmip5/. Gridded precipitation and temperature data for Northern Italy were downloaded from the site https://www.arcis.it.

CONFLICT OF INTEREST

The authors declare there is no conflict.

REFERENCES

Aven, T. & Renn, O. 2015 An evaluation of the treatment of risk and uncertainties in the IPCC reports on climate change. *Risk Analysis* **35**, 701–712. doi:10.1111/risa.12298.

Beven, K. 2021 Issues in generating stochastic observables for hydrological models. *Hydrological Processes* **35**, e14203. doi:10.1002/ hyp.14203.

Boé, J., Terray, L., Habets, F. & Martin, E. 2007 Statistical and dynamical downscaling of the Seine basin climate for hydro-meteorological studies. *International Journal of Climatology* **27**, 1643–1655. doi:10.1002/joc.1602.

Braca, G., Bussettini, M., Ducci, D., Lastoria, B. & Mariani, S. 2019 Evaluation of national and regional groundwater resources under climate change scenarios using a GIS-based water budget procedure. *Rendiconti Lincei. Scienze Fisiche e Naturali* **30**, 109–123. doi:10.1007/ s12210-018-00757-6.

Buser, C. M., Künsch, H. R., Lüthi, D., Wild, M. & Schär, C. 2009 Bayesian multi-model projection of climate: bias assumptions and interannual variability. *Climate Dynamics* **33**, 849–868. doi:10.1007/ s00382-009-0588-6.

Ciccarelli, N., von Hardenberg, J., Provenzale, A., Ronchi, C., Vargiu, A. & Pelosini, R. 2008 Climate variability in north-western Italy during the second half of the 20th century. *Global and Planetary Change* **63**, 185–195. doi:10.1016/j.gloplacha.2008.03.006.

Collins, M., Booth, B. B. B., Harris, G. R., Murphy, J. M., Sexton, D. M. H. & Webb, M. J. 2006 Towards quantifying uncertainty in transient climate change. *Climate Dynamics* **27**, 127–147. doi:10.1007/s00382-006-0121-0.

Déqué, M., Rowell, D. P., Lüthi, D., Giorgi, F., Christensen, J. H., Rockel, B., Jacob, D., Kjellström, E., de Castro, M. & van den Hurk, B. 2007 An intercomparison of regional climate simulations for Europe: assessing uncertainties in model projections. *Climatic Change* **81**, 53–70. doi:10.1007/ s10584-006-9228-x.

Drusch, M., Wood, E. F. & Gao, H. 2005 Observation operators for the direct assimilation of TRMM microwave imager retrieved soil moisture. *Geophysical Research Letters* **32**. doi:10.1029/2005GL023623.

Eyring, V., Bony, S., Meehl, G. A., Senior, C. A., Stevens, B., Stouffer, R. J. & Taylor, K. E. 2012 Overview of the coupled model intercomparison project phase 6 (CMIP6) experimental design and organization. *Geoscientific Model Development* **9**, 1937–2016.

Furrer, R., Knutti, R., Sain, S., Nychka, D. & Meehl, G. A. 2007 Spatial patterns of probabilistic temperature change projections from a multivariate Bayesian analysis of climate model output. *Geophysical Research Letters* **34**. doi:10.1029/2006GL027754.

Giorgi, F. & Francisco, R. 2000 Uncertainties in regional climate change prediction: a regional analysis of ensemble simulations with the HADCM2 coupled AOGCM. *Climate Dynamics* **16**, 169–182. doi:10.1007/PL00013733.

Giorgi, F. & Mearns, L. O. 2003 Probability of regional climate change calculated using the reliability ensemble average (REA) method. *Geophysical Research Letters* **30**, 1629–1632. doi:10.1029/2003GL017130.

Gneiting, T., Balabdaoui, F. & Raftery, A. E. 2007 Probabilistic forecasts, calibration and sharpness. *Journal of the Royal Statistical Society: Series B (Statistical Methodology)* **69**, 243–268. doi:10.1111/j.1467-9868.2007.00587.x.

Grazzini, F., Craig, G. C., Keil, C., Antolini, G. & Pavan, V. 2020 Extreme precipitation events over northern Italy. Part I: a systematic classification with machine-learning techniques. *Quarterly Journal of the Royal Meteorological Society* **146**, 69–85. doi:https://doi.org/ 10.1002/qj.3635.

Hamill, T. M. & Whitaker, J. S. 2006 Probabilistic quantitative precipitation forecasts based on reforecast analogs: theory and application. *Monthly Weather Review* **134**, 3209–3229.

Hersbach, H., Bell, B. & Berrisford, P. E. A. 2020 The ERA5 global reanalysis. *Quarterly Journal of the Royal Meteorological Society*. doi:10.1002/qj.3803.

Hirsch, R. M. 1982 A comparison of four streamflow record extension techniques. *Water Resources Research* **18**, 1081–1088.

Ines, A. V. M. & Hansen, J. W. 2006 Bias correction of daily GCM rainfall for crop simulation studies. *Agricultural and Forest Meteorology* **138**, 44–53.

IPCC 2013 Climate change 2013: the physical science basis. In: *Contribution of Working Group I to the Fifth Assessment Report of the Intergovernmental Panel on Climate Change*. Cambridge University Press, Cambridge, UK; New York, NY, USA. doi:10.1017/ CBO9781107415324.

IPCC 2021 Climate change 2021: summary for policymakers. In: *Climate Change 2021: The Physical Science Basis. Contribution of Working Group I to the Sixth Assessment Report of the Intergovernmental Panel on Climate Change*. Cambridge University Press, Cambridge, UK; New York, NY, USA. doi:10.1017/9781009157896.001.

IPCC 2022 Climate change 2022: impacts, adaptation and vulnerability. In: *Contribution of Working Group II to the Sixth Assessment Report of the Intergovernmental Panel on Climate Change*. Cambridge University Press, Cambridge, UK; New York, NY, USA. doi:10.1017/ 9781009325844.001. Available from: www.climatechange2013.org.

ISPRA 2021 *Il Bilancio Idrologico GIS-Based a Scala Nazionale su Griglia Regolare – BIGBANG. Technical Report 339*. Istituto Superiore per la Protezione e la Ricerca Ambientale.

Jones, P. W. 1999 First- and second-order conservative remapping schemes for grids in spherical coordinates. *Monthly Weather Review* **127**, 2204–2210. doi:10.1175/1520-0493(1999)127 < 2204:FASOCR > 2.0.CO;2.

Knutti, R., Furrer, R., Tebaldi, C., Cermak, J. & Meehl, G. A. 2010 Challenges in combining projections from multiple climate models. *Journal of Climate* **23**, 2739–2758. doi:10.1175/2009JCLI3361.1.

Knutti, R., Sedláček, J., Sanderson, B. M., Lorenz, R., Fischer, E. M. & Eyring, V. 2017 A climate model projection weighting scheme accounting for performance and interdependence. *Geophysical Research Letters* **44**, 1909–1918. doi:10.1002/2016GL072012.

Koutsoyiannis, D. & Montanari, A. 2022 Climate extrapolations in hydrology: the expanded BlueCat methodology. *Hydrology* **9**. doi:10.3390/hydrology9050086.

Levantesi, S. 2022 . *Italy Must Prepare for A Future of Chronic Drought*. Springer Nature, Italy. doi:10.1038/d43978-022-00089-y.

Lorenz, E. N. 1963 Deterministic non-periodic flow. *Journal of the Atmospheric Sciences* **20**, 130–141.

Mahler, H. C. & Dean, C. G. 2001 Chapter 8: credibility. In: *Foundations of Casualty Actuarial Science*, 4th edn. Casualty Actuarial Society, Arlington, VA, pp. 525–526.

Meehl, G. A., Boer, G. J., Covey, C., Latif, M. & Stouffer, R. J. 2000 The coupled model intercomparison project (CMIP). *Bulletin of the American Meteorological Society* **81**, 313–318.

Meehl, G. A., Covey, C., Delworth, T., Latif, M., McAvaney, B., Mitchell, J. F. B., Stouffer, R. J. & Taylor, K. E. 2007 The WCRP CMIP3 multimodel dataset: a new era in climate change research. *Bulletin of the American Meteorological Society* **88**, 1383–1394.

Musolino, D., Vezzani, C. & Massarutto, A. 2018 Chapter 3.11. In: *Drought Management in the Po River Basin, Italy*. John Wiley & Sons, Ltd, pp. 201–215. doi:10.1002/9781119017073.ch11.

Padulano, R., Lama, G. F. C., Rianna, G., Santini, M., Mancini, M. & Stojiljkovic, M. 2020 Future rainfall scenarios for the assessment of water availability in Italy. In: *2020 IEEE International Workshop on Metrology for Agriculture and Forestry (MetroAgriFor)*, 4–6 November 2020. IEEE, New York. pp. 241–246. doi:10.1109/MetroAgriFor50201.2020.9277599.

Palmer, T. N. 2000 Predicting uncertainty in forecasts of weather and climate. *Technical Reports on Progress in Physics* **63**. IOP Science. Available from: https://iopscience.iop.org/article/10.1088/0034-4885/63/2/201.

Palmer, T. N. & Räisänen, J. 2002 Quantifying the risk of extreme seasonal precipitation events in a changing climate. *Nature* **415**, 512–514. doi:10.1038/415512a.

Pavan, V., Antolini, G., Barbiero, R., Berni, N., Brunier, F., Cacciamani, C. & Cagnati, A. 2019 High resolution climate precipitation analysis for north-central Italy, 1961–2015. *Climate Dynamics* **52**, 3435–3453. doi:10.1007/ s00382-018-4337-6.

Pieri, L., Rondini, D. & Ventura, F. 2017 Changes in the rainfall – streamflow regimes related to climate change in a small catchment in northern Italy. *Theoretical and Applied Climatology* **129**, 1075–1087.

Raftery, A. E., Gneiting, T., Balabdaoui, F. & Polakowski, M. 2005 Using Bayesian model averaging to calibrate forecast ensembles. *Monthly Weather Review* **133**, 1155–1174.

Reggiani, P., Todini, E., Boyko, O. & Buizza, R. 2021 Assessing uncertainty for decision-making in climate adaptation and risk mitigation. *International Journal of Climatology* **41**, 2891–2912.

Reggiani, P., Talbi, A. & Todini, E. 2022 Towards informed water resources planning and management. *Hydrology* **9** (8), 1–18. doi:10.3390/hydrology9080136.

Rougier, J. 2007 Probabilistic inference for future climate using an ensemble of climate model evaluations. *Climatic Change* **81**, 247–264. doi:10.1007/s10584-006-9156-9.

Schulzweida, U. 2021 *CDO User Guide. Technical Report*. Max Planck Institute. doi:10.5281/zenodo.5614769.

Stoner, A. M. K., Hayhoe, K., Yang, X. & Wuebbles, D. J. 2013 An asynchronous regional regression model for statistical downscaling of daily climate variables. *International Journal of Climatology* **33**, 2473–2494. doi:10.1002/joc.3603.

Strzepek, K., McCluskey, A., Boehlert, B., Jacobsen, M. & Fant IV, C. 2011 *Climate Variability and Change: A Basin Scale Indicator Approach to Understanding the Risk to Water Resources Development and Management. Technical Report*. The World Bank, Washington, DC, USA.

Taylor, K. E., Stouffer, R. J. & Meehl, G. A. 2012 Summary of the CMIP5 experiment design. *Bulletin of the American Meteorological Society* **93**, 485–498.

Tebaldi, C. & Sansò, B. 2009 Joint projections of temperature and precipitation change from multiple climate models: a hierarchical Bayesian approach. *Journal of the Royal Statistical Society* **172A**, 83–106.

Themeßl, M. J., Gobiet, A. & Leuprecht, A. 2011 Empirical-statistical downscaling and error correction of daily precipitation from regional climate models. *International Journal of Climatology* **31**, 1530–1544. doi:10.1002/joc.2168.

Tramblay, Y. & Somot, S. 2018 Future evolution of extreme precipitation in the Mediterranean. *Climatic Change* **151**, 289–302. doi:10.1023/B:CLIM.0000044622.42657.d4.

Vidale, P. L., Lüthi, D., Frei, C., Seneviratne, S. I. & Schär, C. 2003 Predictability and uncertainty in a regional climate model. *Journal of Geophysical Research: Atmospheres* **108**. doi:10.1029/2002JD002810.

Wood, A. W., Maurer, E. P., Kumar, A. & Lettenmaier, D. P. 2002 Long-range experimental hydrologic forecasting for the eastern United States. *Journal of Geophysical Research* **107**, 1–13. doi:10.1029/2001JD000659.

First received 7 June 2022; accepted in revised form 16 September 2022. Available online 25 October 2022

NHF British Hydrological Society IWA PUBLISHING

doi: 10.2166/nh.2022.201

Multilayer blue-green roofs as nature-based solutions for water and thermal insulation management

Elena Cristiano [a,*], Antonio Annis [b], Ciro Apollonio [c], Dario Pumo [d], Salvatore Urru [a], Francesco Viola [a], Roberto Deidda [a], Raffaele Pelorosso [c], Andrea Petroselli [e], Flavia Tauro [f], Salvatore Grimaldi[f], Antonio Francipane [d], Francesco Alongi [d], Leonardo Valerio Noto [d], Olivier Hoes [g], Friso Klapwijk[h], Brian Schmitt[h] and Fernando Nardi [b,i]

[a] Dipartimento di Ingegneria Civile, Ambientale e Architettura, Università degli Studi di Cagliari, Cagliari, Italy
[b] WARREDOC, Università per Stranieri di Perugia, Perugia, Italy
[c] Dipartimento DAFNE, Università degli Studi della Tuscia, Viterbo, Italy
[d] Dipartimento di Ingegneria, Università degli Studi di Palermo, Palermo, Italy
[e] Dipartimento DEIM, Università degli Studi della Tuscia, Viterbo, Italy
[f] Dipartimento DIBAF, Università degli Studi della Tuscia, Viterbo, Italy
[g] Water Management Department, Delft University of Technology, Delft, the Netherlands
[h] Metropolder Company, Rotterdam, the Netherlands
[i] Institute of Water & Environment, Florida International University, Miami, USA
*Corresponding author. E-mail: elena.cristiano@unica.it

EC, 0000-0002-0725-3014; AA, 0000-0001-6162-4691; CA, 0000-0003-3576-7052; DP, 0000-0002-4274-6316; SU, 0000-0003-4724-1998; FV, 0000-0003-1716-192X; RD, 0000-0001-5469-0199; RP, 0000-0003-1996-5921; AP, 0000-0003-4943-0928; FT, 0000-0002-5176-3492; AF, 0000-0001-7811-8023; FA, 0000-0001-8769-3812; LVN, 0000-0002-3280-2898; OH, 0000-0003-0676-3167; FN, 0000-0002-6562-3159

ABSTRACT

Green roofs have been widely recognized as sustainable nature-based solutions to mitigate floods in urban areas, which, in the last decades, are increasing due to the combination of intense worldwide urbanization and climate change. Besides flood mitigation, green roofs provide additional benefits for the urban environment (e.g., reducing the urban heat island and ensuring energy saving for the underneath building). Moreover, green roofs facilitate the increase of urban biodiversity, attracting different species of small animals, and upgrade the city aesthetic value. Among the different types of green roofs, multilayer blue-green roofs present an additional layer to store water during rainfall events. As part of the Polder Roof field lab project, prototypes of multilayer blue-green roof developed by the Dutch company Metropolder were installed in four Italian cities: Cagliari, Palermo, Perugia, and Viterbo. The four prototypes and the experimental set up are described and the potential benefits of this innovative solution are discussed. Preliminary analyses, from December 2020 to December 2021, enable to estimate runoff reduction and thermal properties of multilayer blue-green roofs, underlying the high potential of this nature-based solution, which allows to retain most of the rainfall events and to mitigate the daily temperature variability.

Key words: green roofs, nature-based solutions, resilience, thermal insulation, water management

HIGHLIGHTS

- Description of multilayer blue-green roofs installed in four Italian cities.
- Analysis of the multilayer blue-green roof retention capacity.
- Analysis of the multilayer blue-green roof thermal insultation properties.

GRAPHICAL ABSTRACT

1. INTRODUCTION

Modern society is dealing with several challenges and global changes that are reshaping the urban environment. Population is growing fast and moving from rural to urban areas: projections show that the global population will reach 9.7 billion by 2050 and two-thirds of it will be living in cities (UN 2018), and a consequent increase in impermeabilization of natural surfaces is expected. At the same time, climate changes are expected to increase the global average temperature, with longer dry periods, alternated by short and highly intense rainfall events (IPCC 2007; Masson-Delmotte *et al.* 2018) also in the Mediterranean region (Treppiedi *et al.* 2021). The interplay of climate and urbanization changes is projected to facilitate increased pluvial flood generation and intensify urban heat islands (Arnone *et al.* 2018). Moreover, the COVID-19 pandemic and the consequent socio-economic crisis are strongly influencing and reshaping the way of thinking and living in the urban environment, and, in particular, the need for private green spaces is emerging (Hanzl 2020; Honey-Rosés *et al.* 2020; Jenkins 2020; Xie *et al.* 2020; Pelorosso *et al.* 2021a).

Different solutions have been proposed to mitigate and adapt to climate changes and population densification, creating smart, sustainable, and resilient cities, aiming to achieve the Sustainable Development Goals (SDGs) proposed in the Sustainable Agenda 2030. A solution to cope with urban rainwater excess is represented by rainwater harvesting systems. Such systems have been developed through the centuries in Mediterranean areas to collect and store water during rainfall events, making it available during drought periods (Beckers *et al.* 2013). Besides this potential usage, rainwater harvesting systems are nowadays mainly used to mitigate extreme rainfall events (Boers & Ben-Asher 1982; Zhang & Hu 2014; Teston *et al.* 2018; Freni & Liuzzo 2019; Akter *et al.* 2020; Cristiano *et al.* 2021b). Thanks to their water storage capacity, rainwater harvesting tanks can reduce the pressure on the drainage systems, detaining a fraction of the rainfall during the events and releasing it in a later moment (Bocanegra-Martínez *et al.* 2014; Chao-Hsien *et al.* 2014; Sample & Liu 2014; Akter & Ahmed 2015; Campisano & Modica 2015; Campisano & Lupia 2017; Adugna *et al.* 2018; Cipolla *et al.* 2018; Abas & Mahlia 2019; Palermo *et al.* 2020). If properly stored and treated, this water can be reused for different domestic purposes, such as home-garden irrigation and flushing toilets, potentially reducing potable water consumption.

Besides these traditional solutions, which are often costly and not flexible, other engineering solutions, known as nature-based solutions (NBSs), have also been investigated to mitigate the impact of pluvial floods and to increase urban resilience to climate and demographic changes (Pelorosso *et al.* 2018; Oral *et al.* 2020; Recanatesi & Petroselli 2020; Calheiros & Stefanakis 2021). NBSs (e.g., green roofs (GRs), rainwater collection parks, permeable pavements, and wetlands) are solutions inspired and supported by nature that can provide multiple environmental, social, and economic benefits (i.e., ecosystem services) through locally adapted, resource-efficient, and systemic interventions (Faivre *et al.* 2017; Nesshöver *et al.* 2017; Oral *et al.* 2020). Among the different NBSs, GRs are one of the most important technologies to deal with the urban challenges. They have been mainly investigated for their capacity to mitigate pluvial floods, retaining a fraction of the rainfall event (Getter *et al.* 2007; Brandão *et al.* 2017; Cristiano *et al.* 2020, 2021a; Liu *et al.* 2020), and to thermal insulate the buildings, reducing the energy consumption (Lazzarin *et al.* 2005; Castleton *et al.* 2010; Coma *et al.* 2016). Moreover, GRs present multiple benefits: they guarantee biodiversity, help lowering the urban temperature and reducing the urban heat island (Takebayashi & Moriyama 2007; Susca *et al.* 2011; Santamouris 2014; Muhammad & Reeho 2017; Solcerova *et al.* 2017), retain contaminants in the soil layer, contribute to the CO_2 sequestration (Yang *et al.* 2008; Gregoire & Clausen 2011; Rowe 2011; Speak *et al.* 2012; Abhijith *et al.* 2017), and add aesthetic values to urban environments (van den Bosch & Ode Sang 2017; Kumar *et al.* 2019).

Multilayer blue-green roof (MGR) is an innovative technology that combines all the potential benefits of traditional GRs with the large storage capacity of rainwater harvesting systems (Shafique *et al.* 2016a, 2016b; Muhammad & Reeho 2017; Andenæs *et al.* 2018; Cristiano *et al.* 2021a). Compared to traditional GRs, this technology presents an additional layer below the soil layer, which can store the water leaked from the soil, thus, when the soil moisture reaches the leakage triggering point, water starts to percolate and is collected in the storage layer. The water level within the storage layer can be regulated by means of a remotely controlled valve, which can be also opened or closed manually if necessary.

As part of the Polder Roof field lab, a project supported by the European Climate KIC programme, four MGR prototypes, called Polder Roofs and developed by the Dutch company Metropolder, have been installed in four Italian cities (Cagliari, Palermo, Perugia, and Viterbo) during Spring 2019, with the aim of exploring the potential benefits of this tool in Mediterranean climate regions. This work has the goal to present some preliminary results obtained during the first period of observation, comparing the potential of the four prototypes, which differ for their setup and installation properties (e.g., soil and vegetation type) and are influenced by different climatological conditions. Two specific benefits of the MGR installation are here investigated: the flood mitigation capacity and the thermal insulation.

The paper is structured as follows. Section 2 provides a general presentation of the MGR system proposed by the Metropolder company, with a detailed description of the four MGRs installed in Italy, focusing on the different characteristics that could influence the MGR retention capacity and temperature mitigation. Moreover, this section illustrates all the installed sensors and the observed data available for each study case. The methodology followed to assess preliminary analysis to evaluate the potential runoff mitigation capacity and thermal insulation potential of this tool are presented in Section 3, while results are illustrated in Section 4. Section 5 presents the potential benefits of MGR, discussing the role of the gate in the outflow generation and how this element can improve the urban drainage management compared to traditional GR. This section also includes an overview of the future research directions. Section 6, finally, summarizes the key points of the project and highlights the main conclusions of this preliminary analysis.

2. MGRS AND DESCRIPTION OF THE CASE STUDIES

2.1. Polder Roof by Metropolder: the blue-green roof technology

The Polder Roof is an innovative system at a Technical Readiness Level (TRL) of 7, designed by the Metropolder Company, to maximize the water retention capacity of a roof. Currently, there are more than 25 functioning Polder Roof systems in the

Netherlands and many more at the engineering stage. The system buffers and holds water while functioning as a foundation of an urban surface most commonly a GR, garden, or park. Though green is the current top layer, there is sustained interest in placing systems on hardened surfaces such as parking areas, streets, and pathways.

The Polder Roof structure is composed of multiple layers (Figure 1): from the top there is the vegetation, the soil substrate, a geotextile filter fabric to enable the water to flow holding the soil particles, a buffer where the water can be stored and a waterproof membrane to protect the roof. The buffer is made of Permavoid, a modular interlocking polypropylene void former. The storage capacity of the Polder Roof can be regulated by a smart remotely controlled valve, which can be opened or closed depending on the user's needs (Busker *et al.* 2022). For example, as shown in Figure 1, respectively, the user could decide to open the valve when heavy rain is forecast (Figure 1(a)) and to keep it closed during sunny periods (Figure 1(b)), with the aim to use the stored water for irrigation.

Using Metropolder's patent pending Sm^rtMILL (Figure 1(c)), the Polder Roof functions as a dynamic water buffer. The Sm^rtMILL contains sensors which measure rainfall, temperature, water height, and wind speed. Based on the levels of each of these components, the water held on the roof can be maintained or released through a valve connected to the roof drain. The Sm^rtMILL transmits its data to a personal dashboard which displays real-time conditions and allows for remote adjustments. Remotely controlled, fine-grained management, and distribution of stored water is the Polder Roofs contribution to urban resilience. The Polder Roof fits building regulations, policy aims, and is easily applicable and well scalable. The innovations regarding urban flooding, building heating and cooling, insulation, groundwater infiltration, grey water re-use, and urban agriculture are what set the Polder Roof apart from any rooftop system seen before (van der Meulen 2019; Pelorosso *et al.* 2021b).

2.2. The four Italian case studies

In this section, the characteristics of the prototypes located in Cagliari, Palermo, Perugia, and Viterbo are described, with a particular focus on the peculiarities that could influence the water retention capacity and the thermal insulation properties of

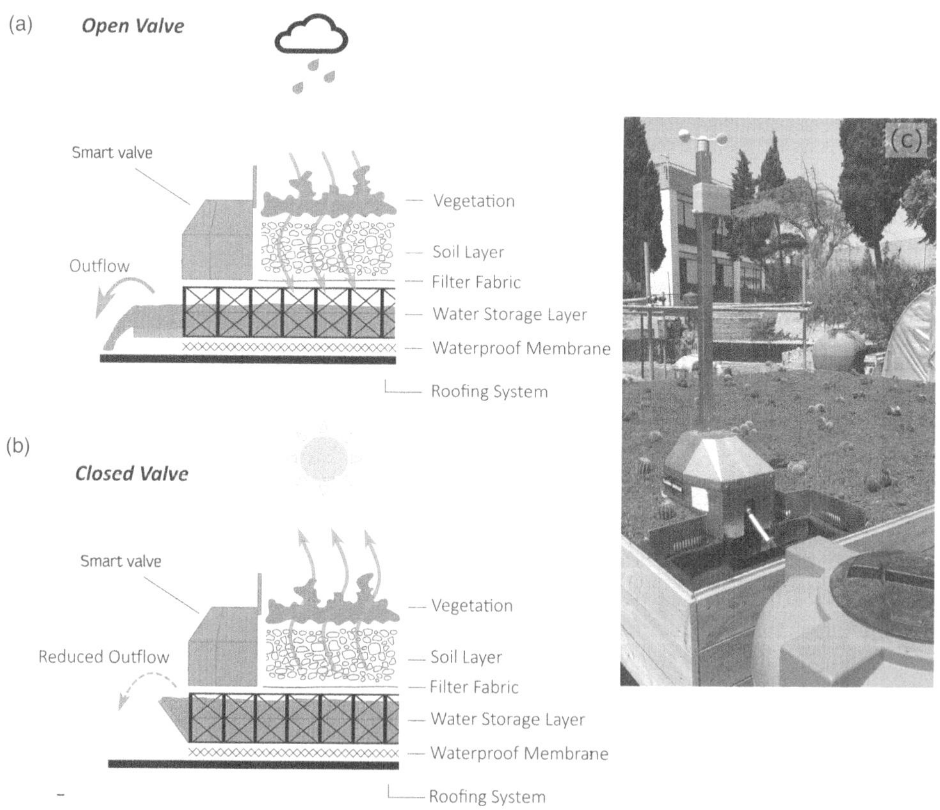

Figure 1 | The blue-green roof technology proposed by Metropolder. Schematization of the Polder Roof structure with a (a) open smart valve and (b) closed smart valve. (c) Example of the smart valve installed on the prototypes.

this system, such as soil and vegetation type. In some locations, the presence of a benchmark roof (either a bare roof or a traditional GR) will allow a comparison with the prototype to highlight the benefits of this innovative system. For each test case, Table 1 resumes the main characteristics of the system, the installed sensors, and the climate of the site where it is installed (e.g., mean annual rainfall depth and temperature reference values, derived from local historical time series).

The four Italian Polder Roofs are equipped with a standard monitoring station, which provides the real-time observation of essential climatic variables, and additional local instrumentation to pursue specific objectives of each case study. In particular, the standard Polder Roof monitoring station includes a tipping-bucket rain gauge, two thermometers for air and water temperature, an anemometer, and a sensor to monitor water level in the storage layer. The Metropolder monitoring station is connected to an online platform, which enables to easily read the different measurements and to regulate the water level in the storage layer, varying the opening degree of the valve.

To measure the runoff generation from the MGR, the outflow from the storage layer is directed into a rain barrel. A Baro-Diver® produced by the Van Essen Instruments company is placed in the rain barrel to measure the variation of pressure and, consequently, the water level. The temporal resolution of the water level measurements can be decided by the user: in Cagliari, for example, a time step of 1 min has been set, while in Palermo and Viterbo a 5-min time step is used. An external barometer allows compensating atmospheric pressure variations in real-time. Water heights are then converted into water volume by a specific calibration curve. The outflow from the MGR is then estimated by a mass conservation equation. Besides the described standard equipment, each Polder Roof has additional sensors that will be presented in the following sections. Figure 2(b)–2(e) illustrates the prototypes installed in the four case studies.

With the aim to better understand the differences among the case studies, Figure 2(a) shows the empirical complementary cumulative distribution function of the daily rainfall of historical time series in the four locations (circles) and the corresponding fit with a generalized Pareto distribution (thick lines), by the Multiple Threshold Method (Deidda 2010). Historical rainfall time series present different lengths for each case study: Cagliari (1947–1997), Palermo (2002–2020), Perugia (1990–2021), and Viterbo (1916–2015). The shape parameter k of the generalized Pareto distribution and the annual average rainfall m are reported in the legend of Figure 2(a). Higher values of the shape parameter k highlight a higher probability of observing extreme values with respect to the local average in the considered location.

2.2.1. MGR in Cagliari

The Polder Roof located in Cagliari (Figure 2(b)) is installed 50 cm above the ground on a wooden structure within the garden of the Department of Hydraulics and Hydrology of the Civil Engineering Faculty, University of Cagliari. The MGR has a surface of 16 m^2 (4 m × 4 m) and it is constituted by an 8 cm storage layer and an 8 cm of soil, classified as sand. The soil is planted with common cactus plants (*Cactaceae*), which represent one of the local species of the vegetation. These plants are characterized by a Crassulacean Acid Metabolism (CAM), that enables them to close the stoma during the day, avoiding water dispersion. The evapotranspiration efficiency of CAM vegetation is lower with respect to other plants, but it ensures the survival of the species even during very hot and dry periods. Moreover, CAM vegetation requires very low maintenance, with consequently low costs (Cristiano *et al.* 2020). For these reasons, this type of vegetation is particularly suitable for regions characterized by long hot and dry summers, such as the Mediterranean areas. The MGR outflow is directed into a 350-l rain barrel equipped with a Baro-Diver®.

In addition to the sensors included in the standard MGR monitoring station provided by Metropolder (see Sections 2.1 and 2.2), four thermometers (MX2203, produced by HOBO) have been installed to analyse the temperature in the soil, and beneath and on the wooden structure, on February 25, 2021. The four thermometers have been set to record the surface temperature with a 30-min resolution. Rainfall data have been provided by the Agency for the Environmental Protection of the Sardinia Region (Agenzia Regionale per la Protezione dell'Ambiente della Sardegna – ARPAS), which manages a dense rain gauge network in the region, with a weather station close to the prototype location. Rainfall data are available for the entire investigated period with a 1-min resolution.

2.2.2. MGR in Palermo

The MGR installed at the University of Palermo (Figure 2(c)) covers a rectangular area of about 32.0 m^2 (i.e., 4.30 m × 7.45 m) and is characterized by a green upper layer composed by two sub-areas with different depths; a central rectangular area, with an extension of about 18 m^2, is characterized by a soil thickness of 20 cm, while the remaining border region, whose extension is 14.0 m^2, has a shallower soil thickness of 10 cm. All sub-areas have been filled with a fertile soil consisting of a mixture of volcanic

Table 1 | Summary of the characteristics of the study cases

		Cagliari	Palermo	Perugia	Viterbo
Local climate characteristics	Average annual rainfall	425.3 mm/year	733.1 mm/year	896.4 mm/year	796.6 mm/year
	Average annual temperature	17.5 °C	17.0 °C	13.2 °C	14.4 °C
Polder Roof characteristics	Total surface	16 m^2 (4 m × 4 m)	32 m^2 (4.30 m × 7.45 m)	55 m^2 (11 m × 5 m)	16 m^2 (4 m × 4 m)
	Soil type	Common soil for Cactaceae, classified as *Sand* (granulometric analysis)	Volcanic materials (about 90% of volcanic lapillus and about 10% of pumice with a grain size of 0–10 mm)	Lightweight soil (40% pumice)	70% pumice, 20% peat, 10% sand
	Soil thickness	80 mm	1st area (14 m^2): 100 mm 2nd area (18 m^2): 200 mm	100 mm	100 mm
	Vegetation type	Cactus plants (*Cactaceae*)	1st area: mixture of sedum and meadow species: *Sedum album Stipa capensis, etc.* 2nd area: mixture of shrub species: *rosmarinus officinalis, teucrium fruticans, teucrium flavum, etc.*	*Sedum, Frankincense (Plectranthus), Aloe Vera, Delospermum*	*Sedum album, Sedum acre*
	Storage layer	80 mm	80 mm	80 mm	80 mm
Measurement tools	Rainfall	Metropolder station, from October 2019, with some gaps + ARPAS rain gauge close to the University (long historical data)	Metropolder station, from June 2019, with some gaps + disdrometer, weighing rain gauge, weather station equipped with a tipping rain gauge very close to the Polder Roof (~ 65 m).	Metropolder station, from October 2019, with some gaps + Official pluviometer from the Regional Hydrographic Service (~1 km from the Polder Roof).	provided by Metropolder station, from July 2019, with several gaps. + Rain gauges in the experimental site (Grimaldi *et al.* 2018)
	Water level	Metropolder station, from October 2019, with some gaps	Metropolder station, from January 2021	Metropolder station, from October 2019, with some gaps	N/A
	Outflow	One Rain barrel (350 l) with Baro-Diver® for the water level since December 2020 ($\Delta t = 1$ min).	Two Rain barrels (2 × 1,000 l) with Baro-Diver® for the water level since December 2020 ($\Delta t = 5$ min)	A rain barrel will be placed soon	Two Rain barrels (2 × 1,000 l) with Baro-Diver® for the water level since October 2020 ($\Delta t = 5$ min) +two additional rain barrels (2 × 1,000 l) installed in November 2021

(Continued.)

Table 1 | Continued

		Cagliari	Palermo	Perugia	Viterbo
	Temperature	Four HOBO sensors since February 2021: 2 below the roof, 1 over the soil, 1 aside north facing on the wooden structure (Figure 4(a))	Four HOBO sensors since December 2020: 1 below and 1 above the unaltered roof, 1 below the Polder Roof, 1 in the soil (~5 cm below the surface) (Figure 4(b)) 1 Lambrecht 8096 thermo-hygrometer sensor to measure the air temperature 2 m above the prototype	Three HOBO sensors installed on February 2020: 1 in the substrate not far from the Sm^rtMILL, 1 under the substrate between the plastic crates, 1 under the roof inside the carpentry (Figure 4(c))	Two HOBO sensors since November 2020: 1 above the Polder Roof and 1 above the benchmark. (Figure 4(d)) 2 additional HOBO thermometers (1 inside and 1 below the Polder Roof) could be included in future monitoring.
Investigated daily events	Number	13 (3 dry season–10 wet season)	50 (9 dry season–41 wet season)	18 (10 dry season–8 wet season)	37 (13 dry season–24 wet season)
	Mean intensity	21.7 mm (11.7 mm dry season–28.0 mm wet season)	14.7 mm (14.8 dry season–14.6 mm wet season)	15.4 mm (20.1 dry season–8.11 wet season)	13.0 mm (12.4 mm dry season–13.3 mm wet season)
	Max intensity	100.8 mm (22.6 mm dry season–100.8 mm wet season)	42.3 mm (42.3 mm dry season–40.6 wet season)	40 mm (40 mm dry season–10.8 mm wet season)	39.4 mm (13.7 dry season–39.4 mm wet season)
Additional notes		Soil moisture sensors have been installed in June 2021	Soil moisture sensors + Eddy Covariance Tower + further temperature sensors will be installed	Soil moisture sensors will be installed	Soil moisture sensors will be installed

Figure 2 | Case studies. (a) Probability distribution of the daily rainfall. Polder Roof prototypes located in (b) Cagliari, (c) Palermo, (d) Perugia, and (e) Viterbo with the geographic location in the upper left corners.

materials (about 90% of volcanic lapillus and about 10% of pumice with a grain size of 0–10 mm). In order to facilitate vegetation growth, a combination of CAM plants and typical local species were selected. More specifically, the first area was installed with a system of shrub species of garrigue and herbaceous species, such as *Rosmarinus officinalis*, *Teucrium fruticans*, *Teucrium flavum*,

Pallenis maritima, Helichrysum rupestre, Trifolium fragiferum, Lavandula spica, Crithmum maritimum, Micromeria graeca, Coridothymus capitatus, Dianthus ruipicola, Origanum heracleoticum, and *Convolvulus cneorum.* The second area, instead, was realized with a mixture of sedum species and typical Mediterranean meadow species, such as *Sedum album, Sedum dasiphyllum, Sedum sediforme, Sedum anopetalum, Stipa capensis, Trifolium angustifolium, Trifolium stellatum, Aegylops geniculata, Silene coeli-rosa.* An almost equal size (36.8 m^2) bare roof area bordering the MGR installation is also monitored for comparison and it is referred to as 'unaltered roof'. Analogously to the installation of Cagliari, a system of two 1,000-l rain barrels and two Baro-Diver® pressure sensors allow to collect and compare the rainwater coming from the MGR and the unaltered roof.

Besides the observations provided by the standard MGR monitoring station (see Sections 2.1 and 2.2), additional data are collected from a weighing rain gauge (OTT Pluvio2-400), a disdrometer (OTT-Parsivel2), and a weather monitoring station equipped with a tipping-bucket rain gauge (DQA130.1 by LSI LASTEM), a thermo-hygrometer sensor (Lambrecht 8096) for the measurement of air temperature and relative humidity, an air pressure sensor (Lambrecht 8128), a global solar radiation sensor (Lambrecht 16131.3), and an ultrasonic wind speed and direction sensor (VENTUS-UBM by Lufft). All the instruments are approximately 65 m from the MGR. Finally, four thermometers (MX2203 HOBO) have been installed, with a couple of sensors placed on both the MGR and the unaltered roof surfaces, and inside the building, in correspondence of the MGR and the unaltered roof.

2.2.3. MGR in Perugia

The MGR in Perugia (Figure 2(d)) is placed on the top of the carpentry of the University for Foreigners of Perugia (Palazzina Orvieto, Perugia). The prototype is 55 m^2 wide and it is constituted by a 10 cm layer of lightened soil (40% pumice) to reduce the weight on the carpentry attic. Different types of plants are included, such as *Sedum, Frankincense (Plectranthus), Aloe Vera, Delospermum.* These plants can resist long dry periods without the need of any water supplies for irrigation in the summer months. The roof has a pathway with white pebbles for recreational activities. The installation of a 1,000-l barrel equipped with a pressure measurement to estimate the water level is in progress.

In addition to the measurement recorded by the standard MGR monitoring station (see Sections 2.1 and 2.2), three temperature sensors (MX2203 HOBO) have been placed, two in the soil layer and one under the roof, inside the carpentry. The temporal frequency of the measurements has been set equal to 5 min.

2.2.4. MGR in Viterbo

The MGR located in Viterbo (Figure 2(e)) presents a structure similar to the prototype of Cagliari. The MGR is located in the hydrological experimental site (www.mechydrolab.org) of Tuscia University. It is placed on top of a wooden structure, 90 cm above the ground, and has a total surface of 16 m^2 (4 m × 4 m). It is formed by an 8 cm height storage layer and 10 cm of soil. The vegetation is characterized by perennial camephytic plants adapted to survive in arid climates (*S. album* and *S. acre*). The MGR outflow is directed into a 1,000-l rain barrel equipped with a Baro-Diver®. A second 16 m^2 bare area, representing a benchmark steel roof, was installed a few meters far from the MGR and is monitored for comparison with another 1,000-l rain barrel and Baro-Diver® sensor. Two additional 1,000-l rain barrels have been installed in November 2021 to increase the time between two emptying and to reduce the risk of overflow in case of intense precipitation.

Besides the standard MGR monitoring station (see Sections 2.1 and 2.2), the climatic variables are monitored by additional sensors in the nearby Macroraingauge experimental installation (Grimaldi *et al.* 2018): four additional rain gauges, one thermometer for standard air temperature (placed in a shaded area), one pyranometer, and one additional anemometer are present. Two thermometers (HOBO) measure the temperatures just above the MGR and the benchmark roof. Two additional thermometers (MX2203 HOBO) have been placed on November 9, 2020 to measure the temperature above the MGR and the benchmark roof, with a time step of 5 min.

3. METHODOLOGY

Among the multiple benefits of the MGR installation on buildings, this study focuses on the potential effects of the MGR prototypes on the runoff generation reduction and their thermal insulation properties. With this aim, this section introduces quantitative indices to evaluate the performances of MGR in terms of water and temperature control.

3.1. Runoff reduction

Thanks to the soil and storage layers, the MGR can retain and store rainwater, thus reducing the runoff generation and consequently contributing to mitigate the pluvial flood risk. The MGR runoff reduction performances have been evaluated through two indices which have been applied at the daily scale, with the aim to evaluate the retention capacity of both the soil layer and the system as a whole.

The first index is the MGR Index of retention R_{MGR}, which is defined as:

$$R_{MGR,i} = 1 - \frac{O_i}{P_i} \quad (1)$$

where O_i and P_i are the depths of daily outflow from the MGR and of daily rainfall at the ith day, respectively. $R_{MGR,i}$ can range between 0, when soil is saturated and storage layer is full and, consequently, no rainwater is retained, and 1, when there is no outflow generation. This index enables to evaluate the retention capacity of the entire MGR structure, but it does not allow to identify whether the rainfall is stored in the soil layer or in the storage layer.

The second index proposed in this study is the GR index of retention R_{GR}, which has been introduced to evaluate the retention capacity of only the soil layer component of the MGR and can be calculated as:

$$R_{GR,i} = 1 - \frac{\Delta h_i + O_i}{P_i} \quad (2)$$

where Δh_i is the daily difference of water level in the storage layer. $R_{GR,i}$ can vary between 0, when the soil is already saturated and no rainwater is retained by the soil layer, and 1, when all the rainfall is stored in the soil layer. This index has been already used in the literature to estimate the retention capacity of traditional GRs (Viola *et al.* 2017; Hellies *et al.* 2018; Cristiano *et al.* 2021b). Both $R_{GR,i}$ and $R_{MGR,i}$ indices provide complementary information to properly describe the behaviour of a MGR.

Rainfall data are available for each location at high temporal resolution (ranging between 1 and 5 min depending on the case study) and for long time windows, covering the period December 2020–December 2021. Water level in the storage layer and outflow measurements, instead, are not available yet in some locations; for this reason, it was not possible to estimate both $R_{GR,i}$ and $R_{MGR,i}$ for all the four installations.

With the aim to investigate the retention capacity of the MGRs under critical conditions, light rainfall events have been excluded from this study. Following Alpert *et al.* (2002), that investigated the increase of extreme daily rainfall events in Mediterranean regions and identified rainfall thresholds, we defined as '*light*' an event with daily rainfall depth lower than 4 mm. Hence, only events with a daily rainfall intensity equal to or higher than 4 mm have been considered in this study to assess $R_{GR,i}$ and $R_{MGR,i}$. The main characteristics of the selected daily events used to estimate the retention capacity of the MGR in each case study are summarized in Table 1.

3.2. Thermal properties

MGRs help reducing the roof surface temperature contributing to the thermal insulation of the building. To investigate this potential benefit and to quantify the temperature mitigation, four indices, proposed by Bevilacqua *et al.* (2017), have been here used. As for the runoff mitigation, this investigation focuses on data available during the period from December 2020 to December 2021. Temperature data are recorded with a time step of 5 min (HOBO sensors in Perugia and Viterbo), 10 min (Metropolder sensors), and 30 min (HOBO sensors in Cagliari and Palermo), while the indices are evaluated at the daily scale.

The four indices compare the temperature T^G in a certain gauge G to the temperature T^* in a reference sensor, which is identified as 'reference' to represent the unaltered conditions, without the installation of the MGR. Since each prototype presents a unique configuration of the temperature sensor placement, in some cases different sensors have been chosen as a benchmark, depending on the scope of the analyses and on the peculiarities of the study case.

The first proposed index is the average daily surface temperature ratio $STR_{av,i}$ at the ith day, defined as the ratio between the average daily temperature $T^G_{av,i}$ recorded at the selected gauge G and the average daily temperature $T^*_{av,i}$ observed by the temperature sensor chosen as benchmark in the analysis:

$$STR_{av,i} = \frac{T^G_{av,i}}{T^*_{av,i}} \quad (3)$$

The $STR_{av,i}$ presents values higher than 1 when the selected location presents an average daily temperature higher than the benchmark and lower than 1 for the opposite situation. $STR_{av,i}$ values close to 1 highlight that there are no significant differences between the two observations.

The second index is the maximum daily surface temperature ratio $STR_{max,\,i}$, defined as:

$$STR_{max,\,i} = \frac{T^{G}_{max,i}}{T^{*}_{max,i}} \tag{4}$$

where $T^{*}_{max,i}$ and $T^{G}_{max,i}$ are the maximum temperatures observed during the ith day at the benchmark gauge and at the selected gauge G, respectively. Similar to the $STR_{av,i}$, $STR_{max,i}$ presents values close to 1 when the difference between the maximum temperature at the selected gauge and at the benchmark gauge is very small.

Following the same approach, the minimum daily surface temperature ratio $STR_{min,i}$ has been defined as:

$$STR_{min,i} = \frac{T^{G}_{min,i}}{T^{*}_{min,i}} \tag{5}$$

where $T^{*}_{min,i}$ and $T^{G}_{min,i}$ are the minimum temperatures during the ith day at the benchmark gauge and at the selected gauge G, respectively. This index has been estimated for the days with minimum daily temperature higher than zero.

The last index is the daily temperature excursion ratio TER_i, which is defined as:

$$TER_i = \frac{T^{G}_{max,i} - T^{G}_{min,i}}{T^{*}_{max,i} - T^{*}_{min,i}} \tag{6}$$

This last index evaluates the effects of the MGR installation on both minimum and maximum temperature, in terms of daily temperature excursion.

4. RESULTS

This section presents and discusses the results obtained from the analysis of the collected data in the four case studies for a 1-year period, from December 2020 to December 2021. The benefits of the MGRs are analysed and evaluated in terms of rainwater retention capacity and the thermal insulation properties, using the indexes introduced in Section 3.

4.1. Runoff mitigation

As explained in Section 3.1, the retention capacity is evaluated through two indices, the R_{MGR} and R_{GR}, only for significant daily rainfall events, with a mean intensity higher than 4 mm/day. Depending on the available measurements, for each location it was possible to estimate R_{MGR} and R_{GR} in order to investigate the retention capacity of the entire MGR and the soil layer, respectively. Rainfall, outflow, and water level data were recorded at different time resolutions and then aggregated and analysed at the daily scale.

In Cagliari, the periods between June 1, 2021 and August 1, 2021, and between October 1 and 30, 2021, have been excluded for this case study, due to sensors malfunctioning. Although the Baro-Diver installed in the rain barrel in Cagliari was not recording, no outflow has been observed during the investigated period. In Perugia, the rain barrel with the Baro-Diver is not recording data yet, and for this reason, also in this case, only the R_{GR} has been estimated and evaluated for events that did not generate runoff. Finally, for the MGRs located in Palermo and Viterbo it was possible to estimate the R_{MGR} for the entire year, while the R_{GR} has been evaluated until September 2021 for the case of Palermo and not analysed for the case of Viterbo, since the sensors that measure the water lever in the storage layer did not work properly over the entire year. The gate has been always kept closed during the investigated period, except for Cagliari, where it has been periodically open for maintenance during the dry periods.

4.1.1. Retention capacity

Results are illustrated in Figure 3, where the boxplots show the annual variability of the two investigated indices R_{GR} and R_{MGR} for the different cases (Figure 3(a) and 3(b), respectively). The solid line indicates the median value of the index, while the diamond represents the average.

As illustrated in Figure 3(a), Cagliari and Palermo present similar R_{GR}, with an average retention capacity of the soil layer of 0.46 and 0.42, respectively, suggesting that most of the investigated rainfall events are only partially retained by the soil layer,

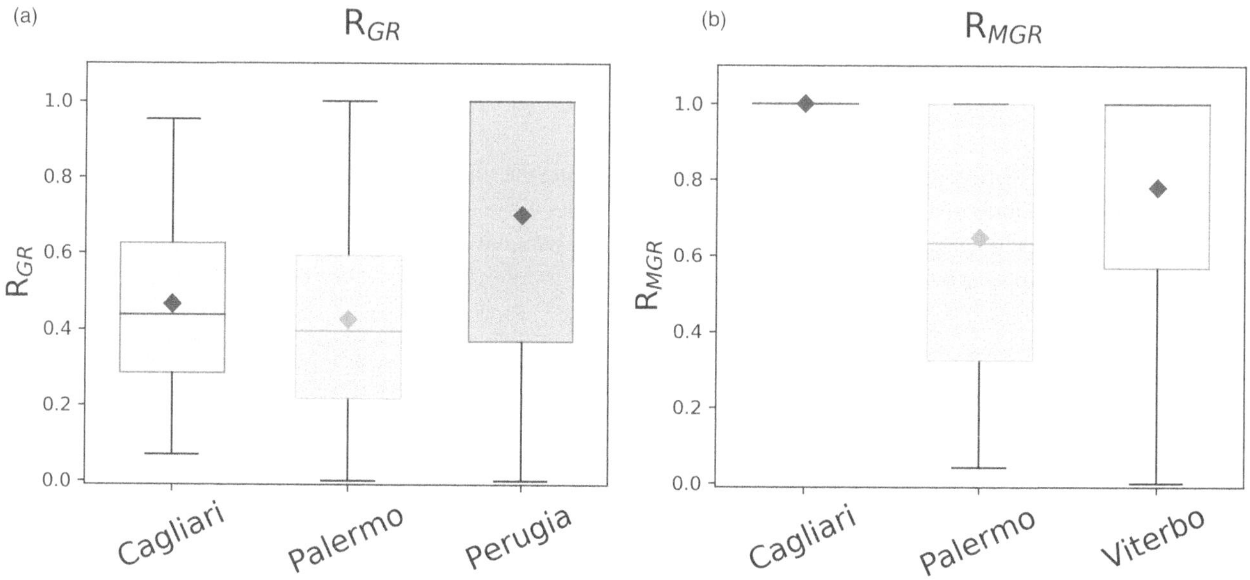

Figure 3 | R_{GR} (a) and R_{MGR} (b) estimated for the investigated locations. The solid line indicates the median value of the index, whereas the diamond represents its average.

and hence would have been only partially retained by a traditional GR, whose retention function is exclusively due to the soil layer retention capacity. The MGR of Perugia shows a higher performance in retaining the rainwater, probably due to the fact that, differently from the other two cases, it was not possible to measure the MGR outflow and, consequently, the events generating runoff, which usually are the most intense, has been excluded from the analysis.

The total retention capacity R_{MGR}, ensured by the MGR prototypes, can be observed in Figure 3(b) for the case study of Cagliari, Palermo, and Viterbo. Thanks to the contribution of the additional storage layer, the performance in this case is higher than the corresponding performance of a traditional GR, with an average R_{MGR} equal to 1 for Cagliari, to 0.65 for Palermo and to 0.75 for Viterbo. The plot shows that a significant number of rainfall events are completely retained by the systems, especially in Cagliari, where all events were fully retained, and in Viterbo, more than 50% of the events presented a R_{MGR} equal to 1.

In Palermo, the MGR retention capacity presents a high variability, with R_{MGR} mostly varying between 0.35 and 1. The high variability is determined by the different initial conditions that characterize the rainfall events. The antecedent conditions of soil water content and water level in the storage layer, in fact, strongly influence the retention capacity of a MGR.

The differences between the two indices, R_{MGR} and R_{GR}, and consequently the additional retention capacity provided by the storage layer in a MGR compared to a traditional GR, are well evident in the case of Palermo and Cagliari. In Palermo, when only the soil layer capacity is evaluated, the average R_{GR} is equal to 0.42, while, when the system as a whole is considered, the average index of retention R_{MGR} rises up to 0.65 and some of the events that were only partially retained by the soil layer, were instead fully retained by the entire system. In Cagliari, where the average annual rainfall depth is lower than in the other case studies (Table 1 and Figure 2), the MGR enabled to completely retain all the events, highlighting the high potential performance of the MGR installation in Mediterranean cities.

4.1.2. Seasonality effects

Results presented in this section show the seasonal performance of the MGRs, analysing separately the retention capacity during a warm and dry season, from March 22 to September 21, 2021, and a cold and rainy season, covering the complementary months of the year. The seasonal analysis emphasizes the high impact that antecedent soil moisture in the soil layer and water level in the storage layer have on the runoff mitigation.

The Spring-Summer rainfall events have been almost fully retained by the MGR in Palermo, with an average R_{MGR} equal to 0.87, and seven out of nine events completely retained by the structure. The two not fully retained events are consecutive and happened on March 22 and 23, 2021: in this case, the soil was already partially saturated and the storage layer partially full before the events, especially for the second day where conditions close to saturation were reached after the previous rainy day and the R_{MGR} dropped to 0.3. In Viterbo, all the 13 Spring-Summer events have been fully retained by the MGR, confirming the importance that this tool can have on the runoff generation reduction.

The retention capacity estimated during the Autumn-Winter period is quite lower compared to the Spring-Summer seasons, due to the higher frequency and intensity of the rainfall events: the average R_{MGR} drops to 0.59 in Palermo and to 0.65 in Viterbo. Moreover, the temperature that characterizes this season is consistently lower than in the Spring-Summer period, limiting the evapotranspiration losses from the soil layer.

4.2. Thermal mitigation properties

This section analysed the potential thermal benefits that the installation of a MGR on a roof can provide, to contribute to the creation of a smart and resilient environment. The four thermal indices described in Section 3.2, are here estimated, and used to discuss the potential benefits of the installation of MGRs in terms of thermal impacts. For each location, multiple sensors have been installed in different points of the system, as illustrated in Figure 4.

In Cagliari, temperatures recorded from four gauges have been used to evaluate the potential thermal impacts of the MGR. The sensor position is schematized in Figure 4(a). Two types of sensors have been used for the analysis. The 'Air' sensor is provided by the Metropolder monitoring stations and refer to the air temperature above the wooden structure. Three HOBO sensor records have been investigated: one is installed few centimetres beneath the soil surface ('Soil'), and other two sensors are situated beneath the prototype, at diagonally-opposite corners ('Under1' and 'Under2'). The two sensors situated beneath the prototype are used to help understand the effects on the hypothetical underneath building. An average of the two-time series is plotted in Figure 4(e). The temperature time series for the Cagliari case study are available from February 25th to December 21, 2021 and are plotted in Figure 4(e).

Four HOBO temperature sensors have been installed in Palermo to evaluate the thermal insulation properties of the MGR, as illustrated in Figure 4(b). One sensor has been placed in the soil layer ('Soil'), a few centimetres below the surface, one has been located on the 'Unaltered Roof' surface, and other two sensors have been situated inside the building: one in correspondence of the MGR sensor ('Room1'), and the other one in correspondence of the 'Unaltered Roof' ('Room2'). Moreover, a Lambrecht 8096 thermo-hygrometer sensor is set to measure the air temperature 2 m above the roof ('Air'). In Palermo, temperature observations are available from December 2020 to December 2021 and are plotted in Figure 4(f).

In Perugia, temperature measurements from three gages are available: the 'Air temperature', provided by the Metropolder monitoring stations and other two HOBO sensors placed at 5 cm under the substrate, under the substrate lean on the plastic crate. In Perugia, the temperature time series are available from August 2021 and are illustrated in Figure 4(g).

Measurements from the two HOBO sensors have been evaluated in Viterbo for the entire 2021 and are presented in Figure 4(h). As schematized in Figure 4(d), one sensor has been set to measure the air temperature ('Air'), while the other two gauges have been installed in the soil layer of the MGR ('Soil') and on the unaltered roof ('Unaltered Roof'). In this way, it was possible to estimate the impact of the MGR installation on the surface temperature.

Although a direct comparison among all the case studies is not possible due to differences that characterize each location, some representative analyses may illustrate the thermal behaviour of a MGR and its thermal impact.

4.2.1. MGR vs unaltered roof

In order to investigate the thermal properties of a MGR, the first step is to compare the soil temperature with benchmark values referring to an unaltered roof. For this purpose, in Palermo and in Viterbo, a temperature sensor has been installed on an unaltered roof close by the MGR prototype (see Figure 4(b) and 4(d)). The four indices, proposed in Section 3.2, are here used to facilitate the visualization of the MGR thermal properties, comparing the soil temperature with the surface temperature of the unaltered roof. Figure 5(a) investigates the differences in the average daily temperature through the index STR_{av}, and it shows that on average the MGR daily temperature is 5–7% lower than the surface temperature of the unaltered roof for both the cases.

When comparing the differences in the minimum daily temperatures through the STR_{min} (Figure 5(b)), the MGR temperature is overall higher than the unaltered roof performance, suggesting that the prototype can ensure a good thermal inertia, increasing minimum temperature usually occurring during the night. In Palermo, for example, the MGR temperature is on average 40% higher than the unaltered roof temperature.

Following the same approach with the STR_{max}, plotted in Figure 5(c), it is clear that, in most of the cases, the presence of the MGR provides a benefit for the peak, usually diurnal, daily surface temperature, since the MGR temperature is lower than the

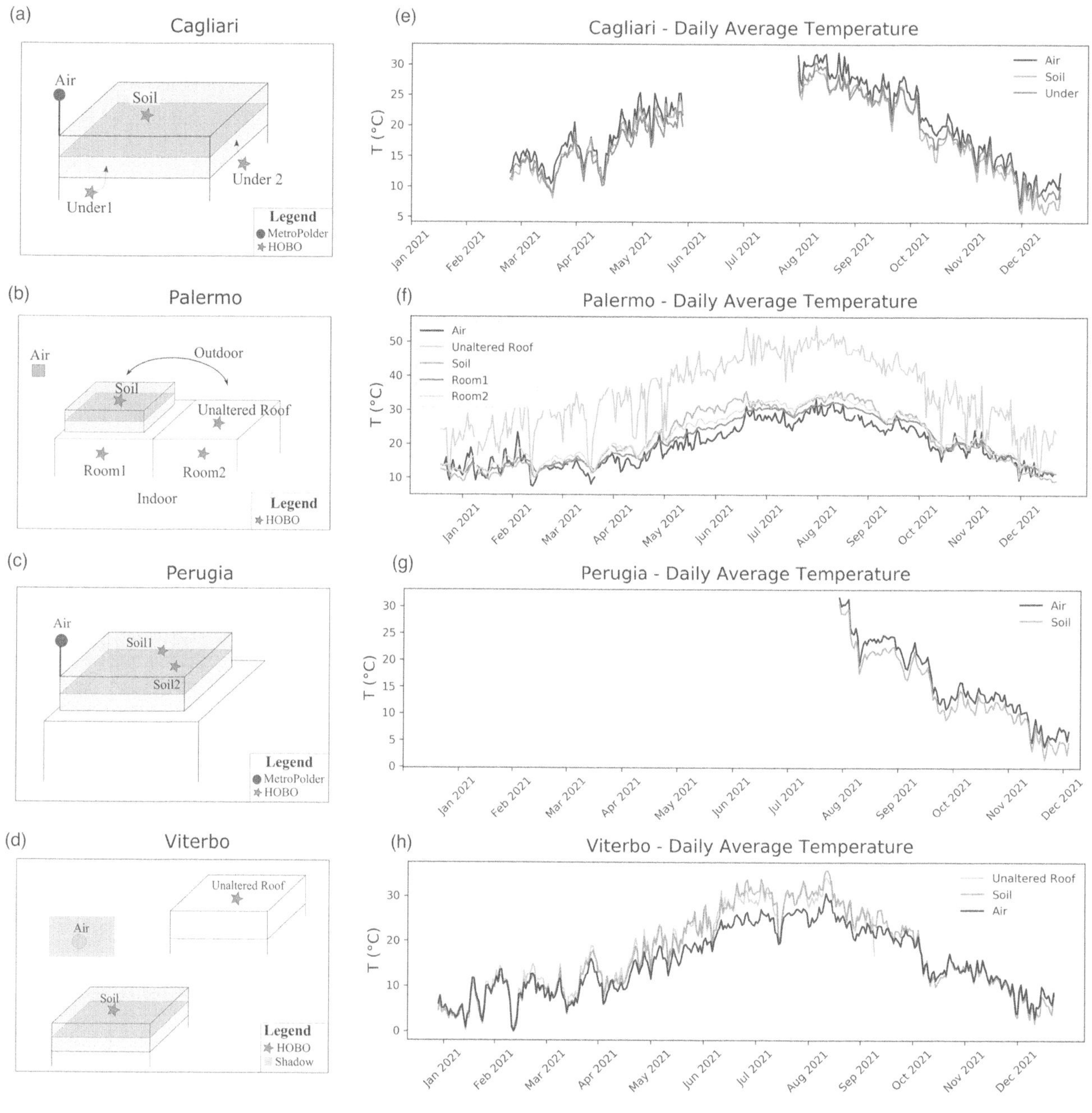

Figure 4 | Schematization of the MGR structure with the locations of the installed temperature sensors for each case study: (a) Cagliari, (b) Palermo, (c) Perugia and (d) Viterbo and related daily average temperature time series (e–h).

unaltered roof temperature. The effects are stronger in Palermo, where the daily peak air temperature, especially during the Summer, are usually higher than in Viterbo and, in fact, the maximum daily temperature on the MGR is on average reduced by about 40% compared to the corresponding unaltered values, while in Viterbo the average reduction is about 10%.

Finally, the variability of the TER, which describes the thermal excursion of the MGR with respect to that of the Unaltered Roof, is plotted in Figure 5(d). In both locations, the MGR shows a lower daily thermal excursion than the Unaltered Roof, with an average TER equal to 0.15 in Palermo and to 0.85 in Viterbo. Beyond the different climate conditions, the observed marked difference in performance between the two locations can be also addressed to the different materials of the Unaltered Roofs and the different vegetation coverage of the MGRs. Nevertheless, all the indices analysed here in both the case studies are concordant and highlight overall high performances of the MGR in mitigating the surface temperature of the roof,

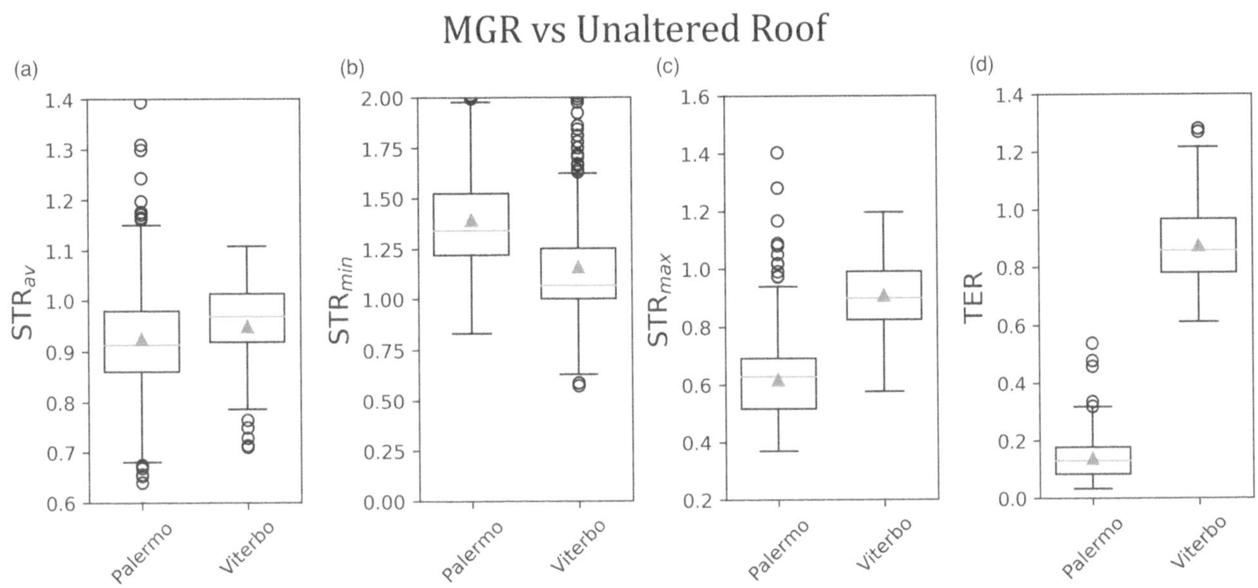

Figure 5 | Comparison between the MGR soil temperature and the surface temperature of the Unaltered Roof in Palermo and Viterbo. Boxplots for the indices (a) STR_{av}, (b) STR_{min}, (c) STR_{max}, and (d) *TER*.

suggesting an increased thermal inertia of the external surfaces, whose temperature results less influenced by the air temperature values and their daily excursions.

4.2.2. MGR and air temperature

The relationship between the surface temperature of a MGR and the air temperature is investigated through the four indices presented in Section 3.2, using the air temperature as the reference and the MGR soil temperature as the main gauge. Results are illustrated in Figure 6, where the boxplots represent the variability of STR_{av}, STR_{min}, STR_{max} and TER for the locations of Cagliari and Perugia.

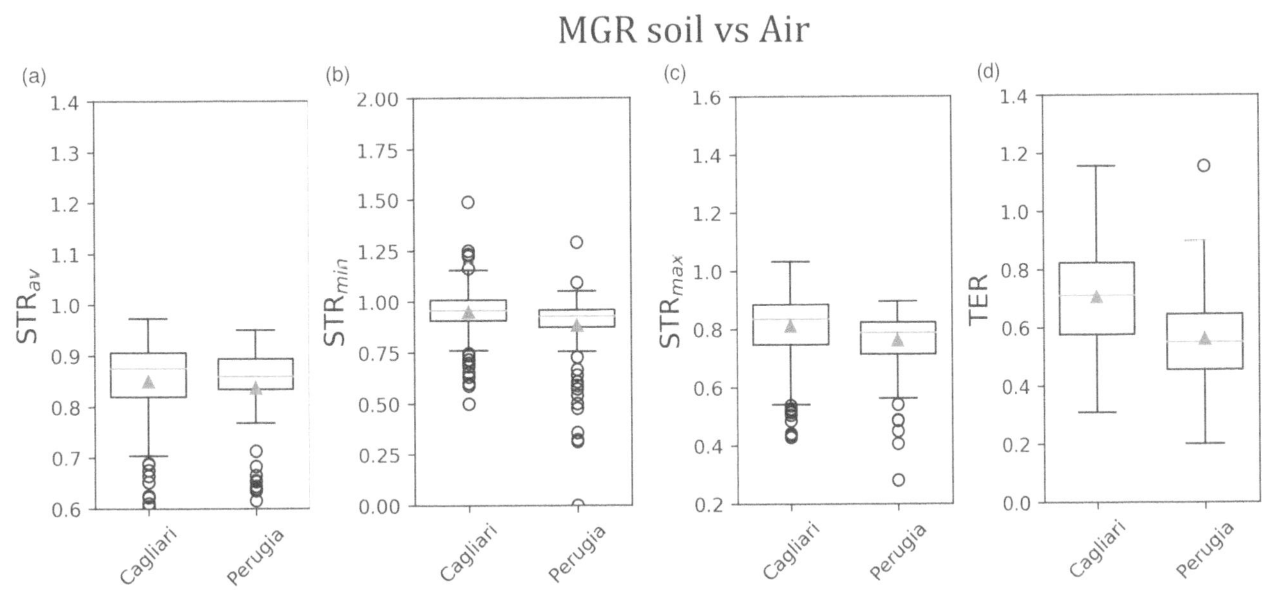

Figure 6 | Comparison between the MGR soil temperature and the air temperature in Cagliari and Perugia. Boxplots for the indices (a) STR_{av}, (b) STR_{min}, (c) STR_{max}, and (d) *TER*.

The STR_{av} analysis shows that the average MGR daily temperature is always lower than the air temperature, with an average reduction of about 15–20% (Figure 6(a)). The minimum and maximum daily temperatures for the MGR also presents overall values lower than the corresponding air temperature; more specifically, the differences between the minimum daily temperatures are rather low, with an average STR_{min} close to 1 in the Cagliari case study (Figure 6(b)), while a more significant lowering (on average reduction by the 20%) of the MGR temperatures compared to the air temperatures can be noticed for the maximum values (Figure 6(c)). More relevant differences can be observed between the daily thermal excursion of the MGR and the air (Figure 6(d)): the average TER is equal to 0.7 in Cagliari and to 0.55 in Perugia, confirming the high MGR capability to limit the daily temperature variability.

4.2.3. Potential thermal insulation for the building

The last analysis aims to directly investigate the thermal insulation properties of a MGR, comparing the temperature underneath the prototype with the soil temperature, used as reference value for the thermal indices. As shown in Figure 4, in the Cagliari case study, two HOBO temperature sensors have been placed beneath the prototype. The average between their measurements has been here used to derive the four thermal indices. In Palermo, a HOBO sensor has been installed inside the building, in correspondence with the prototype. Although the direct comparison between the two case studies is not possible, since in Palermo the roof structure influences the temperature inside the building, it is still possible to derive some conclusions regarding the thermal insulation properties of the MGR.

The STR_{av} varies around the unit, with average and median values over the entire observation period corresponding to about 1 year close to 1 for both locations (Figure 7(a)). However, analysing the values at seasonal scale (not reported in figure), it can be notice how STR_{av} assumes values above the unit prevalently during the Autumn-Winter season and below 1 during the Summer-Spring season, implying temperatures under the MGR (indoor for the case of Palermo) higher than the temperature on the soil surface during the colder period of the year and, on the contrary, lower temperature under the MGR during the hotter period, with a potential enhancement of the indoor internal comfort.

Different results can be achieved when investigating the minimum and maximum daily temperatures (Figure 7(b) and 7(c), respectively): in these cases, in fact, STR_{min} values prevalently higher than 1 and STR_{max} values lower than 1 over the entire 1-year period confirm that the temperature under the MGR is highly mitigated by the presence of the prototype itself. This can be inferred also by the analysis of the TER (Figure 7(d)), which reports for both location a thermal excursion under the MGR and inside the building constantly lower than that for soil surface of the MGR.

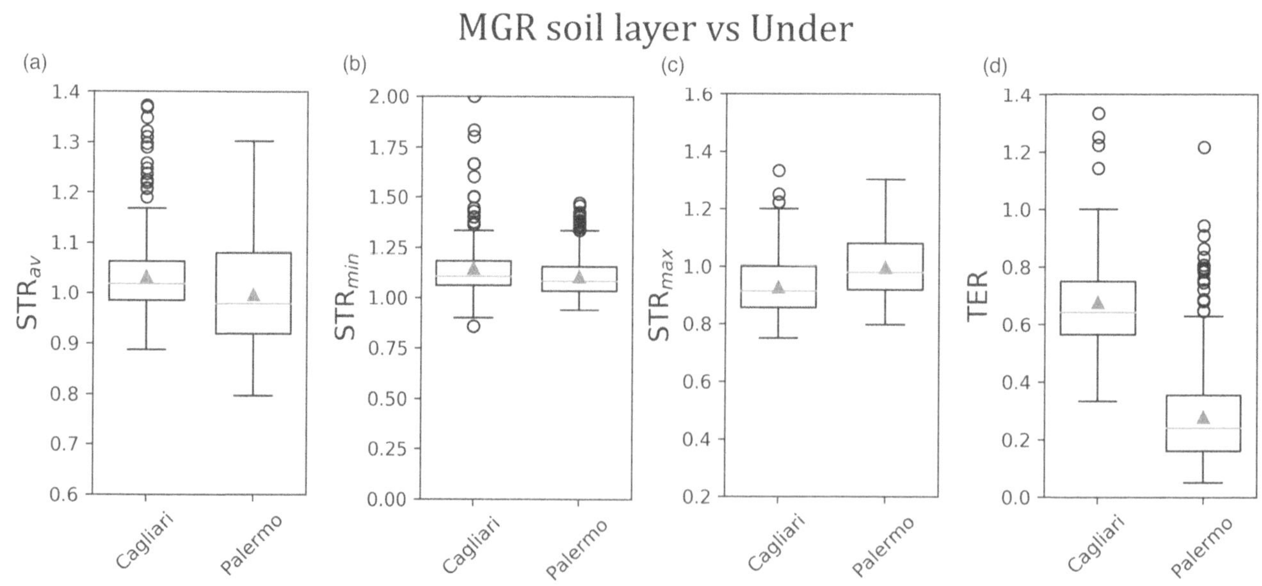

Figure 7 | Comparison between the MGR soil temperature and the temperature beneath the MGR structure in Cagliari and Palermo. Boxplots for the indices (a) STR_{av}, (b) STR_{min}, (c) STR_{max}, and (d) TER.

The roof configuration in Palermo enables to investigate the influence of the MGR on the temperature inside the building, comparing the recorded measures in Room 1, in correspondence of the prototype, and in Room 2, below the unaltered roof (see Figure 4(b) for the sensor configuration in Palermo and Figure 4(f) for the average daily time series). Daily temperatures in Room 1 are on average 1.3 °C lower than in Room 2 during the investigated period, while during the dry period, the difference increases up to a maximum of 3.7 °C. Moreover, the average TER between the temperatures recorded in Room 1 and in Room 2 mostly varies between 0.38 (25%-ile) and 0.95 (75%-ile), suggesting that the MGR contributes not only in reducing the average temperature inside the building, but also in limiting the daily excursion.

5. DISCUSSION

5.1. MGR performance

This study confirms the high potential of MGRs installed in Mediterranean areas in mitigating the runoff generation and in thermal insulating the underneath building. Traditional GRs guarantee an average rainfall reduction generally higher than the 50% of medium and intense rainfall events (Palla *et al.* 2010; Brandão *et al.* 2017; Palermo *et al.* 2019), with a strong variability connected to the structure features (such as soil type and thickness, vegetation type) and weather characteristics. These results have been confirmed by this study, where the analysis of the retention capacity guaranteed by the soil layer, represented by the R_{GR}, reported an average rainfall reduction between 42 and 70%, for the investigated case studies. Thanks to the additional layer, the MGRs ensure a higher volume to store rainwater, with an average retention index that reaches 0.65 in Palermo, 0.75 in Viterbo, and 1 in Cagliari.

These results support the findings presented by Busker *et al.* (2022), who investigated the hydrological performance of a MGR in the Netherlands. The different climatological environment leads to a different retention capacity, slightly lower in The Netherlands than in Mediterranean areas. As shown in Section 4, in fact, the retention and storage capacity of the MGRs installed in the Mediterranean areas enable to retain all the rainfall events in Spring-Summer. The high retention capacity in this case is facilitated by the low frequency of the events and by the high evapotranspiration rates, which ensures a higher probability that a rainfall event occurs when the soil is completely dry, and the storage layer is empty.

The retention capacity of the system, especially during the Autumn-Winter season, when rainfall events present higher intensity and frequency in Mediterranean areas, could be increased using a thicker storage layer. In many cases, this solution, however, might not be feasible, especially for already existing building, where the roof might not be able to sustain the weight of a large additional storage layer. An easier solution is given by a functional management of the gate that can optimize the water level in the storage layer, increasing the volume available to store rainwater.

Besides the high retention capacity, the preliminary analysis presented in this work, also emphasizes the additional benefits of a MGR, especially regarding thermal insulation. Compared to a traditional roof, the MGRs have shown lower maximum daily surface temperatures and higher minimum ones, and a lower thermal excursion during the day. These preliminary results suggest a high thermal insulation capacity for the underneath building, which can be translated in a potential save of energy for heating and cooling (Castleton *et al.* 2010; Coma *et al.* 2016).

5.2. The role of the gate to control the outflow

In this preliminary analysis, where the main goal was to present the potential benefits of the MGR prototypes under Mediterranean climate, the influence of the gate management has not been evaluated, since the gate was always closed. However, it cannot be ignored that the retention capacity and the thermal insulation properties of a MGR are strongly affected by the water level in the storage layer.

The retention capacity of a MGR can be largely increased through a wise management of the gate that regulates the water storage layer. As shown in Figure 1, the gate allows to regulate the water level in the blue layer, with the aim to accumulate rainfall and release it as runoff when it is more convenient, depending on the user's needs. If the main goal is to mitigate runoff, for example, the gate should be closed when it is raining, and the harvested water should be released when the rainfall event is over, and the urban drainage system is not stressed anymore. On the other hand, if there is the need to re-use the harvested water for irrigation or other domestic purposes, the gate should be kept closed, to accumulate a larger volume of water. It is therefore clear that, depending on the goals that need to be achieved with the MGR installation, a different management of the gate should be evaluated to optimize the benefits of the system. Busker *et al.* (2022) recently investigated the potential improvement of the retention capacity of a MGR very similar to those tested in this work, when integrating the gate management with the weather forecasts. This choice enables to empty the storage layer before intense rainfall events and could largely reduce the peak runoff and increases the peak delay. The study calculates that the MGR in the Netherlands can reach a capture rate of 90% for total precipitation when using weather forecast to manage the gate opening. In arid and semi-arid climates, an optimized gate management can allow not only to mitigate the runoff generation, but also ensures water storage, that can be used, especially in Mediterranean cities, to limit additional irrigations and to reduce the pressure on the water supply system. Future analyses will be, hence, dedicated to understanding how the system may alter water quality to evaluate the potential reuses of water collected by the MGR for different purposes.

5.3. Urban scale MGR installation

Recently, Busker *et al.* (2022) assessed that converting all the potentially suitable roofs in Amsterdam to the MGR would determine 11% of extreme precipitation (>20 mm/hour) to be captured when using proper gate management with the weather forecasts. However, the different climate and typology of Mediterranean cities call for specific and local evaluations. Although the preliminary analysis presented in this work already highlights the potential benefits of MGRs as a sustainable solution for the development of the urban environment, further investigations need to be carried out to fully understand the behaviour of this innovative system in Mediterranean climate and to emphasize the potential impacts and benefits for society. From this point of view, an evaluation of the potential impacts of the MGR at the urban scale would be desirable, i.e., assuming the installation of this system over a great number of suitable roofs in a large neighbourhood or at city scale.

Future applied research will aim at implementing and testing modelling tools to identify best solution and localization of MGR for the mitigation of hydraulic risk and runoff reduction in performance-based planning and design contexts (La Rosa & Pappalardo 2020; Pelorosso 2020). An evaluation at the urban catchment scale will then allow flood mitigation capacity of NBS-based scenarios to be tested considering different costs and levels of interventions. Further analyses will be then dedicated to the definition of the criteria for the identification of potential buildings for the MGR retrofit and the creation of smart and resilient Mediterranean cities. Following this approach, it will also be possible to assess other ecosystem services, as the heat island attenuation or the pollution retention, for a more holistic evaluation of the MGR technology.

6. CONCLUSIONS

This work describes the installation of MGRs prototypes in four Italian cities (Cagliari, Palermo, Perugia, and Viterbo) as part of the Polder Roof field lab project. Compared to traditional GRs, MGRs present an additional water storage layer that increases the rainwater retention capacity of the system. The four installed prototypes are equipped with a standard monitoring station, provided by Metropolder, and differ for their setting, geometry and extension, vegetation and soil types, and some additional monitoring instruments. Moreover, although all cities are located in the Mediterranean area, each site is characterized by different climatological conditions. This work presents preliminary analyses that investigate the retention capacity and the thermal insulation properties of MGRs, comparing measurements and results from the four prototypes during the period from December 2020 to December 2021.

To estimate the rainwater retention capacity of this system, two indices have been proposed, focusing on the performance of the entire structure or only of the soil layer. Results show how the MGR structure ensures a retention capacity higher than a traditional GR and enables to retain most of the high-intensity rainfall events, especially in the warm season. This suggests the need for a weather-based optimization of the MGR gate to control water storage in the wet season for Mediterranean climate.

Four indices have been introduced to evaluate the potential implications of the MGR installation on the temperature of the surrounding environment and in the buildings underneath the multilayer roof. Since each site presents a peculiar configuration of the MGR and different locations of the temperature sensors, it was not possible to make a direct comparison among all study cases. However, in all locations it was possible to observe a positive impact of the MGR, which helps to mitigate the daily thermal excursion in all sites. The high potential of the MGR is clearly visible in the Palermo case study, where the daily temperature excursion of the room directly beneath the MGR is lower than that measured at the room beneath the unaltered roof.

On top of the exploratory hydrological and energetical analysis, this work offers a panoramic view of the future research outlines and directions that needs to be followed in order to fully understand the potential benefits of MGR in Mediterranean Climate. Further analysis will focus on the following:

- the implementation of an ecohydrological model for reproducing the MGR behaviour, from the soil moisture dynamics to the evapotranspiration and outflow;

- the optimal management practices for the gate control, to facilitate the regulation of the water storage, based on the specific purposes;
- the water quality of the MGR outflow, in order to evaluate possible re-uses and plan adequate treatments; and
- the assessment of the potential impact of the MGRs at the urban scale, identifying suitable locations for the NBS installation in each of the four cities.

ACKNOWLEDGEMENTS

Funding received from the European Union Climatic KIC programme with grant number EIT SGA 2018 supporting project 'Polder Roof Fieldlabs (180522)'. This work has also been partially funded by the Florisa Melone Award, promoted by the Italian Hydrological Society (IHS).

DATA AVAILABILITY STATEMENT

All relevant data are included in the paper or its Supplementary Information.

CONFLICT OF INTEREST

The authors declare there is no conflict.

REFERENCES

Abas, P. E. & Mahlia, T. 2019 Techno-economic and sensitivity analysis of rainwater harvesting system as alternative water source. *Sustainability* **11** (8), 2365.

Abhijith, K. V., Kumar, P., Gallagher, J., McNabola, A., Baldauf, R., Pilla, F., Broderick, B., Di Sabatino, S. & Pulvirenti, B. 2017 Air pollution abatement performances of green infrastructure in open road and built-up street canyon environments – a review. *Atmospheric Environment* **162**, 71–86. https://doi.org/10.1016/j.atmosenv.2017.05.014.

Adugna, D., Jensen, M. B., Lemma, B. & Gebrie, G. S. 2018 Assessing the potential for rooftop rainwater harvesting from large public institutions. *International Journal of Environmental Research and Public Health* **15** (2). doi:10.3390/ijerph15020336.

Akter, A. & Ahmed, S. 2015 Potentiality of rainwater harvesting for an urban community in Bangladesh. *Journal of Hydrology* **528**, 84–93. doi:10.1016/j.jhydrol.2015.06.017.

Akter, A., Tanim, A. H. & Islam, M. K. 2020 Possibilities of urban flood reduction through distributed-scale rainwater harvesting. *Water Science and Engineering* **13** (2), 95–105. https://doi.org/10.1016/j.wse.2020.06.001.

Alpert, A., Ben-Gai, T., Baharad, A., Benjamini, Y., Yekutieli, D., Colacino, M., Diodato, L., Ramis, C., Homar, V., Romero, R., Michaelides, S. & Manes, A. 2002 The paradoxical increase of Mediterranean extreme daily rainfall in spite of decrease in total values. *Geophysical Research Letters* **29** (11), 31-1–31-4. https://doi.org/10.1029/2001GL013554.

Andenæs, E., Kvande, T., Muthanna, T. M. & Lohne, J. 2018 Performance of blue-green roofs in cold climates: a scoping review. *Buildings* **8** (4), 55.

Arnone, E., Pumo, D., Francipane, A., La Loggia, G. & Noto, L. V. 2018 The role of urban growth, climate change, and their interplay in altering runoff extremes. *Hydrological Processes* **32** (12), 1755–1770.

Beckers, B., Berking, J. & Schütt, B. 2013 Ancient water harvesting methods in the drylands of the Mediterranean and Western Asia. *Journal for Ancient Studies* **2**, 145–164.

Bevilacqua, P., Mazzeo, D., Bruno, R. & Arcuri, N. 2017 Surface temperature analysis of an extensive green roof for the mitigation of urban heat island in southern mediterranean climate. *Energy and Buildings* **150**, 318–327. https://doi.org/10.1016/j.enbuild.2017.05.081.

Bocanegra-Martínez, A., Ponce-Ortega, J. M., Nápoles-Rivera, F., Serna-González, M., Castro-Montoya, A. J. & El-Halwagi, M. M. 2014 Optimal design of rainwater collecting systems for domestic use into a residential development. *Resources, Conservation and Recycling* **84**, 44–56. https://doi.org/10.1016/j.resconrec.2014.01.001.

Boers, T. M. & Ben-Asher, J. 1982 A review of rainwater harvesting. *Agricultural Water Management* **5** (2), 145–158. https://doi.org/10.1016/0378-3774(82)90003-8.

Brandão, C., Cameira, M. d. R., Valente, F., Cruz de Carvalho, R. & Paço, T. A. 2017 Wet season hydrological performance of green roofs using native species under Mediterranean climate. *Ecological Engineering* **102**, 596–611. https://doi.org/10.1016/j.ecoleng.2017.02.025.

Busker, T., de Moel, H., Haer, T., Schmeits, M., van den Hurk, B., Myers, K., Cirkel, D. G. & Aerts, J. 2022 Blue-green roofs with forecast-based operation to reduce the impact of weather extremes. *Journal of Environmental Management* **301**, 113750. DOI:https://doi.org/10.1016/j.jenvman.2021.113750.

Calheiros, C. S. C. & Stefanakis, A. I. 2021 Green roofs towards circular and resilient cities. *Circular Economy and Sustainability*. doi:10.1007/s43615-021-00033-0.

Campisano, A. & Lupia, F. 2017 A dimensionless approach for the urban-scale evaluation of domestic rainwater harvesting systems for toilet flushing and garden irrigation. *Urban Water Journal* **14** (9), 883–891. doi:10.1080/1573062X.2017.1279192.

Campisano, A. & Modica, C. 2015 Appropriate resolution timescale to evaluate water saving and retention potential of rainwater harvesting for toilet flushing in single houses. *Journal of Hydroinformatics* **17** (3), 331–346. doi:10.2166/hydro.2015.022.

Castleton, H. F., Stovin, V., Beck, S. B. M. & Davison, J. B. 2010 Green roofs; building energy savings and the potential for retrofit. *Energy and Buildings* **42** (10), 1582–1591. https://doi.org/10.1016/j.enbuild.2010.05.004.

Chao-Hsien, L., En-Hao, H. & Yie-Ru, C. 2014 Designing a rainwater harvesting system for urban green roof irrigation. *Water Supply* **15** (2), 271–277. doi:10.2166/ws.2014.107.

Cipolla, S. S., Altobelli, M. & Maglionico, M. 2018 Systems for rainwater harvesting and greywater reuse at the building scale: a modelling approach. *Environmental Engineering and Management Journal* **17** (10), 2349–2360.

Coma, J., Pérez, G., Solé, C., Castell, A. & Cabeza, L. F. 2016 Thermal assessment of extensive green roofs as passive tool for energy savings in buildings. *Renewable Energy* **85**, 1106–1115.

Cristiano, E., Urru, S., Farris, S., Ruggiu, D., Deidda, R. & Viola, F. 2020 Analysis of potential benefits on flood mitigation of a CAM green roof in Mediterranean urban areas. *Building and Environment* **183**, 107179. https://doi.org/10.1016/j.buildenv.2020.107179.

Cristiano, E., Deidda, R. & Viola, F. 2021a The role of green roofs in urban Water-Energy-Food-Ecosystem nexus: a review. *Science of The Total Environment* **756**, 143876. https://doi.org/10.1016/j.scitotenv.2020.143876.

Cristiano, E., Farris, S., Deidda, R. & Viola, F. 2021b Comparison of blue-green solutions for urban flood mitigation: a multi-city large-scale analysis. *PLOS ONE* **16** (1), e0246429. doi:10.1371/journal.pone.0246429.

Deidda, R. 2010 A multiple threshold method for fitting the generalized Pareto distribution to rainfall time series. *Hydrol. Earth Syst. Sci.* **14** (12), 2559–2575. doi:10.5194/hess-14-2559-2010.

Faivre, N., Fritz, M., Freitas, T., de Boissezon, B. & Vandewoestijne, S. 2017 Nature-based solutions in the EU: innovating with nature to address social, economic and environmental challenges. *Environmental Research* **159**, 509–518.

Freni, G. & Liuzzo, L. 2019 Effectiveness of rainwater harvesting systems for flood reduction in residential urban areas. *Water* **11** (7), 1389.

Getter, K. L., Rowe, D. B. & Andresen, J. A. 2007 Quantifying the effect of slope on extensive green roof stormwater retention. *Ecological Engineering* **31** (4), 225–231. https://doi.org/10.1016/j.ecoleng.2007.06.004.

Gregoire, B. G. & Clausen, J. C. 2011 Effect of a modular extensive green roof on stormwater runoff and water quality. *Ecological Engineering* **37** (6), 963–969. https://doi.org/10.1016/j.ecoleng.2011.02.004.

Grimaldi, S., Petroselli, A., Baldini, L. & Gorgucci, E. 2018 Description and preliminary results of a 100 square meter rain gauge. *Journal of Hydrology* **556**, 827–834.

Hanzl, M. 2020 Urban forms and green infrastructure – the implications for public health during the COVID-19 pandemic. *Cities & Health* 1–5. doi:10.1080/23748834.2020.1791441.

Hellies, M., Deidda, R. & Viola, F. 2018 Retention performances of green roofs worldwide at different time scales. *Land Degradation & Development* **29** (6), 1940–1952. doi:10.1002/ldr.2947.

Honey-Rosés, J., Anguelovski, I., Chireh, V. K., Daher, C., Konijnendijk van den Bosch, C., Litt, J. S., Mawani, V., McCall, M. K., Orellana, A., Oscilowicz, E., Sánchez, U., Senbel, M., Tan, X., Villagomez, E., Zapata, O. & Nieuwenhuijsen, M. J. 2020 The impact of COVID-19 on public space: an early review of the emerging questions – design, perceptions and inequities. *Cities & Health* 1–17. doi:10.1080/23748834.2020.1780074.

IPCC, I.P.o.C.C 2007 *Climate Change 2007: Impacts, Adaptation and Vulnerability. Contribution of Working Group II to the Fourth Assessment Report of the Intergovernmental Panel on Climate Change.* Cambridge University Press, Cambridge, UK, p. 976.

Jenkins, A. 2020 Biotic systems as a critical urban infrastructure during crisis: learning from the COVID-19 pandemic. *Cities & Health* 1–3. doi:10.1080/23748834.2020.1789821.

Kumar, P., Druckman, A., Gallagher, J., Gatersleben, B., Allison, S., Eisenman, T. S., Hoang, U., Hama, S., Tiwari, A., Sharma, A., Abhijith, K. V., Adlakha, D., McNabola, A., Astell-Burt, T., Feng, X., Skeldon, A. C., de Lusignan, S. & Morawska, L. 2019 The nexus between air pollution, green infrastructure and human health. *Environment International* **133**, 105181. https://doi.org/10.1016/j.envint.2019.105181.

La Rosa, D. & Pappalardo, V. 2020 Planning for spatial equity – a performance based approach for sustainable urban drainage systems. *Sustainable Cities and Society* **53**, 101885. https://doi.org/10.1016/j.scs.2019.101885.

Lazzarin, R. M., Castellotti, F. & Busato, F. 2005 Experimental measurements and numerical modelling of a green roof. *Energy and Buildings* **37** (12), 1260–1267. https://doi.org/10.1016/j.enbuild.2005.02.001.

Liu, L., Sun, L., Niu, J. & Riley, W. J. 2020 Modeling green roof potential to mitigate urban flooding in a Chinese city. *Water* **12** (8), 2082.

Masson-Delmotte, V., Zhai, P., Pörtner, H.-O., Roberts, D., Skea, J. & Shukla, P. R. (eds). 2018 Global warming of 1.5 C. An IPCC Special Report on the impacts of global warming. IPCC, Geneva, pp. 1-9. Available from: https://www.ipcc.ch/site/assets/uploads/sites/2/2019/06/SR15_Full_Report_Low_Res.pdf.

Muhammad, S. & Reeho, K. 2017 Application of green blue roof to mitigate heat island phenomena and resilient to climate change in urban areas: a case study from Seoul, Korea. *Journal of Water and Land Development* **33** (1), 165–170. https://doi.org/10.1515/jwld-2017-0032.

Nesshöver, C., Assmuth, T., Irvine, K. N., Rusch, G. M., Waylen, K. A., Delbaere, B., Haase, D., Jones-Walters, L., Keune, H., Kovacs, E., Krauze, K., Külvik, M., Rey, F., van Dijk, J., Vistad, O. I., Wilkinson, M. E. & Wittmer, H. 2017 The science, policy and practice of nature-based solutions: an interdisciplinary perspective. *Science of the Total Environment* **579**, 1215–1227. https://doi.org/10.1016/j.scitotenv.2016.11.106.

Oral, H. V., Carvalho, P., Gajewska, M., Ursino, N., Masi, F., van Hullebusch, E. D., Kazak, J. K., Exposito, A., Cipolletta, G., Raaschou Andersen, T., Finger, D. C., Simperler, L., Regelsberger, M., Rous, V., Radinja, M., Buttiglieri, G., Krzeminski, P., Rizzo, A., Dehghanian,

K., Nikolova, M. & Zimmermann, M. 2020 A review of nature-based solutions for urban water management in European circular cities: a critical assessment based on case studies and literature. *Blue-Green Systems* **2** (1), 112–136. doi:10.2166/bgs.2020.932.

Palermo, S. A., Turco, M., Principato, F. & Piro, P. 2019 Hydrological effectiveness of an extensive green roof in Mediterranean climate. *Water* **11** (7), 1378.

Palermo, S. A., Talarico, V. C. & Pirouz, B. 2020 Optimizing rainwater harvesting systems for non-potable water uses and surface runoff mitigation. In: *Numerical Computations: Theory and Algorithms* (Sergeyev, Y. D. & Kvasov, D. E., eds). Springer International Publishing, Cham, pp. 570–582.

Palla, A., Gnecco, I. & Lanza, L. G. 2010 Hydrologic restoration in the urban environment using green roofs. *Water* **2** (2), 140–154.

Pelorosso, R. 2020 Modeling and urban planning: a systematic review of performance-based approaches. *Sustainable Cities and Society* **52**, 101867. https://doi.org/10.1016/j.scs.2019.101867.

Pelorosso, R., Gobattoni, F. & Leone, A. 2018 Increasing hydrological resilience employing nature-based solutions: a modelling approach to support spatial planning. In: *Smart Planning: Sustainability and Mobility in the Age of Change* (Papa, R., Fistola, R. & Gargiulo, C., eds). Springer International Publishing, Cham, pp. 71–82. doi:10.1007/978-3-319-77682-8_5.

Pelorosso, R., La Rosa, D., Floris, S. C. & Cerino, N. 2021a Planning accessible urban green infrastructure for healthy and fair historical towns: the study case of Viterbo, Central Italy. In: *Innovation in Urban and Regional Planning. INPUT 2021. Lecture Notes in Civil Engineering*, Vol. 146 (Rosa, D. L. & Privitera, R., eds). Springer, Cham.

Pelorosso, R., Petroselli, A., Apollonio, C. & Grimaldi, S. 2021b Blue-green roofs: hydrological evaluation of a case study in Viterbo, Central Italy. In: *Innovation in Urban and Regional Planning. INPUT 2021. Lecture Notes in Civil Engineering*, Vol. 146 (Rosa, D. L. & Privitera, R., eds). Springer, Cham.

Recanatesi, F. & Petroselli, A. 2020 Land cover change and flood risk in a peri-urban environment of the metropolitan area of Rome (Italy). *Water Resources Management* **34** (14), 4399–4413. doi:10.1007/s11269-020-02567-8.

Rowe, D. B. 2011 Green roofs as a means of pollution abatement. *Environmental Pollution* **159** (8), 2100–2110. https://doi.org/10.1016/j.envpol.2010.10.029.

Sample, D. J. & Liu, J. 2014 Optimizing rainwater harvesting systems for the dual purposes of water supply and runoff capture. *Journal of Cleaner Production* **75**, 174–194. https://doi.org/10.1016/j.jclepro.2014.03.075.

Santamouris, M. 2014 Cooling the cities–a review of reflective and green roof mitigation technologies to fight heat island and improve comfort in urban environments. *Solar Energy* **103**, 682–703.

Shafique, M., Kim, R. & Lee, D. 2016a The potential of green-blue roof to manage storm water in urban areas. *Nature Environment and Pollution Technology* **15** (2), 715.

Shafique, M., Lee, D. & Kim, R. 2016b A field study to evaluate runoff quantity from blue roof and green blue roof in an urban area. *International Journal of Control and Automation* **9** (8), 59–68.

Solcerova, A., van de Ven, F., Wang, M., Rijsdijk, M. & van de Giesen, N. 2017 Do green roofs cool the air? *Building and Environment* **111**, 249–255. https://doi.org/10.1016/j.buildenv.2016.10.021.

Speak, A. F., Rothwell, J. J., Lindley, S. J. & Smith, C. L. 2012 Urban particulate pollution reduction by four species of green roof vegetation in a UK city. *Atmospheric Environment* **61**, 283–293. https://doi.org/10.1016/j.atmosenv.2012.07.043.

Susca, T., Gaffin, S. R. & Dell'Osso, G. 2011 Positive effects of vegetation: urban heat island and green roofs. *Environmental Pollution* **159** (8–9), 2119–2126.

Takebayashi, H. & Moriyama, M. 2007 Surface heat budget on green roof and high reflection roof for mitigation of urban heat island. *Building and Environment* **42** (8), 2971–2979.

Teston, A., Teixeira, C. A., Ghisi, E. & Cardoso, E. B. 2018 Impact of rainwater harvesting on the drainage system: case study of a condominium of houses in Curitiba, Southern Brazil. *Water* **10** (8), 1100.

Treppiedi, D., Cipolla, G., Francipane, A. & Noto, L. 2021 Detecting precipitation trend using a multiscale approach based on quantile regression over a Mediterranean area. *International Journal of Climatology* **41**, 5938–5955.

UN 2018 *United Nations Final Report on World Urbanization Prospects 2018.*

van den Bosch, M. & Ode Sang, Å. 2017 Urban natural environments as nature-based solutions for improved public health – a systematic review of reviews. *Environmental Research* **158**, 373–384. https://doi.org/10.1016/j.envres.2017.05.040.

van der Meulen, S. H. 2019 Costs and benefits of green roof types for cities and building owners. *Journal of Sustainable Development of Energy, Water and Environment Systems* **7** (1), 57–71.

Viola, F., Hellies, M. & Deidda, R. 2017 Retention performance of green roofs in representative climates worldwide. *Journal of Hydrology* **553**, 763–772. doi:10.1016/j.jhydrol.2017.08.033.

Xie, J., Luo, S., Furuya, K. & Sun, D. 2020 Urban parks as green buffers during the COVID-19 pandemic. *Sustainability* **12** (17), 6751.

Yang, J., Yu, Q. & Gong, P. 2008 Quantifying air pollution removal by green roofs in Chicago. *Atmospheric Environment* **42** (31), 7266–7273. https://doi.org/10.1016/j.atmosenv.2008.07.003.

Zhang, X. & Hu, M. 2014 Effectiveness of rainwater harvesting in runoff volume reduction in a planned industrial park, China. *Water Resources Management* **28** (3), 671–682. doi:10.1007/s11269-013-0507-9.

First received 22 September 2021; accepted in revised form 25 July 2022. Available online 8 August 2022

© 2022 The Authors
doi: 10.2166/nh.2022.126

The limits of watershed delineation: implications of different DEMs, DEM resolutions, and area threshold values

Srijon Datta[a], Shyamal Karmakar [a,*], Symon Mezbahuddin[b,c,d], Mohammad Mozaffar Hossain[a], B. S. Chaudhary [e], Md. Enamul Hoque[f], M. M. Abdullah Al Mamun [a,g] and Tarit Kumar Baul[a]

[a] Institute of Forestry and Environmental Sciences, University of Chittagong, Chattogram - 4331, Bangladesh
[b] Natural Resource Management Branch, Alberta Agriculture and Forestry, AB, Edmonton, Canada
[c] Department of Renewable Resources, University of Alberta, AB, Edmonton, Canada
[d] 4S Analytics and Modelling Ltd, AB, Edmonton, Canada
[e] Department of Geophysics, Kurukshetra University, Kurukshetra, Haryana 136119, India
[f] Department of Oceanography, University of Chittagong, Chattogram - 4331, Bangladesh
[g] Aerospace Information Research Institute, Chinese Academy of Sciences, Beijing, China
*Corresponding author. E-mail: shyamal.karmakar@cu.ac.bd

SK, 0000-0001-7282-0479; BSC, 0000-0003-3789-8443; MMA, 0000-0002-7210-930X

ABSTRACT

Identifying and demarcating watershed areas provides a basis for designing and planning for water resources. In this study, DEMs-based estimates of watershed characteristics of three rivers of Bangladesh – Halda, Sangu, and Chengi – were derived using eight Digital Elevation Models (DEMs) of 30 m, 90 m, and 225 m resolution in the Soil and Water Assessment Tool (SWAT). We have assessed watershed characteristics concerning DEMs, resolutions, and Area Threshold Values (ATVs). Though the elevation data differed, high correlation values among DEMs and resolutions confirm the negligible effect of elevation in the watershed delineation. However, the slope and watershed delineation vary for different DEMs and resolutions. The 90 m DEMs estimated larger areas for Halda and Chengi and lower perimeter values for all three rivers. In watershed delineation, the area near the mouth and flat terrain did not coincide with DEMs. The common intersected area by DEMs can be used as the focal area of watershed management. ATV $\leq$ 40 km^2 significantly influences sub-basin counts and stream network extraction for these watershed areas. Though watershed size and shape were independent of the different ATVs, the DEM-based watershed delineation process in SWAT needs optimum ATV values to represent the stream network precisely.

Key words: catchment, DEM resolution, hydrological model, network, stream, SWAT

HIGHLIGHTS

- DEM source influences the slope, and DEM resolution affects the basin perimeter of a watershed.
- The flat terrain area influences watershed delineation.
- The sub-basin counts and stream network extraction should consider an optimum ATV value (here, $\leq$40 km^2).
- The watershed size and shape were independent of the ATVs.
- The common intersected area, independent of the DEMs and resolutions, can be used as the watershed management area.

GRAPHICAL ABSTRACT

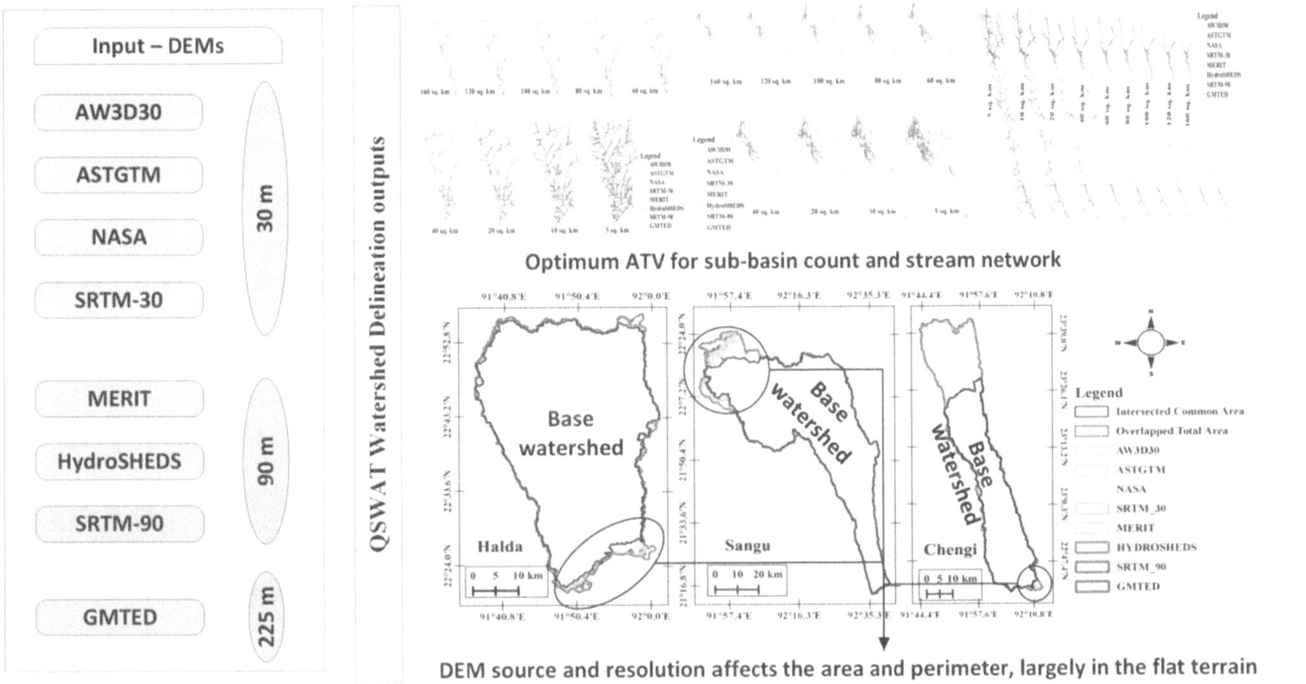

1. INTRODUCTION

A watershed is a natural geo-hydrological unit where water flows downhill, moves through a common outlet by a system of streams and distributes to rivers, reservoirs, or lakes (Paranjape *et al.* 1998; Sowmya *et al.* 2020). It consists of a natural water divider separating drainage basins and offering a natural boundary for planning, management, and development (LaMoreaux *et al.* 2008; Kumar & Dhiman 2014). Watershed delineation is a prerequisite in hydrologic and environmental evaluation of a watershed, such as runoff estimation, water quality modelling, flood assessment, and disaster risk assessment (Luo *et al.* 2011; Giridhar *et al.* 2015; Ray 2018). Before the advancement of remote sensing and Geographical Information System (GIS) technologies, watershed delineation was predominantly manual drawings of the watersheds and stream networks using visual interpretations and digitization of the topographic and contour maps (Ehsani *et al.* 2010; Salih & Hamid 2017). With recently advanced GIS technology, watershed delineations are done using remotely sensed terrain features processed by different digital elevation models (DEMs) at various spatial resolutions (Munoth & Goyal 2019).

Most hydrological models use DEM to account for geomorphological and hydrological variables in a watershed spatio-temporal modelling. The spatial resolution of a DEM can affect the outputs for elevation, slope, stream network, sub-watersheds, watershed area, sediment and nutrient loads (Chaubey *et al.* 2005; Zhang *et al.* 2014; Reddy & Reddy 2015; Buakhao & Kangrang 2016; Xu *et al.* 2016; Wu *et al.* 2017; Ray 2018; Munoth & Goyal 2019). A DEM with a coarser resolution can simulate larger basin areas with shorter streams, flatter slopes and insignificant differences in altitude, which reduces heterogeneity and accuracy (Chaplot 2005; Chaubey *et al.* 2005; Liu *et al.* 2010). Conversely, a finer resolution DEM can delineate a more accurate river network but less accurate watershed boundary (Liu *et al.* 2010; Ray 2018). Moreover, DEMs vary in horizontal and vertical outputs, primarily affecting stream network and delineation in relatively flatter topography (Li & Wong 2010; Luo *et al.* 2011; Al-Khafaji & Al-Sweiti 2017; Shafiq *et al.* 2020). The outcomes of GIS-based hydrological models are greatly influenced by the spatial resolution and accuracy of DEMs (Piwowar & LeDrew 1990; Moore *et al.* 1991; Wolock & Price 1994; Wolock & McCabe 1995; Cho & Lee 2001). Hence, a DEM selection is critical in the watershed delineation process and subsequent catchment solute transport study. The use of two or more DEMs was suggested for efficient hydrogeological research and watershed management of an area (Ray 2018), though the use of a higher number of DEMs has not been reported in the literature until now.

Apart from DEM resolution, Area Threshold Values (ATVs) also control the quality of hydrological model outputs (Munoth & Goyal 2019). A decrease in ATV detects minute differences within the extracted drainage parameters by different DEMs that produce a proportionately large number of drainage channels and, in turn, affects hydrological model outputs such as sub-basin count and surface runoff (Reddy *et al.* 2018; Gautam *et al.* 2019; Munoth & Goyal 2019). However, none of these researches suggested any optimum ATV for a specific range of watersheds in any hydrological modelling study. A hydrological modelling software, such as the Soil and Water Assessment Tool (SWAT), derived drainage channel delineation affected by DEM resolution and ATVs (Kalcic *et al.* 2015).

SWAT is a widely-used physically-based basin-scale distributed continuous hydrologic model (Arnold *et al.* 1998). In SWAT, the stream burning process modifies DEM elevation data to overlap the surface drainage patterns with existing stream network locations (Wang *et al.* 2011; Lindsay 2016). The SWAT-generated stream network depends on ATVs and allows users to divide the basin into sub-basins based on topography to incorporate spatial details (Munoth & Goyal 2019). Since a lower ATV gives a higher sub-basin count and subsequent surface runoff reduction, an ATV significantly influences basin hydrological characterization and modelling.

In Bangladesh, hydrological processes, which influence aquatic habitat management and flood control, have been modelled using SWAT-based hydrological models with geomorphologic inputs from open-source DEMs (Raihan *et al.* 2020). Such modelling helps adapt water resource management and monitoring systems for a river like Halda, a significant aquatic habitat in Bangladesh (Raihan *et al.* 2020). SRTM (Shuttle Radar Topography Mission) DEM of 30 m resolution was used in the SWAT model to study the effects of hydrological structures on the flow of the Halda and the sustainability of the aquatic habitat (Saha *et al.* 2019a) and susceptibility to flash floods to the Karnaphuli and the Sangu River basins (Adnan *et al.* 2019). Evaluation of the hydrological modelling using SRTM 90 m DEM in the flat terrain of twelve catchments in Bangladesh showed that the slope parameter affected river network delineation (Rahman *et al.* 2010). However, only SRTM DEM and one ATV were used for specific watershed areas in these studies, which apparently could not represent the variation in outputs using different sources and qualities of DEMs in different terrains.

In this study, we aimed to examine the challenges in watershed delineation using eight different sources of DEM, varying resolutions (30 m, 90 m, and 225 m), and for nine ATVs in delineating three watersheds, i.e. Sangu, Chengi, and Halda with an area between 1,200 km^2 and 3,900 km^2 located in south-eastern Bangladesh (Figure 1). These rivers created complex stream networks at their estuaries. Though these basin areas lie within Bangladesh country boundary and possess a considerable management implication in water resources, to date, the discharge or sediment load was not reported in research. Moreover, the effects of DEMs on watershed hydrological parameters were compared for the first time for any river basin in Bangladesh. We hypothesized that the watershed output of three rivers (Halda, Sangu, and Chengi) basins from different DEMs would produce varied results in hydrological attributes or water resource management studies. Moreover, attributed ATV should show resembling trends in sub-basin count, stream length, and drainage density for the different river basins, where the elevation and slope of the DEMs are sensitive to DEM resolution. Additionally, we will evaluate the performance of all freely available DEM products for the generalized utility (viz., basin hypsometry and slope). Finally, a generalized ATV for the catchment size during stream network processing would be possible to suggest from this study.

2. MATERIALS AND METHODS

2.1. Catchment areas of the rivers

Halda River catchment: Halda originated from the Badnatali Hills range in Ramgarh, a sub-district of Khagrachari District, and ended in Karnaphuli River as one of the major tributaries, about 35 km away from the Bay of Bengal (Saha *et al.* 2019b). The watershed of Halda is bounded by 91 °36.6′E–92 °1.2′E longitude and 22 °56.4′N–22 °20.4′N latitude (Figure 1). The river basin lies inside the political boundary of Bangladesh and is well known as the only natural carp-breeding habitat in South-east Asia (Azadi 2005; Correspondent 2014; Saha *et al.* 2019a; 2019b). The Bangladesh government decided to declare the river as the Ecologically Critical Area to protect the sanctuary of the carp fishes (Hussain 2016). It is declared the 'Bangabondu Fisheries Heritage' site (Report 2020) that prohibits all water abstraction, fishing and sand quarrying activities on this river.

Sangu river catchment: Sangu River originates in the North Arakan Hills of Myanmar and enters Bangladesh from the east of Remarki, Thanchi, a sub-district of Bandarban district. The watershed of Sangu is bounded between 91 °48′E–92 °42′E and 22 °24.6′N–21 °14.4′N (Figure 1). Sangu supports irrigation, fisheries, and navigation in the region.

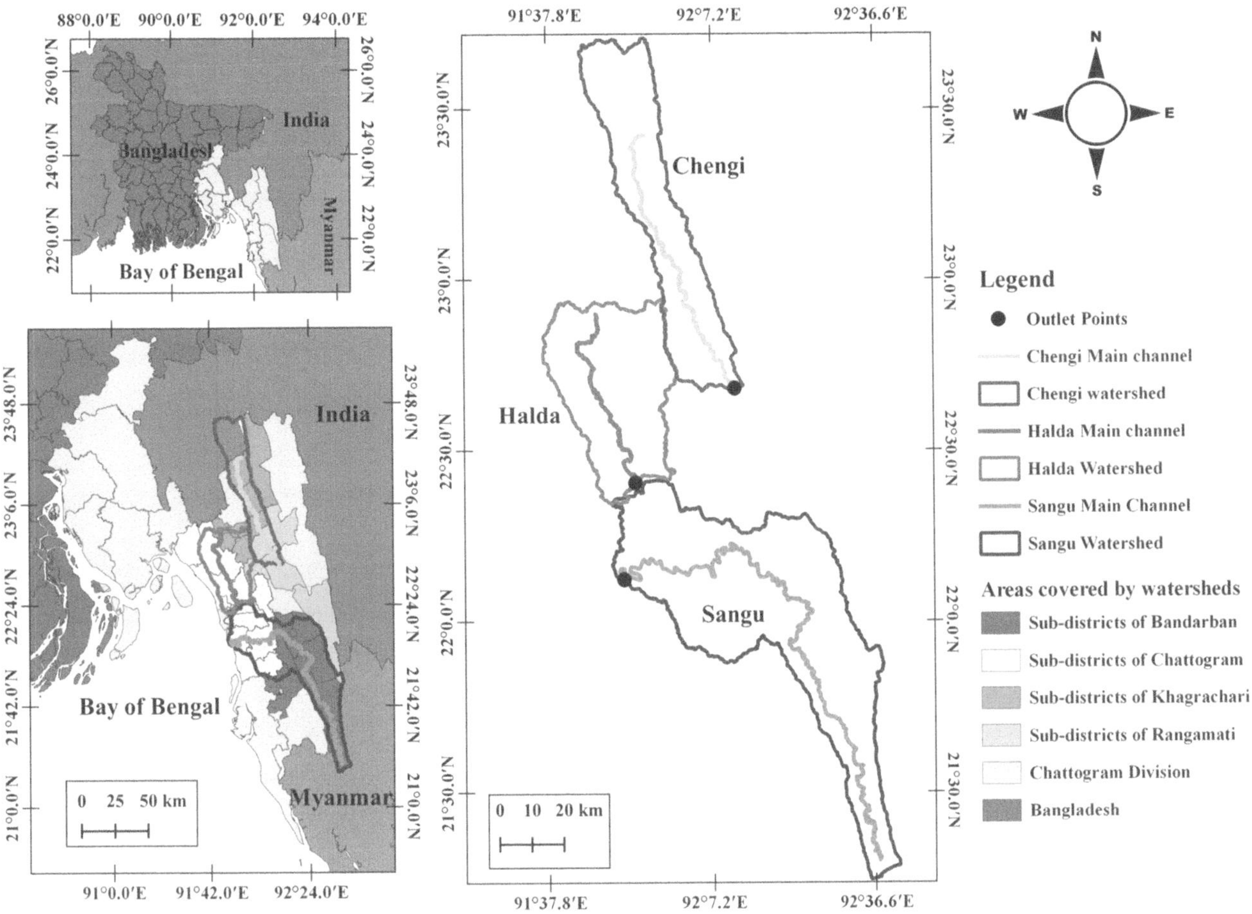

Figure 1 | Study area map.

Chengi river catchment: Chengi spreads through 3 sub-districts of Khagrachari district and 2 sub-districts of Rangamati district, starting from Gandacherra, India. The watershed of Sangu is bounded between 91 °43.2′E–92 °12′E and 23 °42.4′N–22 °39′N (Figure 1). In hilly areas of the Khagrachari district, Chengi is the longest and one of the most potent water sources for ethnic local communities. People use the river water for domestic, agricultural, navigation, extraction, and transport of forest resources and recreational purposes. The water quality of this river has been deteriorating lately due to the disposal of untreated industrial effluents like rubber factories, municipal sewage, rubber dam, riverbank erosion, and overuse of fertilizer (Latifa *et al.* 2019).

2.2. Watershed delineation in SWAT

We have used the QGIS (v.2.6.1-Brighton) interface of SWAT (QSWAT v. 1.9) for watershed delineation. QSWAT uses TauDEM (Terrain Analysis Using Digital Elevation Models) to perform this process. Uniform spatial projection UTM 46 N is used for DEM and vector files of outlet points and observed stream networks (Figure 2). The user defines an ATV for creating sub-basins and extracts stream networks in the watershed. The stream input dataset creates continuous lines that go through wetlands and hydraulic structures but avoids crossing the DEM edge. The outlet points were selected within the defined snap threshold distance (300 m) to ensure proper determination of the stream network.

2.2.1. Source of different DEMs and primary data

In this study, AW3D30 (Advanced Land Observation Satellite World 3D 30 m), ASTGTM (Advanced Space-borne Thermal Emission and Reflection Radiometer Global Digital Elevation Model, Version 3), NASA (National Aeronautics and Space Administration, USA), and SRTM (Shuttle Radar Topographic Mission) DEMs of ~30 m horizontal resolution; Hydro-SHEDS (Hydrological data and maps based on Shuttle Elevation Derivatives at multiple Scales), MERIT (Multi-Error-

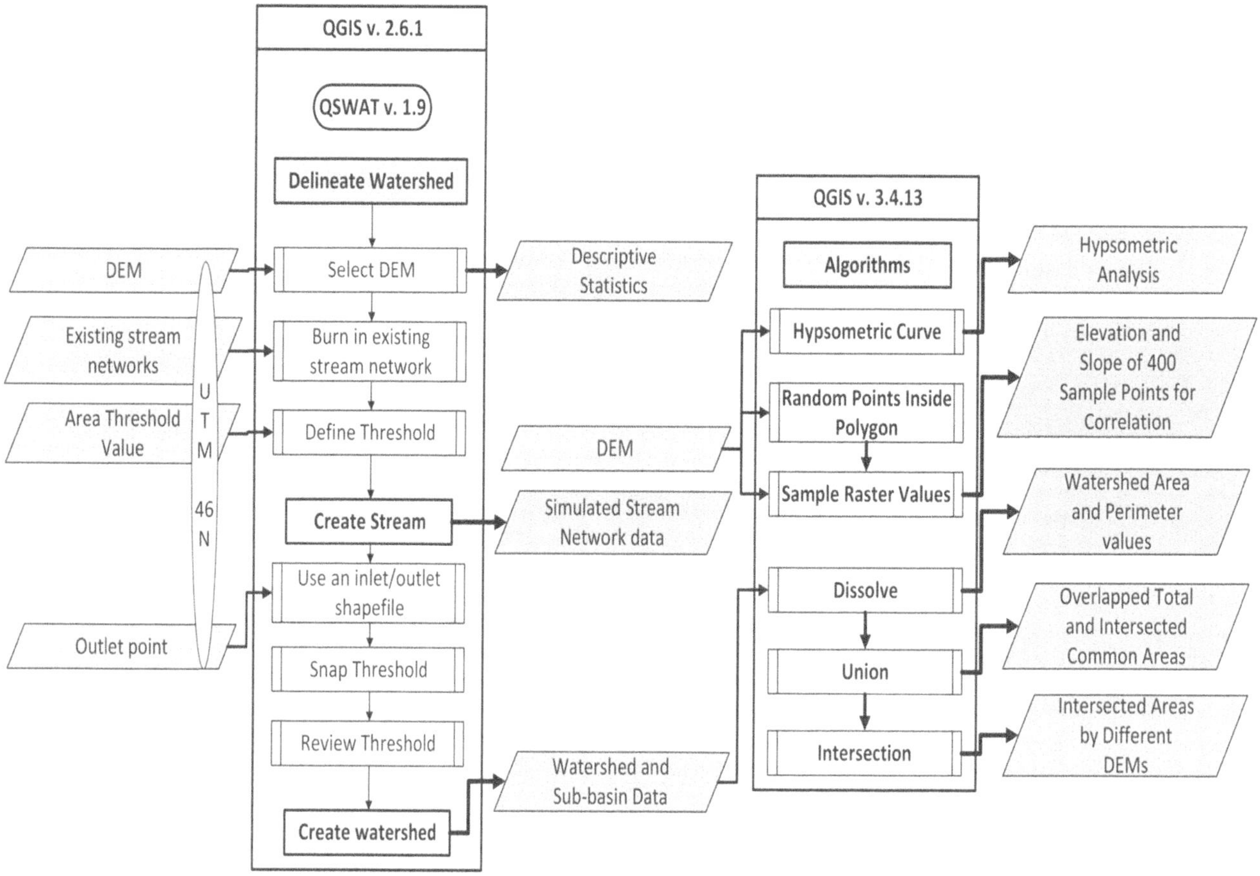

Figure 2 | Stepwise watershed delineation process in QSWAT and data arrangement and analysis in QGIS.

Removed Improved-Terrain), SRTM DEMs of ~90 m resolution, and GMTED (Global Multi-resolution Terrain Elevation Data) DEM of ~225 m resolution were used to carry out the analysis (Table 1 and SI – Fig. S1).

2.2.2. Stream network extraction

The observed stream network may differ from the extracted ones for a DEM due to different resolutions and data sources or modification of a river course by river training (Reddy & Reddy 2015). To define the accurate position of the main streams, a digitized and georeferenced main channel (polyline vector file) from Google Earth was used for each of the Halda, Sangu, and Chengi river basins during the stream network extraction process. We conducted a field survey and communicated with water management authorities to affirm the accuracy of the stream burn-in method and outlet points. Sub-basin creation followed the 8 DEMs for the threshold of 5, 10, 20, 40, 60, 80, 100, 120, and 160 km^2 areas, giving 72 experiments for the stream networks.

2.2.3. Data analysis: hypsometry, correlation, watershed and impact of ATVs

We have performed geospatial data organization, analyses, and visualizations in QGIS v. 3.4.13 (Madeira). Hypsometric curves from different DEMs were constructed by plotting relative areas (a/A) on the x-axis (%) and relative heights (h/H) compared to mean sea level (MSL) on the y-axis (%). The hypsometric integral (HI) was calculated as the ratio of elevation differences estimated in DEMs (Equation(1)). The HI, along with the hypsometric curves, describes the stage of geological development of a watershed and respective landmass distribution (Kusre 2013; Golekar *et al.* 2015). The elevation-relief ratio (E) is defined as equivalent to the HI (Pike & Wilson 1971), as follows

$$E \approx HI = \frac{\text{Elevation}_{\text{mean}} - \text{Elevation}_{\text{minimum}}}{\text{Elevation}_{\text{maximum}} - \text{Elevation}_{\text{minimum}}} \quad (1)$$

Table 1 | Information about DEMs

DEMs	AW3D30	ASTGTM	NASA	SRTM-30	HydroSHEDS	MERIT	SRTM-90	GMTED
Spatial resolution/ Posting interval	1 arc second (approximately. 30 m)	1 arc second (approximately. 30 m)	1 arc second (approximately. 30 m)	1 arc-second (approximately. 30 m)	3 arc second (approximately. 90 m)	3 arc second (approximately. 90 m)	3 arc second (approximately. 90 m)	7.5 arc second (approximately. 225 m)
Publication date and acquired version (V)	April 2020, V 3	13 June 2019, V 3	24 May 2018, V 1	23 September 2014, V 3	2006	15 October 2018, v 1.0.3	November 2018, V 4	11 November, 2010
Observation period or temporal extent	2006–2011	March 1, 2000– November 30, 2013	11–21 February, 2000	11–21 February, 2000	2006	Variable	11–21 February, 2000	Not specified
Coordinate System or Projection	ITRF97	Geographic	Geographic	Geographic	Geographic	Geographic	Geographic	Geographic
Datum (ellipsoid/ geoid)	GRS80/EGM96	WGS84/EGM96	WGS84/EGM96	WGS84/EGM96	WGS84/EGM96	WGS84/EGM96	WGS84/EGM96	WGS84/EGM96
Global vertical accuracy	5 m	16.8 m	Not specified	≤16 m	Not specified	2 m	≤16 m	26–30 m
Data format	GeoTIFF (.tif)	GeoTIFF (.tif)	HGT (.hgt)	HGT (.hgt)	Header (.hdr)	GeoTIFF (.tif)	GeoTIFF (.tif)	GeoTIFF (.tif)
Publisher/ generating agency	EORC, JAXA	LP DAAC, EOSDIS NASA, and USGS	LP DAAC, EOSDIS NASA, and USGS	LP DAAC, EOSDIS NASA, and USGS	HydroSHEDS database, WWF, and USGS	Dai YAMAZAKI, Institute of Industrial Sciences, The University of Tokyo	CGIAR-CSI SRTM	EROS, USGS
Download source	https://www.eorc.jaxa.jp/ALOS/en/aw3d30/data/index.htm	https://earthdata.nasa.gov/	https://earthdata.nasa.gov/	https://earthexplorer.usgs.gov/	https://www.hydrosheds.org/downloads	http://hydro.iis.u-tokyo.ac.jp/~yamadai/MERIT_DEM/	http://srtm.csi.cgiar.org/SELECTION/inputCoord.asp	https://earthexplorer.usgs.gov/
References	(JAXA 2020)	(NASA/METI/AIST/Japan Space Systems 2019)	(NASA 2020)	(NASA 2013)	(Lehner *et al.* 2008)	(Yamazaki *et al.* 2017)	(Jarvis *et al.* 2008)	(Danielson & Gesch 2011)

To evaluate the correlation among DEMs, 400 random points were extracted from across the watershed using the Random Points inside Polygon and Sample Raster Values algorithms (SI – Table S1-S3). The efficiency of DEMs in delineating a watershed area can be compared easily by visual inspection and observing intersected areas between DEMs (Buakhao & Kangrang 2016). The dissolve, union, and intersection tools of QGIS were used to extract the overlapping segments of the DEM layers to determine the intersected area derived from DEMs. Furthermore, a quantitative relationship was derived between the ATV and sub-basin count for a specific DEM for the individual river by a power function $y = ax^b$. Here, the sub-basin number (y) of a watershed for an individual DEM (constant a and exponent factor b) can be predicted if a specific ATV ($x > 0$) changes.

3. RESULTS

3.1. Watershed physiographic characteristics defined in DEMs

3.1.1. Elevation, slope, and hypsometry estimated in different DEMs

Elevation of a watershed characterizes its climatic, vegetation, and runoff attributes. For Halda, Sangu, and Chengi watersheds, about 68% of the area lies 43.62 m, 188.36 m, and 57.48 m from the mean elevation of 40.02 m, 173.76 m, and 102.34 m, respectively (Table 2). This indicates that the Halda watershed has a higher proportion of flat areas than Sangu and Chengi (Figure 3, SI – Figure S1). AW3D30 produced the lowest elevation for the Halda (−18 m) and Sangu (−30 m) watershed and the lowest mean value of elevation for the Sangu and Chengi rivers (Table 2). GMTED, ASTGTM, and MERIT for the Halda watershed; ASTGTM and HydroSHEDS for the Chengi watershed overestimated the lowest average elevation (Table 2). The 90 m resolution DEMs underestimate the maximum average elevation for the Sangu and the Chengi river watershed. HydroSHEDS and SRTM-90 show relatively congruous values of all parameters for the watershed areas (Table 2). The DEMs elevation parameters are varied for all the studied watersheds.

Table 2 | Descriptive statistics of elevation (m) and slope (degree) values of Halda, Sangu and Chengi river watershed from different DEM sources and resolutions

	Range (Minimum-maximum)			Mean ± Standard deviation		
DEM (resolution)	**Halda**	**Sangu**	**Chengi**	**Halda**	**Sangu**	**Chengi**
Elevation (m)						
AW3D30 (30 m)	−18–529	−30–970	26–520	41.47 ± 44.24	170.21 ± 185.93	91.81 ± 58.52
ASTGTM (30 m)	0–540	0–1,001	15–513	39.31 ± 44.11	172.80 ± 189.57	102.55 ± 58.55
NASA (30 m)	−10–519	−5–1,018	25–508	39.18 ± 43.48	178.46 ± 187.58	103.80 ± 58.54
SRTM-30 (30 m)	−15–515	−3–1,044	24–520	41.16 ± 43.63	174.73 ± 188.58	106.37 ± 58.13
MERIT (90 m)	−5.61–524.55	−0.69–937.97	27.31–471.86	36.50 ± 41.35	171.50 ± 192.40	97.40 ± 56.08
HydroSHEDS (90 m)	−10–535	0–902	20–419	40.51 ± 43.76	176.25 ± 188.52	105.00 ± 55.98
SRTM-90 (90 m)	−12–535	−1–901	30–419	40.88 ± 43.87	173.98 ± 188.00	105.55 ± 55.89
GMTED (225 m)	−3–519	−2–1,020	27–521	41.19 ± 44.52	172.17 ± 186.33	106.20 ± 57.86
Average elevation	−9.20–527.07	−5.21–974.25	24.49–486.48	40.02 ± 43.62	173.76 ± 188.36	102.34 ± 57.48
Slope (Degree)						
AW3D30 (30 m)	0–51.80	0–60.27	0–46.22	7.20 ± 6.77	12.41 ± 10.09	10.93 ± 7.8
ASTGTM (30 m)	0–45.00	0–60.50	0–45.66	5.79 ± 5.25	11.91 ± 9.50	8.29 ± 5.98
NASA (30 m)	0–49.22	0–69.79	0–45.19	4.76 ± 4.78	11.78 ± 9.64	7.50 ± 5.83
SRTM-30 (30 m)	0–50.50	0–67.53	0–49.14	5.00 ± 5.02	11.68 ± 9.63	7.83 ± 6.17
MERIT (90 m)	0–45.36	0–55.67	0–31.59	1.91 ± 3.23	8.50 ± 8.44	4.06 ± 4.65
HydroSHEDS (90 m)	0–45.72	0–54.09	0–32.46	2.54 ± 3.25	8.68 ± 8.06	4.74 ± 4.54
SRTM-90 (90 m)	0–45.19	0–50.57	0–32.24	2.46 ± 3.19	8.42 ± 8.00	4.61 ± 4.50
GMTED (225 m)	0–28.27	0–41.78	0–27.55	1.30 ± 1.99	5.94 ± 6.23	2.57 ± 2.80
Average slope	0–45.13	0–57.52	0–38.76	3.87 ± 4.19	9.91 ± 8.70	6.32 ± 5.28

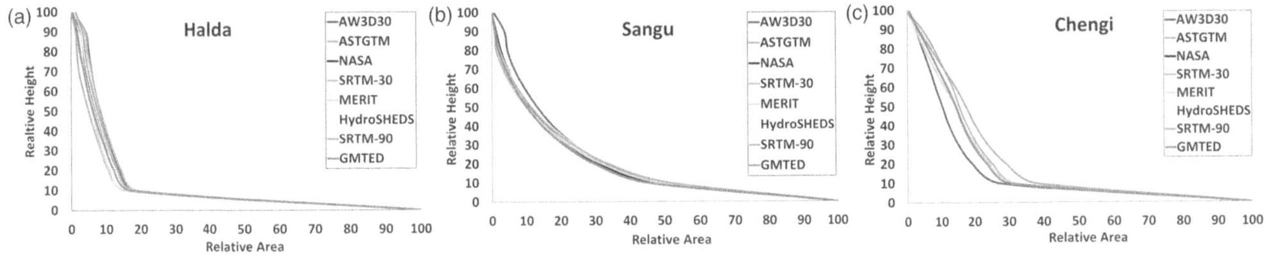

Figure 3 | Hypsometric curves of three river watersheds derived from different DEMs.

Variation of slope among 90 m DEMs is relatively lower for Sangu and Chengi rivers than 30 m DEMs for all the rivers (Table 2). Mean and standard deviation values are highest for AW3D30, and GMTED has the lowest values for all the river watershed parameters (Table 2). A maximum slope of 69.79 degrees is estimated for Sangu from NASA DEM (Table 2). The minimum slope derived from different DEMs is zero degrees. In contrast, slopes' maximum, mean, and standard deviation are higher for 30 m resolution and lower for coarser resolutions (Table 2).

The hypsometric curves of the watersheds were concave-up (Figure 3). About 17% of the Halda watershed was steep sloping with more than the elevation of 83 m, and the remaining areas were relatively flat (Figure 3). Though Chengi had similar attributes, it had a steeper slope (about 32% of the area) and different curve positions for DEMs in high-altitude areas. Unlike the two watersheds, Sangu had a relatively smooth sloping pattern at the gradual increment of the area, and around 50% of the area had more than 105 m (average of all DEMs) elevation (Figure 3). Halda watershed had lower HI values (ranged between 0.07 and 0.11 with an average of 0.09) than Sangu (ranged between 0.17 and 0.20 with an average of 0.18) and Chengi (ranged between 0.13 and 0.21 with an average of 0.17) for all the DEMs.

3.1.2. Correlation of different DEMs' elevation and slope data

Correlation values among the elevation data are between 0.99 and 1.00 for Halda and Sangu River and 0.98 and 1.00 for Chengi river (SI – Table S2). GMTED correspond least to other DEMs data for any river. NASA and SRTM-30; HydroSHEDS, and SRTM-90 DEMs are mostly correlated for all the rivers. All the DEMs show higher similarity to the elevation data among the studied rivers for the Sangu River. A higher correlation is observed in 90 m DEMs for the Sangu River watershed. The correlation among 30 m DEM is higher in every river watershed. MERIT DEM (90 m) correlates with all 30 m DEMs than coarser DEMs in the Chengi watershed. Additionally, all DEMs show a robust correlation (≥98%) for the elevation.

Among the different DEM slope data, correlation value varies from 0.62 to 0.95, 0.68–1.00, and 0.41 to 1.00 for Halda, Sangu, and Chengi river, respectively (SI – Table S3). HydroSHEDS and SRTM-90 are the most correlated DEMs, and GMTED has the lowest correlation value with all other DEMs for all the rivers, especially the lowest with ASTGTM in Halda and AW3D30 in Sangu and Chengi river watersheds. ASTGTM has lower correlation values with all coarser resolution (90 m and 225 m) DEMs in the Halda watershed, but AW3D30 has the same for other river watersheds. NASA and SRTM-30 have the highest correlation among all other 30 m resolution DEMs. The correlation among 90 m DEMs is higher than among finer DEMs (30 m).

3.2. DEMs sources and resolutions on watersheds morphometric parameters

3.2.1. Watershed area

The largest area was delineated from SRTM-90 (1,769.52 km^2), AW3D30 (3,820.29 km^2), and MERIT (1,836.78 km^2), while ASTGTM (1,750.13 km^2), NASA (3,591.15 km^2), and AW3D30 (1,253.98 km^2) delineated the smallest area for Halda, Sangu and Chengi river watershed, respectively (Figure 4). For Halda and Chengi watersheds, 90 m resolution DEMs generated the larger watershed area compared to 30 m and 225 m DEMs. However, no clear relationship is observed between the watershed area and DEM resolutions for the Sangu watershed. The largest perimeter value of the Halda, Sangu and Chengi watershed is from ASTGTM (345.04 km, 616.51 km, and 438.51 km, respectively). The lowest perimeter value (340.47 km) of the Chengi river watershed is from AW3D30; however, for the other two river watersheds, GMTED delineated the

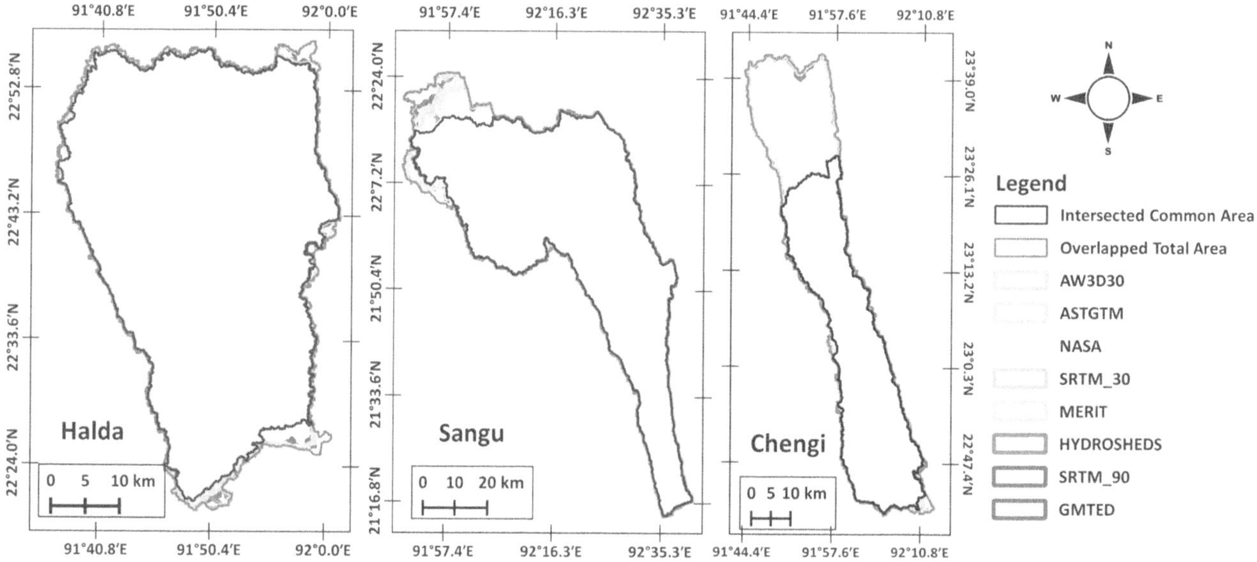

DEMs	AW3D30	ASTGTM	NASA	SRTM-30	MERIT	HydroSHEDS	SRTM-90	GMTED
Resolutions	30 m	30 m	30 m	30 m	90 m	90 m	90 m	225 m
				Sub-Basin area (km^2)				
Halda	1756.46	**1750.13**	1765.48	1762.10	1766.32	1768.91	**1769.52**	1763.13
Sangu	**3820.29**	3709.80	**3591.15**	3744.22	3705.67	3682.00	3770.86	3752.38
Chengi	**1253.98**	1825.60	1828.02	1829.11	**1836.78**	1836.73	1835.38	1822.79
				Perimeter (km)				
Halda	319.41	**345.04**	323.58	315.60	298.43	303.20	299.67	**284.94**
Sangu	612.30	**616.51**	562.04	602.65	567.10	566.78	569.16	**550.92**
Chengi	**340.47**	**438.51**	428.94	428.21	407.93	409.32	410.20	398.60

Figure 4 | Watershed size and perimeter of the three river watersheds delineated by different DEMs.

lowest perimeter values (Halda – 284.94 km and Sangu – 550.92 km). The overall perimeter of the watersheds is larger in 30 m resolution DEMs (except AW3D30 in Chengi) for all the watersheds (Figure 4). The results indicated that DEMs resolution related to the watershed perimeter inversely, independent of the watershed area in watershed delineation.

3.2.2. Watershed delineation

The size and shape of the Halda, Sangu, and Chengi river watersheds are different as computed by the DEMs. Halda, Sangu, and Chengi rivers intersected common areas have 6.79, 12.38, and 0.03% differences with the overlapped total area, respectively, except AW3D30 DEM (53.75%) for the Chengi watershed (Figure 4). The delineated shapes of Halda, Sangu, and Chengi river watersheds for different DEMs are shown in Figure 4 (and SI – Figure S1).

Northeast and south of the Halda watershed showed minor differences at every bend of the river, which may be caused by loss of information at lower elevations by different DEMs. The variation of the boundary in the east and west parts was inferior. The shape differences for the Sangu River watershed were seen on the northwest side, mainly the coastal area with low elevation. Though a low deviation is observed in the southeast, delineated watersheds are typical for all the DEMs, except AW3D30 DEM in the Chengi river watershed. In agreement with the mountainous Gilgit watershed in Pakistan, SRTM, ASTER, and GTOPO30 DEM showed close agreement in delineating the watershed, but SRTM (30 m) was more accurate than others in ridge demarcation (Shafiq *et al.* 2020).

3.2.3. Intersected areas by DEMs

The intersected areas between the defined watersheds from various DEMs are extracted. The intersected areas by different DEMs are 97.62–99.73%, 93.08–99.92%, and 67.80–99.92% for Halda, Sangu, and Chengi rivers, respectively (Figure 4 and SI – Table S4). The difference in the intersected areas is small for Halda and larger for Chengi. The ASTGTM is comparatively less intersected by other DEMs (97.62%) among the delineated watersheds for Halda. But the highest value (99.73%) for SRTM-30 when AW3D30 is intersected by it. For Sangu, the lowest intersection area (93.08%) is found between AW3D30 and NASA. It is evident from Figure 4 that for the Chengi, AW3D30 intersects the least with other DEMs and has the lowest value (67.80%) for intersecting GMTED.

3.3. ATVs, source and resolution of DEM effects on sub-basin counts and stream network

3.3.1. Watershed sub-basin counts

Extracted sub-basin counts rapidly decrease with the 5–40 km^2 ATVs. However, the sub-basin counts decreased between 40 km^2 and 60 km^2 for the Chengi, 40 km^2 and 80 km^2 for the Halda, and 40 km^2 and 100 km^2 for the Sangu watershed (Figure 5(a)). No significant improvement in sub-basin count could be achieved beyond the mentioned ATVs ranges. And the change in the number of sub-basins counts at different ATVs for sources and resolutions of DEM is not varied

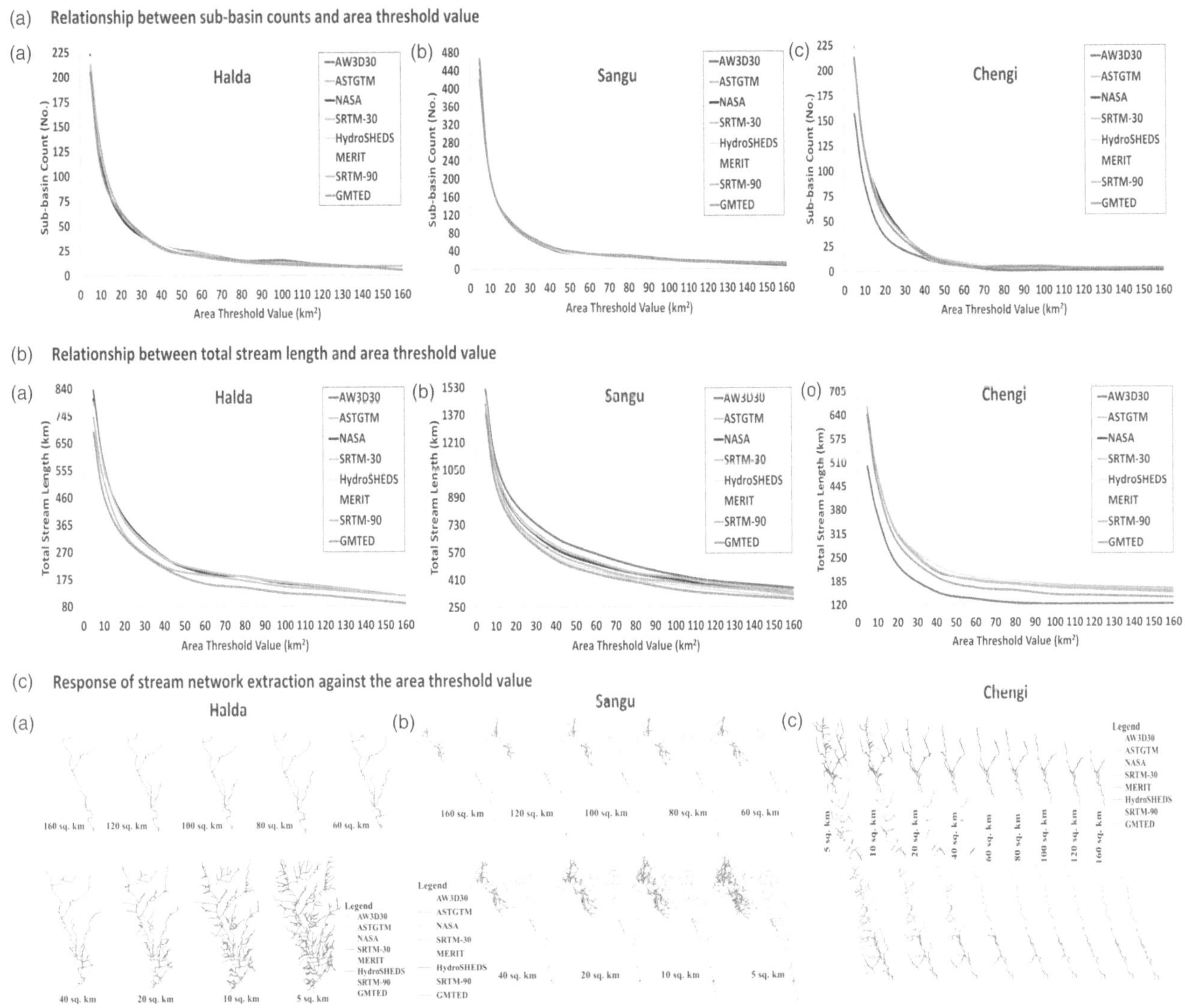

Figure 5 | Impact of different ATVs on (a) sub-basin counts, (b) total streams length and (c) extracted stream networks using different DEMs for three river watersheds.

significantly, which might be due to deviation of simulation range accepted by inherent differences among sources and resolutions of DEM (Xu *et al.* 2016). GMTED showed the lowest, and AW3D30 showed the highest sub-basin count for all ATVs. The sub-basin numbers were closely varied for different DEMs, with 7 (NASA in 160 km^2) – to 467 (ASTGTM in 5 km^2) for various ATVs in the Sangu River watershed. For the Chengi river watershed, AW3D30 had the lowest number of sub-basins with different ATVs due to the lowest watershed area delineated for this DEM.

SWAT follows the power function as $y = ax^b$, in sub-basin and drainage network extraction based on an ATV (Wu *et al.* 2017). The 'a' values range from 1,045.60 to 1,517.60, 2,252.50 to 2,699.50, and 2,344.90 to 3,731.20 for Halda, Sangu, and Chengi river watersheds, respectively, where R^2 is >0.9473 (Figure 5(a) and Table 3). Variations in 'b' value show that Halda has the lowest range (−0.97∼−1.07) values following the Sangu (−1.03∼−1.09) and Chengi has the highest values (−1.36∼−1.69). These curve fitting values are varied for different DEMs having different resolutions, but these variations are basin-specific.

3.3.2. Total stream lengths of watershed

Around 85.86, 77.19, and 76.83% of diminution happened in stream length derivation for all DEMs at ATV 5–160 km^2 for Halda, Sangu, and Chengi river watersheds, respectively (Figure 5(b)). The extracted stream had increased networking, and obscure streams were more visible when the imposed ATVs decreased, irrespective of DEM sources and resolutions (Figure 5(c)). Usually, the extracted total stream length from the finer-resolution (30 m) DEMs (except AW3D30 in Halda and Chengi) are longer than from the coarser resolutions (SI – Table S5). GMTED shows the lowest total stream length among all other DEMs for different ATVs, and AW3D30 shows the most extensive stream length (834.61 km) for Halda. The minimum and maximum sum of stream lengths for the Sangu river watershed were 291.61 km (GMTED), 1,519.85 km (AW3D30) at ATVs 160 km^2 and 5 km^2, respectively. As AW3D30 delineated the smallest watershed (1,253.98 km^2) for Chengi, it had the smallest stream network and the lowest total stream length among all the DEMs (Figures 4 and 5(c) and SI – Table S5).

4. DISCUSSION

4.1. Sensitivity of elevation and slope to different DEMs and resolutions

Considering hypsometric curves, 9, 40, and 17% of landmass elevations are above mean elevations (40.02 m, 173.76 m, and 102.34 m) for the Halda, Sangu, and Chengi rivers watersheds, respectively. These basins are at the monadnock stage, concave up with low integrals (HI < 0.35), usually formed from isolated bodies of resistant rock of major hills (monadnock) that are found above the subdued surface and old and deeply dissected landscapes (Strahler 1952; Kusre 2013). Three river basins have varied estuary connections and topography. All the lowest elevation values of the watershed are near the stream network.

ASTGTM shows an unlikely lower bound value (0 m, identical to MSL), and AW3D30 shows a meagre elevation value (−30 m or −18 m) on stream surface area for the Halda and Sangu watersheds which does not physically exist (observed during field investigation). Halda connects with Karnaphuli about >35 km before the estuary of the Bay of Bengal. Here,

Table 3 | Fitted curves for sub-basins counts for different ATVs concerning DEMs for Halda, Sangu, and Chengi River watersheds

Rivers	Halda		Sangu		Chengi	
DEMs	Formula	R^2	Formula	R^2	Formula	R^2
AW3D30 (30 m)	$y = 1{,}130.0x^{-0.972}$	0.9937	$y = 2{,}283.3x^{-1.026}$	0.9966	$y = 3{,}731.2x^{-1.693}$	0.9473
ASTGTM (30 m)	$y = 1{,}045.6x^{-0.968}$	0.9952	$y = 2{,}477.0x^{-1.064}$	0.9990	$y = 2{,}686.3x^{-1.404}$	0.9792
NASA (30 m)	$y = 1{,}517.6x^{-1.069}$	0.9888	$y = 2{,}550.1x^{-1.087}$	0.9820	$y = 2{,}507.3x^{-1.369}$	0.9781
SRTM-30 (30 m)	$y = 1{,}234.9x^{-1.018}$	0.9932	$y = 2{,}327.3x^{-1.036}$	0.9972	$y = 2{,}365.3x^{-1.361}$	0.9813
MERIT (90 m)	$y = 1{,}211.4x^{-1.009}$	0.9956	$y = 2{,}252.5x^{-1.036}$	0.9934	$y = 2{,}820.7x^{-1.414}$	0.9765
HydroSHEDS (90 m)	$y = 1{,}173.3x^{-1.001}$	0.9946	$y = 2{,}409.8x^{-1.057}$	0.9984	$y = 2{,}712.5x^{-1.397}$	0.9774
SRTM-90 (90 m)	$y = 1{,}137.6x^{-0.993}$	0.9919	$y = 2{,}290.8x^{-1.036}$	0.9880	$y = 2{,}815.0x^{-1.431}$	0.9611
GMTED (225 m)	$y = 1{,}150.7x^{-1.009}$	0.9857	$y = 2{,}699.5x^{-1.083}$	0.9929	$y = 2{,}344.9x^{-1.367}$	0.9714

MERIT (90 m) shows a comparable elevation value to other DEMs elevation ranges. These elevation discrepancies are derived from the different acquisition and update time and calculation errors of DEMs (Stefanescu *et al.* 2012; Xu *et al.* 2016), resulting in high-value depressions, which do not exist. However, ≥98% correlation values among the DEMs for elevation data confirm that DEM sources and resolutions have minimal effect on elevation data for specific watershed types and topographical variation.

We observed that the slope data vary with different DEMs and resolutions which is in agreement with Buakhao & Kangrang (2016). However, coarse-resolution DEMs (90 m) show lower slope parameter variation (e.g., maximum). Hence, coarser-resolution DEMs ignore topographic information details (Liu *et al.* 2010; Chen 2013; Xu *et al.* 2016). In topographic parameter extraction, DEM resolutions have more impact than DEM sources (Xu *et al.* 2016), indicating that elevation and slope data are more sensitive to the resolution of DEMs (Dixon & Earls 2009; Tan *et al.* 2015). Though the variation in elevation is not significant, these complex DEM data are processed (filled, conditioned) to extract water flow direction and channel properties such as channel length, width, depth, slope, and watershed separation (Liu *et al.* 2010; Rao & Yang 2010). For these reasons, the effect of DEM resolution is minimal in streamflow simulation in SWAT (Zhang *et al.* 2014; Gautam *et al.* 2019; Munoth & Goyal 2019). On the contrary, TOPMODEL, Hydrological Simulation Program-FORTRAN (HSPF) streamflow showed sensitivity against DEM resolution compared to DEM sources and different resampling techniques (Suliman *et al.* 2019; Roostaee & Deng 2020) or in SWAT (Rocha *et al.* 2020).

4.2. Watershed physiography from different DEMs and resolutions

4.2.1. Watershed area variation

In the study, Halda, Sangu, and Chengi river watersheds differed in size and shape for all the DEMs. We estimated larger areas for Halda and Chengi and lower perimeter values for all three rivers from 90 m DEMs, indicating that coarser-resolution DEMs represent lower information than finer resolutions DEMs. This is contrary to the findings of Al-Khafaji & Al-Sweiti (2017). They inferred inconspicuous trends in the watershed area with DEM resolutions. A larger delineated area (3,842.82 km^2) for the Sangu river watershed is reported using SRTM-30 DEM (Adnan *et al.* 2019), where the outlets of the streams vary in our study. Akhter (2015) found a 1,727.06 km^2 area by SRTM-90 DEM using HEC-GeoHMS, and Chowdhury *et al.* (2020) found a 1,756.91 km^2 area by ASTER DEM for the Halda river. Area differences in various studies could be due to the different outlet/discharge point selections for the main channel and individual streams and the choice of DEMs (source and resolution) or software packages used in watershed delineation. Considering minor deviation, the SRTM 90 m DEM delineated a higher watershed area than other DEMs and resolutions. The boundary delineated by DEM may overlook an area delineated by another (Buakhao & Kangrang 2016). Though primarily the difference is derived from the local sink-related problem for all hydrological tools, SWAT produces a minimum error in watershed delineation than others (Ray 2018). Using high-resolution DEM in the automated watershed delineation process provides no significant benefit (Buakhao & Kangrang 2016), though it extracts more accurate streams (Li & Wong 2010). However, coarser-resolution DEMs perform more precisely than higher resolution DEMs in the delineation process (Ray 2018) due to fewer topographic variations at boundaries (Xu *et al.* 2016).

4.2.2. DEMs vertical projection and common area demarcation

One of the reasons for the difference between the delineated overlapped total area and intersected common area by the DEMs could be the difference in vertical datum projection like ITRF97 - GRS80 in AW3D30 than other DEMs – WGS84 or converting them to UTM 46 N. Moreover, the MSL elevations of the reflective surfaces of any features are elevated above the bare earth by AW3D30 DSM. However, it cannot be confirmed for the Halda and Sangu River watersheds. There has been no demarcated watershed boundary for the management of these rivers. Hence, for initiating such approaches, the common intersected area, independent of the DEMs and their resolutions, can be used as a starting point to define basin boundary by keeping the total overlapped area as a buffer or boundary zones for the watersheds.

4.2.3. Correlation among intersected areas by DEMs

The correlation values among delineated watersheds by different DEMs and resolutions indicate a close association among DEMs. ASTGTM and NASA, AW3D30 showed lower intersected areas than all the studied DEMs (SI – Table S4), reflecting the correlation values of slope (SI – Table S3). DEMs accurately delineate watersheds surrounded by mountains (Shafiq *et al.* 2020). Hence, along the flood plain, coastal areas, or flat topography near the discharge points of the catchment, Figure 4 reveals a significant variation in the size and shape of the catchment. Hence DEM selection in flat areas may have more significant implications on discharge and pollution load estimation in any hydrological study.

4.3. ATVs' effects on watershed characteristics

4.3.1. Sub-basin counts

DEM data with different resolutions have little influence on the extraction results if ATV lies between 5 and 100 km^2 (Figure 5(a)). However, at ATV $\leq 40\ km^2$, the DEM resolution greatly influences sub-basin counts. The area and perimeter of a catchment are independent of the ATVs. Wu *et al.* (2017) also reported that the values of a and b of the ATVs to subbasin count equation, $y = ax^b$, were varied for different watersheds but not for the resolutions of a specific or single DEM. The inverse relationship (as exponent b is negative) between ATVs and sub-basin counts (Munoth & Goyal 2020) become more insignificant for the watersheds for different DEM sources and resolutions at $x < 60\ km^2$ for Chengi, $x < 80\ km^2$ for Halda, and $x < 100\ km^2$ for Sangu River. An efficient and rational sub-basin number is needed in hydrological modelling (Chang 2009). This is achieved here at ATV $< 40\ km^2$ for the watershed area 1,200 km^2–4,000 km^2. DEM resolution influences sub-basin count and stream length (Figure 5(b) and 5(c)). Additionally, the processing time by the software package also increases with higher resolution DEMs, which agrees with Munoth and Goyal's observation (2019). However, Reddy & Reddy (2015) reported that the accuracy of the sub-basin decreases with the lower resolution of DEM. Hence sub-basin count operation affects the processing time of DEM in delineation, and the input ATVs command it.

4.3.2. Stream network

Figure 5(c) reveals that to detect a detail stream network, $\leq 40\ km^2$ ATV is needed. Though lowest ATVs cannot be suggested for an acceptable stream network extraction. The complex network of small streams, reservoirs, lakes, and depressions of the flat area would cause difficulty in the identification of braided streams and the natural routing process controlled by water structures or other anthropogenic changes in stream networks (Wilson & Gallant 2000; Turcotte *et al.* 2001; Luo *et al.* 2011). However, this threshold would be feckless for a larger watershed ($>$4,000 km^2) because large stream density may increase the computation cost and complexity of the model. Luo *et al.* (2011) suggested that manual editing after automated delineation of sub-basin boundary and stream layers to adjust the location, range, and hydrologic connection could bring better results in the hydrologic model setup. The field survey in the catchment is essential while doing the manual edit for the studied river basins.

4.3.3. Drainage density

Depending on the ATV, the number of streams and stream lengths will vary in a watershed. A smaller ATV will increase stream length and a denser stream network. Hence, the drainage density, which is defined as total stream lengths (km)/ basin area (km^2), directly depends on total stream length and follows the same trend of sub-basin for a specific watershed. Cartosat-1 and ASTER DEM showed 0.05 km^2 as the optimum threshold value to obtain a refined and higher drainage density for the Neri watershed, Maharashtra, India, where the watershed area (about 31.96 km^2) was small (Reddy *et al.* 2017). Therefore, threshold value selection is crucial and directly influences the basin characteristics and stream network (Gopinath *et al.* 2014) as well as the hydrologic model performance (Gautam *et al.* 2019).

5. CONCLUSION

This study offers a crucial answer to the question of available, open-ended, and space-borne DEMs, their resolutions, and user-defined ATVs' influence on the physiographic and morphometric characteristics of different river watersheds. DEMs' sources and resolutions influence different outputs of automatic watershed delineation because of the watershed features such as size and shape differences, stream network, watershed area, and sub-watershed size and shape. The studied watersheds also observed the influence of different ATVs in stream networks and sub-basin for a specific river watershed. This study suggests depending on the watershed size, slope, and topography, DEM-based watershed delineation process in SWAT needs optimum ATV values to capture a detailed and accurate subbasin area, shape and stream network. The detection and delineation of watershed areas in the mountainous part are relatively precise, but the flatter surface creates discrepancies, which suggest the use of the common intersected area of DEMs to confine a river's watershed management area. Results of this study indicate that the choice of input data (here DEMs or topographic features) can be a factor in defining terrain differences and optimum stream complexation covering expected watershed and sub-basin areas. We suggest the comparative study of outputs and operational differences of similar experiments using other geospatial analysis software packages. Moreover, river watersheds similar or different in size can affect the variation or preciseness of the impacts of DEMs, resolution, and ATVs. A base watershed is necessary to suggest a particular DEM source or resolution in watershed delineation. Accuracy assessment of a particular DEM source or resolution for stream network extraction by comparing the base watershed or observed ones is recommended for future study.

ACKNOWLEDGEMENTS

The first and second authors highly appreciate the financial support of the Planning and Development Division of Chittagong University for the research grant on the Halda watershed. The authors highly appreciate the reviewers' suggestions and comments, which significantly improved the manuscript.

DATA AVAILABILITY STATEMENT

All relevant data are included in the paper or its Supplementary Information.

CONFLICT OF INTEREST

The authors declare there is no conflict.

REFERENCES

Adnan, M. S. G., Dewan, A., Zannat, K. E. & Abdullah, A. Y. M. 2019 The use of watershed geomorphic data in flash flood susceptibility zoning: a case study of the Karnaphuli and Sangu river basins of Bangladesh. *Natural Hazards* **99** (1), 425–448. doi:10.1007/s11069-019-03749-3.

Akhter, F. 2015 *Change of Halda River Flow due to Different Water Control Structures and its Impact on Halda Ecosystem. MS in Water Resource Development*, Bangladesh University of Engineering and Technology, BUET, Dhaka. Available from: http://lib.buet.ac.bd:8080/xmlui/handle/123456789/4657

Al-Khafaji, M. S. & Al-Sweiti, F. H. 2017 Integrated impact of digital elevation model and land cover resolutions on simulated runoff by SWAT model. In: *Hydrology And Earth System Sciences Discussions*, pp. 1–26. doi:10.5194/hess-2017-653.

Arnold, J. G., Srinivasan, R., Muttiah, R. S. & Williams, J. R. 1998 Large area hydrologic modeling and assessment part I: model development. *Journal of the American Water Resources Association (JAWRA)* **34** (1), 73–89. https://doi.org/10.1111/j.1752-1688.1998.tb05961.x.

Azadi, M. 2005 *The Halda River, Only Breeding Ground of Major Carps in Bangladesh: Conservation and Management*. National Fish Fortnight Compendium, Directorate of Fisheries, Ministry of Fisheries and Livestock, Bangladesh.

Buakhao, W. & Kangrang, A. 2016 DEM resolution impact on the estimation of the physical characteristics of watersheds by using SWAT. In: *Advances in Civil Engineering*, pp. 1–10. doi:10.1155/2016/8180158.

Chang, C. L. 2009 The impact of watershed delineation on hydrology and water quality simulation. *Environmental Monitoring and Assessment* **148** (1–4), 159–165. doi:10.1007/s10661-007-0147-8.

Chaplot, V. 2005 Impact of DEM mesh size and soil map scale on SWAT runoff, sediment, and NO3-N loads predictions. *Journal of Hydrology* **312** (1), 207–222. https://doi.org/10.1016/j.jhydrol.2005.02.017.

Chaubey, I., Cotter, A. S., Costello, T. A. & Soerens, T. S. 2005 Effect of DEM data resolution on SWAT output uncertainty. *Hydrological Processes* **19** (3), 621–628. doi:10.1002/hyp.5607.

Chen, N. 2013 Influence of resolutions of DEM on the error of slope. *Wuhan Daxue Xuebao (Xinxi Kexue Ban)/Geomatics and Information Science of Wuhan University* **38**, 594–598.

Cho, S.-M. & Lee, M. 2001 Sensitivity considerations when modeling hydrologic processes with digital elevation model. *Journal of the American Water Resources Association* **37** (4), 931–934. https://doi.org/10.1111/j.1752-1688.2001.tb05523.x.

Chowdhury, M., Hasan, M. E. & Abdullah-Al-Mamun, M. M. 2020 Land use/land cover change assessment of Halda watershed using remote sensing and GIS. *The Egyptian Journal of Remote Sensing and Space Science* **23** (1), 63–75. https://doi.org/10.1016/j.ejrs.2018.11.003.

Correspondent, S. 2014 Rubber dams render blow to ecosystem. *The Daily Star*. Available from: https://www.thedailystar.net/rubber-dams-render-blow-to-ecosystem-26148

Danielson, J. J. & Gesch, D. B. 2011 *Global Multi-Resolution Terrain Elevation Data 2010 (GMTED2010) (Vol. Open-File Report 2011–1073)*. US Department of the Interior, US Geological Survey.

Dixon, B. & Earls, J. 2009 Resample or not?! Effects of resolution of DEMs in watershed modeling. *Hydrological Processes* **23** (12), 1714–1724. https://doi.org/10.1002/hyp.7306.

Ehsani, A. H., Quiel, F. & Malekian, A. 2010 Effect of SRTM resolution on morphometric feature identification using neural network-self organizing map. *GeoInformatica* **14** (4), 405–424. doi:10.1007/s10707-009-0085-4.

Gautam, S., Dahal, V. & Bhattarai, R. 2019 Impacts of Dem source, resolution and area threshold values on SWAT generated stream network and streamflow in two distinct Nepalese catchments. *Environmental Processes* **6**, 597–617. https://doi.org/10.1007/s40710-019-00379-6.

Giridhar, M. V. S. S., Ramaraju, A., Rao, S. & Sowmya, P. 2015 DEM processing for watershed delineation using QSWAT. *International Journal of Scientific and Engineering Research* **6** (8), 1670–1674.

Golekar, R. B., Baride, M. V., Patil, S. N. & Adil, S. H. 2015 Altimetric and hypsometric analysis for soil and water conservation: a case study of Anjani and Jhiri river basin, Northern Maharashtra, India. *Earth Sciences Research Journal* **19**, 51–58.

Gopinath, G., Swetha, T. V. & Ashitha, M. K. 2014 Automated extraction of watershed boundary and drainage network from SRTM and comparison with Survey of India toposheet. *Arabian Journal of Geosciences* **7** (7), 2625–2632. doi:10.1007/s12517-013-0919-0.

Hussain, A. 2016 Halda River to be declared as Ecologically Critical Area. *Dhaka Tribune*. Available from: https://www.dhakatribune.com/bangladesh/2016/10/08/halda-river-declared-ecologically-critical-area

Jarvis, A., Reuter, H. I., Nelson, A. & Guevara, E. 2008 Hole-filled SRTM for the globe Version 4. from International Centre for Tropical Agriculture (CIAT). Available from: http://srtm.csi.cgiar.org/srtmdata/

JAXA 2020 ALOS World 3D - 30 m (AW3D30) Version 3.1. from Earth Observation Research Center, Japan Aerospace Exploration Agency. Available from: https://www.eorc.jaxa.jp/ALOS/en/aw3d30/data/index.htm

Kalcic, M. M., Chaubey, I. & Frankenberger, J. 2015 Defining Soil and Water Assessment Tool (SWAT) hydrologic response units (HRUs) by field boundaries. *International Journal of Agricultural and Biological Engineering* **8** (3), 69–80.

Kumar, A. & Dhiman, R. 2014 Manual and automated delineation of watershed boundaries–a case study from Kangra region of western Himalaya, India. *International Journal of Environmental Sciences* **5** (1), 16–22.

Kusre, B. C. 2013 Hypsometric analysis and watershed management of Diyung watershed in north eastern India. *Journal of the Geological Society of India* **82** (3), 262–270. doi:10.1007/s12594-013-0148-x.

LaMoreaux, P. E., Soliman, M. M., Memon, B. A., LaMoreaux, J. W. & Assaad, F. A. 2008 *Environmental Hydrogeology*, 2nd edn. IWA Publishing and CRC Press, USA.

Latifa, G. A., Majumder, A. K., Kabir, M. H. & Chakma, A. 2019 Water quality and fish diversity of Chengi river of Khagrachari district. *Bangladesh Journal of Zoology* **47** (2), 343–353.

Lehner, B., Verdin, K. & Jarvis, A. J. E. 2008 New global hydrography derived from spaceborne elevation data. *Transactions American Geophysical Union* **89** (10), 93–94.

Li, J. & Wong, D. W. S. 2010 Effects of DEM sources on hydrologic applications. *Computers, Environment and Urban Systems* **34** (3), 251–261. https://doi.org/10.1016/j.compenvurbsys.2009.11.002.

Lindsay, J. B. 2016 The practice of DEM stream burning revisited. *Earth Surface Processes and Landforms* **41** (5), 658–668. https://doi.org/10.1002/esp.3888.

Liu, H.-H., Long, Y., Yan, D.-C. & Hu, B. 2010 Accuracy of topographic parameters in three gorges reservoir region from different resolutions of DEMs. *Journal of Yangtze River Scientific Research Institute* **27** (11), 21–24.

Luo, Y., Su, B., Yuan, J., Li, H. & Zhang, Q. 2011 GIS techniques for watershed delineation of SWAT model in plain polders. *Procedia Environmental Sciences* **10**, 2050–2057. https://doi.org/10.1016/j.proenv.2011.09.321.

Moore, I. D., Grayson, R. B. & Ladson, A. R. 1991 Digital terrain modelling: a review of hydrological, geomorphological, and biological applications. *Hydrological Processes* **5** (1), 3–30. https://doi.org/10.1002/hyp.3360050103.

Munoth, P. & Goyal, R. 2019 Effects of DEM source, spatial resolution and drainage area threshold values on hydrological modeling. *Water Resources Management* **33** (9), 3303–3319. doi:10.1007/s11269-019-02303-x.

Munoth, P. & Goyal, R. 2020 Hydromorphological analysis of Upper Tapi River Sub-basin, India, using QSWAT model. *Modeling Earth Systems and Environment* **6** (4), 2111–2127. doi:10.1007/s40808-020-00821-x.

NASA, J. 2013 *NASA Shuttle Radar Topography Mission Global 1 arc Second From NASA EOSDIS Land Processes DAAC*. https://doi.org/10.5067/MEaSUREs/SRTM/SRTMGL1.003.

NASA, J. 2020 *NASADEM Merged DEM Global 1 arc Second V001. From NASA EOSDIS Land Processes DAAC*. https://doi.org/10.5067/MEaSUREs/NASADEM/NASADEM_HGT.001.

NASA/METI/AIST/Japan Space Systems, a. U. S. J. A. S. T. 2019 ASTER Global Digital Elevation Model V003 (Publication no. https://doi.org/10.5067/ASTER/ASTGTM.003). from NASA EOSDIS Land Processes DAAC. Available from: https://search.earthdata.nasa.gov/search?q=C1575726572-LPDAAC_ECS

Paranjape, S., Joy, K. J., Machadeo, T., Varma, A. K. & Swaminathan, S. 1998 *Watershed Based Development: A Source Book*. Bharat Gyan Vigyan Samithi, New Delhi, India.

Pike, R. J. & Wilson, S. E. 1971 Elevation-relief ratio, hypsometric integral, and geomorphic area-altitude analysis. *GSA Bulletin* **82** (4), 1079–1084. doi:10.1130/0016-7606(1971)82[1079:ERHIAG]2.0.CO;2.

Piwowar, J. M. & LeDrew, E. F. 1990 Integrating spatial data: a user's perspective. *Photogrammetric Engineering and Remote Sensing* **56** (11), 1497–1502.

Rahman, M. M., Arya, D. S. & Goel, N. K. 2010 Limitation of 90 m SRTM DEM in drainage network delineation using D8 method – a case study in flat terrain of Bangladesh. *Applied Geomatics* **2** (2), 49–58. doi:10.1007/s12518-010-0020-2.

Raihan, F., Beaumont, L. J., Maina, J., Saiful Islam, A. & Harrison, S. P. 2020 Simulating streamflow in the Upper Halda Basin of southeastern Bangladesh using SWAT model. *Hydrological Sciences Journal* **65** (1), 138–151. doi:10.1080/02626667.2019.1682149.

Rao, M. N. & Yang, Z. 2010 Groundwater impacts due to conservation reserve program in Texas County, Oklahoma. *Applied Geography* **30** (3), 317–328. https://doi.org/10.1016/j.apgeog.2009.08.006.

Ray, L. K. 2018 Limitation of automatic watershed delineation tools in coastal region. *Annals of GIS* **24** (4), 261–274. doi:10.1080/19475683.2018.1526212.

Reddy, A. S. & Reddy, M. J. 2015 Evaluating the influence of spatial resolutions of DEM on watershed runoff and sediment yield using SWAT. *Journal of Earth System Science* **124** (7), 1517–1529.

Reddy, G. P. O., Kumar, N., Sahu, N. & Singh, S. K. 2017 Evaluation of automatic drainage extraction thresholds using ASTER GDEM and Cartosat-1 DEM: a case study from basaltic terrain of Central India. *The Egyptian Journal of Remote Sensing and Space Science* **21** (1), 95–104. https://doi.org/10.1016/j.ejrs.2017.04.001.

Reddy, N. N., Reddy, K. V., Vani, J. S. L. S., Daggupati, P. & Srinivasan, R. 2018 Climate change impact analysis on watershed using QSWAT. *Spatial Information Research* **26** (3), 253–259.

Report, T. 2020 Govt declares Halda as 'Bangabandhu Fisheries Heritage'. *The Business Standard*. Available from: https://www.tbsnews.net/economy/agriculture/govt-declares-halda-bangabandhu-fisheries-heritage-175183

Rocha, J., Duarte, A., Silva, M., Fabres, S., Vasques, J., Revilla-Romero, B. & Quintela, A. 2020 The importance of high resolution digital elevation models for improved hydrological simulations of a Mediterranean forested catchment. *Remote Sensing* **12** (20), 3287.

Roostaee, M. & Deng, Z. 2020 Effects of digital elevation model resolution on watershed-based hydrologic simulation. *Water Resources Management* **34** (8), 2433–2447. doi:10.1007/s11269-020-02561-0.

Saha, P., Islam, M., Oyshi, J. T., Khanum, R. & Nishat, A. 2019a A sustainability study of the flow regulation impacts by dams in a carp breeding river using the hydrodynamic model and building block analysis. *SN Applied Sciences* **1** (1429), 1–20.

Saha, P., Oyshi, J. T., Haider, S., Nipa, S. & Khanum, R. 2019b Analysis on the changing trends of the channel pattern of Halda river and public-private partnership (PPP) as a catalyst for the sustainable development. In *Paper Presented at the Proceedings on International Conference on Disaster Risk Management*, January 12–14.

Salih, A. A. M. & Hamid, A. A. 2017 Hydrological studies in the White Nile State in Sudan. *The Egyptian Journal of Remote Sensing and Space Science* **20**, S31–S38. https://doi.org/10.1016/j.ejrs.2016.12.004.

Shafiq, M., Irfan, M. & Khan, M. 2020 Using multi-mission satellite elevation data for delineation of Gilgit watershed in Pakistan in geographical information technology environment. *International Journal of Economic and Environmental Geology* **11** (2), 19–24.

Sowmya, D. R., Shenoy, P. D. & Venugopal, K. R. 2020 Watershed delineation using SRTM DEM for Bangalore South Region. *International Journal of Engineering Science Invention* **9** (3), 01–09.

Stefanescu, E. R., Bursik, M., Cordoba, G., Dalbey, K., Jones, M. D., Patra, A. K., Pieri, D. C., Pitman, E. B. & Sheridan, M. F. 2012 Digital elevation model uncertainty and hazard analysis using a geophysical flow model. **468** (2142), 1543–1563. doi:10.1098/rspa.2011.0711.

Strahler, A. N. 1952 Hypsometric (area-altitude) analysis of erosional topography. *GSA Bulletin* **63** (11), 1117–1142. doi:10.1130/0016-7606(1952)63[1117:HAAOET]2.0.CO;2.

Suliman, A. H. A., Gumindoga, W., Awchi, T. A. & Katimon, A. 2019 DEM resolution influences on peak flow prediction: a comparison of two different based DEMs through various rescaling techniques. *Geocarto International* **36** (7), 803–819. doi:10.1080/10106049.2019.1622599.

Tan, M. L., Ficklin, D. L., Dixon, B., Ibrahim, A. L., Yusop, Z. & Chaplot, V. 2015 Impacts of DEM resolution, source, and resampling technique on SWAT-simulated streamflow. *Applied Geography* **63**, 357–368. https://doi.org/10.1016/j.apgeog.2015.07.014.

Turcotte, R., Fortin, J. P., Rousseau, A. N., Massicotte, S. & Villeneuve, J. P. 2001 Determination of the drainage structure of a watershed using a digital elevation model and a digital river and lake network. *Journal of Hydrology* **240** (3), 225–242. https://doi.org/10.1016/S0022-1694(00)00342-5.

Wang, J., Li, L., Hao, Z. & Gourley, J. 2011 Stream guiding algorithm for deriving flow direction from DEM and location of main streams. *IAHS-AISH Publication* **346**, 198–205.

Wilson, J. P. & Gallant, J. C. 2000 *Terrain Analysis: Principles and Applications*, 1st edn. John Wiley & Sons, New York, NY.

Wolock, D. M. & McCabe Jr, G. J.Jr 1995 Comparison of single and multiple flow direction algorithms for computing topographic parameters in TOPMODEL. *Water Resources Research* **31** (5), 1315–1324. https://doi.org/10.1029/95WR00471.

Wolock, D. M. & Price, C. V. 1994 Effects of digital elevation model map scale and data resolution on a topography-based watershed model. *Water Resources Research* **30** (11), 3041–3052. https://doi.org/10.1029/94WR01971.

Wu, M., Shi, P., Chen, A., Shen, C. & Wang, P. 2017 Impacts of DEM resolution and area threshold value uncertainty on the drainage network derived using SWAT. *Water SA* **43** (3), 450–462. http://dx.doi.org/10.4314/wsa.v43i3.10.

Xu, F., Dong, G., Wang, Q., Liu, L., Yu, W., Men, C. & Liu, R. 2016 Impacts of DEM uncertainties on critical source areas identification for non-point source pollution control based on SWAT model. *Journal of Hydrology* **540**, 355–367.

Yamazaki, D., Ikeshima, D., Tawatari, R., Yamaguchi, T., O'Loughlin, F., Neal, J. C., Sampson, C. C., Kanae, S. & Bates, P. D. 2017 A high-accuracy map of global terrain elevations. *Geophysical Research Letters* **44** (11), 5844–5853.

Zhang, P., Liu, R., Bao, Y., Wang, J., Yu, W. & Shen, Z. 2014 Uncertainty of SWAT model at different DEM resolutions in a large mountainous watershed. *Water Research* **53**, 132–144. https://doi.org/10.1016/j.watres.2014.01.018.

First received 20 November 2021; accepted in revised form 27 June 2022. Available online 15 July 2022

doi: 10.2166/nh.2022.027

Identification of suitable sites for traditional *pokhari* water harvesting in mountain rural communities of the Himalaya

Eleonora Forzini [a,*], Luigi Piemontese [a], Elena Bresci [a], Blandine Barthod[b], Florian Bielser[b], Marc Sylvestre[b], Nirmal Adhikari[c], Sony Pun[c] and Giulio Castelli [a]

[a] Department of Agriculture, Food, Environment and Forestry (DAGRI), University of Florence, Via San Bonaventura, 13, Firenze 50145, Italy
[b] International Rainwater Harvesting Alliance (IRHA), Chemin de Balexert 9, 1219, Châtelaine, Geneva, Switzerland
[c] Kanchan Nepal (KN), Janapriya Marg-7, Pokhara, Nepal
*Corresponding author. E-mail: eleonora.forzini@unifi.it

EF, 0000-0001-6727-3706; LP, 0000-0002-1600-5450; EB, 0000-0002-1988-1720; GC, 0000-0002-1600-5450

ABSTRACT

Storing runoff during the monsoon season in Himalayan hills is crucial to have enough water to cope with the dry season, especially considering that climate change is changing rainfall intensity and patterns. Traditional Nepalese water ponds, called *pokharis*, are used to store runoff mainly for cattle rearing and rice fields' supplementary irrigation. Local communities are interested in restoring existing *pokharis* and building new ones to improve their economical and living conditions. Selecting the most suitable locations for *pokharis* is of crucial importance; however, scarce information is available for large-scale site selection. A comprehensive analysis of multiple relevant parameters for traditional ponds siting can lead to more efficient rainwater collection and provide a useful water resource management tool. In this work, we propose a methodology for *pokharis*' best-siting analysis, based on geographic information system (GIS), multi-criteria decision-making (MCDM), and participatory research. To our knowledge, this is the first large-scale best-siting analysis for traditional ponds in Nepal. An area of 0.423 km^2 (3.47% of the study area) is classified as highly suitable, with 100% agreement with existing *pokharis*. Despite the low data resolution, which requires a further field inspection for the final site selection, our results provide reliable guidance for *pokharis*' large-scale suitability, supporting water-resilience projects in the area.

Key words: drought, spatial multi-criteria analysis, sustainable water management, water harvesting best siting, water harvesting ponds

HIGHLIGHTS

- Large-scale best-siting analyses of traditional Nepalese ponds (*pokharis*) are lacking despite their importance for water management in Nepal.
- Uphill communities, being more prone to water scarcity, were prioritized through appropriate criteria setting.
- We propose a composite of standard and in-field surveyed criteria for *pokharis*' best siting.
- The study provides useful guidance for building water harvesting ponds in Nepal.

INTRODUCTION

For centuries, farmers have adapted to climate variability by implementing various kinds of practices including soil and water conservation techniques and water harvesting (WH) (Al-Adamat *et al.* 2012; Adhikari 2018). This latter includes a wide range of techniques aimed at storing runoff water in the soil or dedicated structures to use it subsequently for productive purposes (Adhikari *et al.* 2018; Subedi *et al.* 2020). WH has been applied in many areas of the world through low-cost and eco-friendly interventions, developed from the indigenous knowledge of local people (Lal & Verma 2008; Chen *et al.* 2019). In mountainous areas, where water availability is a major concern, WH allows extending the cropping season and increasing food security by providing water while conventional water sources are drying out (Subedi *et al.* 2020). Moreover, it reduces soil erosion and flood risk and can increase water infiltration in soil and groundwater downstream (Lal & Verma 2008; Dile *et al.* 2016). WH has been proven to considerably increase crop yields, soil moisture, and farmers' income; and to reduce soil erosion in various mountainous zones of the world, among which the Himalayan region, including Nepal (Bastakoti *et al.* 2016; Adhikari 2018; Adhikari *et al.* 2018; Subedi *et al.* 2020).

Nepal is a landlocked country in the Himalayan region (UNDP 2020). Moving from North to South, the Nepalese climate ranges from an extremely cold subarctic climate to a monsoon-influenced tropical one, covering the Köppen and Geiger classification codes Dfd, Dwc, Dwa, Cwb, and Cwa. Despite being perceived as a water-rich country, with a mean annual rainfall of 1,344 mm, some areas of Nepal suffer from water stress during the dry season, especially from March to June (Adhikari 2018). The reason is the unbalanced rainfall distribution throughout the year with about 80% of total annual precipitation limited to the monsoon season between June and September (Adhikari *et al.* 2018). Moreover, climate change is altering rainfall patterns and decreasing total precipitation (Adhikari 2018; Subedi *et al.* 2020). Climate change consequences can be particularly severe in Nepal, where around 80% of the population depends on subsistence agriculture, especially in hilly areas, and more than 50% of agricultural land is rainfed (Subedi *et al.* 2020). Nepalese farmers report an increased frequency of droughts, late monsoon onset, rising temperatures, reduction of water availability, and flooding during monsoon as major impacts on their livelihoods (Bastakoti *et al.* 2016). If monsoon onset is delayed, farmers need to irrigate their crops to avoid severe crop losses (Bastakoti *et al.* 2016; Pandey 2019).

Traditional Nepalese ponds called *pokharis* (Bastakoti *et al.* 2016) are used, especially in rural areas, to harvest and store rainwater for supplemental irrigation, cattle rearing, fish farming, fire control, recreational, and sometimes domestic uses (Lal & Verma 2008; Chen *et al.* 2017; Subedi *et al.* 2020). Very similar structures can be found also in India and China (Lal & Verma 2008; Chen *et al.* 2017, 2019). They are hand-dug, shallow (1.2–1.5 m) generally small ponds (extended from a few to 500 m^2), but their dimension can vary substantially (Lal & Verma 2008; Bastakoti *et al.* 2016; Chen *et al.* 2017, 2019). The bottom of the *pokharis* is lined with red loamy soil compacted by animal passage, while lateral walls are covered with stones and red soil to be impermeable. Each pond also presents two stairs to allow animals access to the pond itself. Maintenance is carried out by village communities. *Pokharis* (Figure 1) are built near housing and fields, but they can be found also near roads and forests when cattle rearing is their main use (Lal & Verma 2008). They also have an important socio-cultural value, as people usually meet near ponds to rest during daily activities and socialize (Chen *et al.* 2019).

Recently, *pokharis* are being abandoned due to the development of irrigation facilities in some areas, while many others are degrading because of abandonment of farmlands (Chen *et al.* 2017, 2019). Underutilization and poor management of these traditional structures reduces not only water availability in dry months but also ecosystem services provided by *pokharis* like downhill flood risk reduction and sediment control (Lal & Verma 2008; Bastakoti *et al.* 2016; Chen *et al.* 2019).

Various authors (Lal & Verma 2008; Bastakoti *et al.* 2016; Chen *et al.* 2017; Adhikari 2018), also reporting farmers' opinions, point out that restoring existing *pokharis* and building new ones could foster resilience to climate change through

Figure 1 | Picture of a *pokhari*, taken during a field visit in October 2021 in the study area.

increased water availability. Usually, the construction sites of new *pokharis*, and ponds in general, are selected 'arbitrarily', considering only the proximity to the point where water is needed (Chen *et al.* 2017).

However, a more comprehensive and systematic site selection can lead to more efficient WH structures, which in turn contribute to improving water availability and agricultural productivity (Adham *et al.* 2016; Mugo & Odera 2019). Nevertheless, site selection is a very challenging task especially at large spatial scales, owing to the lack of detailed hydrological and soil data (Singhai *et al.* 2019).

In the last three decades, many studies have attempted to identify the most suitable sites for different types of WH structures, often using geographical information system (GIS) and multi-criteria decision-making (MCDM) (Adham *et al.* 2016). GIS is widely used because it allows us to analyze large areas by integrating spatial physical and socio-economic information in a timely and cost-effective way (Adham *et al.* 2016). On the other hand, MCDM allows us to compare and relate different kinds of data producing a single final output displaying various options (Krois & Schulte 2014).

The literature on the topic reports the need to select some criteria for WH structures' best siting, such as environmental and physical characteristics of the area of interest, and its socio-economic features (Adham *et al.* 2016; Grum *et al.* 2016). Common physical criteria considered for best-siting analysis are slope, soil type, runoff, and land use (Al-Adamat *et al.* 2012; Dile *et al.* 2016). On the other hand, the main socio-economic data are the distance from roads, villages, and fields (de Winnaar *et al.* 2007). In several cases, these latter criteria and participatory activities to assess stakeholders' preferences are neglected despite being necessary to ensure the proper implementation of site-specific effective measures (Adham *et al.* 2016; Grum *et al.* 2016).

In this work, we use MCDM and GIS to produce, to our knowledge, the first best-siting analysis for *pokharis*. While considering similar MCDM parameters as other conventional best-siting studies for WH structures, we include innovative parameters, such as upslope contributing area and elevation.

Finally, we complement the geophysical analysis with community-driven criteria, including indigenous knowledge, to integrate and validate our analysis.

STUDY AREA

The study area is in the north-eastern Kaski district, Nepal. This district lies in the mid-hill ecological region of Nepal, in the central-western part of the country (Figure 2). It is located in the monsoon-influenced tropical climatic zone and the mean annual rainfall in this region, which is the highest in Nepal, is around 3,900 mm, the majority of which (60–90%) occurs during the monsoon season from June to September (Subedi *et al.* 2020). The average annual temperature nearby Pokhara city, at 854 m a.s.l., is 18.3 °C. From a socio-economical point of view, the Kaski district is characterized by high rates of out-migration, both towards other districts or countries and within the district itself, from less productive upland areas towards downhill zones and Pokhara city. This leads to farmland degradation and abandonment, especially in more marginal and poor areas of the Kaski hills (Pandey 2019).

The study area, which is around 1,186 km^2 large, ranges between 28°12′51″N, 84°4′26″E and 28°10′37″N, 84°11′53″E. The elevation ranges from 407 m to 1,443 m a.s.l. South-facing slopes are usually terraced for cultivation, while North-facing slopes of forest and shrublands do not receive enough solar radiation to allow for cropping.

Our analysis focuses on four communities within the study area: Thulakot, Majhkot, Begnas, and Hansapur (hereon 'locations of interest'), considered representative of the local socio-economical setting. The four locations of interest are also characterized by their position on the upslope parts of the watersheds in the hilly mountainous area, located in the northern surroundings of Pokhara city (Figure 2). These locations (i.e. upslope and south-facing) have been selected since they are the most prone to water scarcity. The total area of the location of interest is 12.11 km^2.

DATA

All the datasets used in this work – i.e. soil type, European Space Agency (ESA) Worldcover Land Use/Land Cover (LULC) map and a Digital Elevation Model (DEM) – were downloaded from websites of different public agencies. Flow accumulation (also said upslope contributing area), slope, and elevation maps were derived from the DEM through QGIS software. The lakes and rivers spatial data retrieved by Open Street Map were provided by the International Rainwater Harvesting Alliance (IRHA), a Swiss NGO carrying out a project in this area with its Nepalese local partner NGO Kanchan Nepal. We downloaded a gridded soil map from the Nepalese National Soil Science Research Center (NARC) website. The soil data were

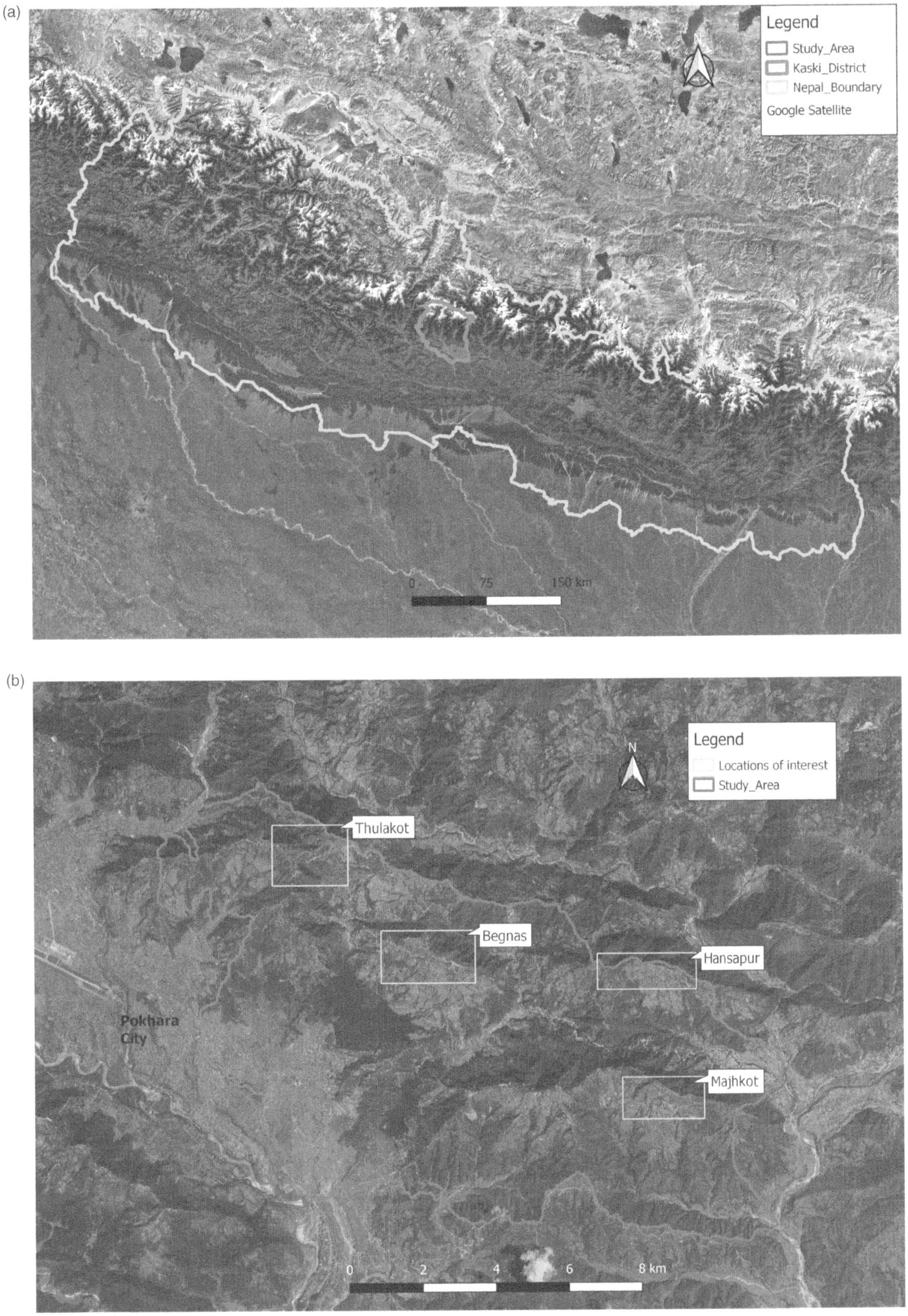

Figure 2 | (a) Study area (red line) within Nepal (green line) and Kaski district (pink line); (b) zoom on the four top hills object of this work within the study area. Please refer to the online version of this paper to see this figure in colour: https://dx.doi.org/10.2166/nh.2022.027.

developed with machine learning and predict soil conditions up to 2 m. Kanchan Nepal provided some GPS locations of existing *pokharis* in the study area as well, while some other locations were gathered during a field visit in the same zone. Data on terraced and private areas, and traditional criteria considered in *pokhari* building, were collected during the field visit. This work was carried out in the framework of IRHA's project. Table 1 reports the sources from which we accessed the data.

METHODS

Methodological framework: multi-criteria decision-making

Spatial MCDM procedures, also known as the weighted overlay process (WOP), are used to combine different thematic maps through a few simple steps to obtain a single final suitability map (Krois & Schulte 2014; Adham *et al.* 2016). A suitability map allows us to visualize multiple options ranked based on preferences expressed through the categorization of various criteria, weighted depending on the influence they have on the overall objective (Al-Adamat *et al.* 2012). The methodological steps of MCDM can be summarized as follows (Grum *et al.* 2016):

1. Selection of relevant suitability criteria for achieving the objective;
2. Establishment of suitability ranges for each criterion (usually on a scale from 1 to 9 where 9 represents the optimal value) and normalization of each criterion on a 1–9 scale;
3. Assignment of a weight to each criterion according to the influence they have on the result – potentially according to experts or local people's opinions. Regardless of their distribution, the sum of all weights must be 1.
4. Combination of suitability values of all weighted criteria to produce the final suitability map.

For each cell of the suitability map, the final suitability score (S) is calculated as follows:

$$S = \sum_i w_i \cdot x_i \qquad (1)$$

where w_i is the weight of factor i and x_i is the suitability score of factor i (Dile *et al.* 2016).

In the present study, the MCDM process was structured in two stages, as shown in the flowchart of the methodology (Figure 3). As a first step, we reviewed various best-siting papers for WH structures applying MCDM and GIS to select the relevant criteria for our work (Literature review). Based on the review, a preliminary set of bio-geophysical criteria and relative weights was defined, which was later used to generate a first preliminary version of the best-siting map. The robustness of the preliminary map was cross-checked with two tests: (a) a sensitivity analysis performed by changing the weights of the MCDM equation, following the approach of Dile *et al.* (2016); and (b) a validation of the most suitable areas, carried out by analysing the value S of the GPS locations of existent *pokharis*. The second step of the analysis allowed a refinement of the suitability map, by maximizing the findings on the on-site survey, aimed at gathering additional information such as terraced and private areas and identifying new socio-economic criteria through a participatory analysis. A new set of

Table 1 | Main information on data sources and format

Data	Data format	Source
Soil	Raster (250 m resolution)	Nepalese National Soil Science Research Center (NARC) website (accessed 21 May 2021)
LULC	Raster (10 m resolution)	ESA Worldcover 10 m 2020 Product (Zanaga *et al.* 2021) (accessed 5 October 2021)
Digital Elevation Model (DEM)	Raster (30 m resolution)	Japan Aerospace Exploration Agency (JAXA), 2021 data (accessed 10 May 2021)
Water sources	Vector	International Rainwater Harvesting Alliance (IRHA) (NGO)
Existing ponds GPS position	Vector	Provided by Kanchan Nepal and Collected during Field visit (21–25 October 2021)
Terraced and private areas	Vector	Collected during Field visit (21–22 and 24–25 October 2021)
Pokhari building traditional criteria	Text	Collected during Field visit (20–22 October 2021)

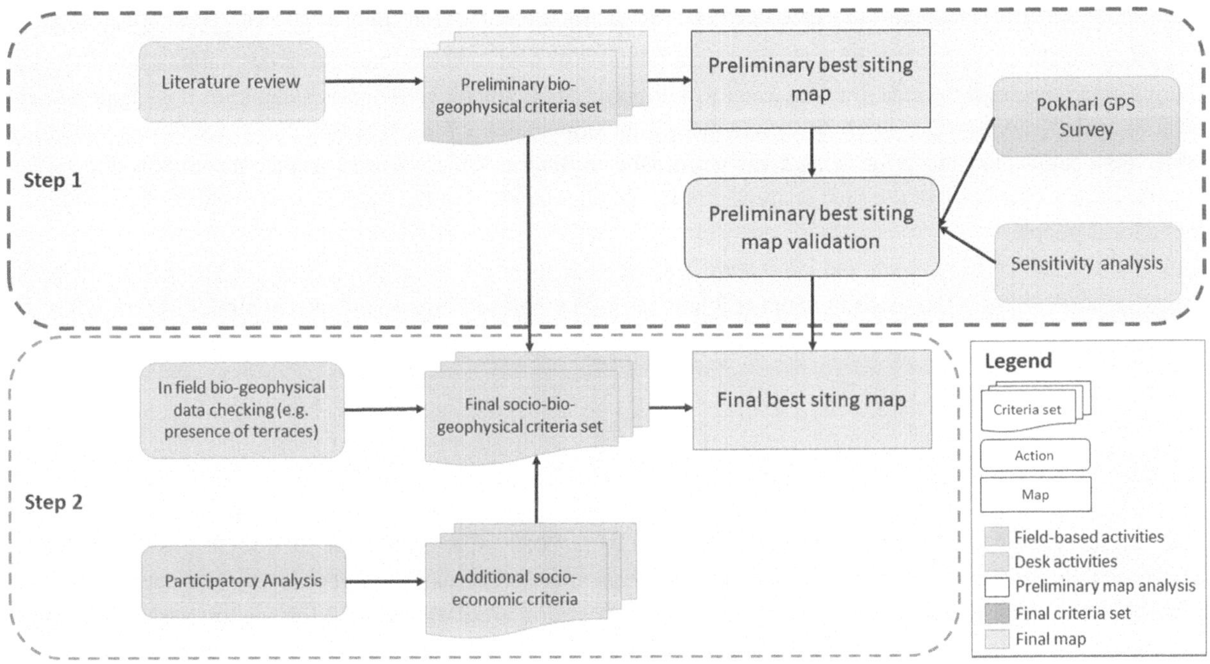

Figure 3 | Flowchart of applied methodology.

socio-bio-geophysical criteria was then obtained merging the criteria of step 1 and of step 2. The additional socio-economic criteria were included by selecting all the areas in the preliminary map falling in the highest quartile of S score (Q4) and removing the ones not meeting the criteria, in order to display only highly suitable areas fitting with socio-economic features. In fact, in our case, these latter do not allow for a suitability range but are rather Boolean criteria, which can only match or not match a given condition.

Literature review and preliminary bio-geophysical criteria set

We reviewed various best-siting works carried out using MCDM and GIS focusing on ponds in different geographic areas and selected the most relevant criteria for the objective of the analysis (de Winnaar *et al.* 2007; Al-Adamat *et al.* 2012; Dile *et al.* 2016; Grum *et al.* 2016). As summarized in the review paper of Adham *et al.* (2016), the most common criteria are slope, soil type, and land use. Other common geophysical criteria are rainfall and runoff, while the main socio-economic criteria are the distance from roads, villages, streams, and fields. The slope affects the quantity of rainfall that infiltrates into the soil or runs off, the distribution of runoff, and the quantity of earthwork needed to build WH reservoirs (Ramakrishnan *et al.* 2009; Dile *et al.* 2016). Moreover, a small slope is needed to convey water towards storage sites, thus gentle slopes are preferred for WH implementation with respect to flat areas (Ward *et al.* 2021). The soil type affects water infiltration and storage in the soil (Mugo & Odera 2019), and consequently, the runoff (Singhai *et al.* 2019). Soil type suitability varies with the type of WH structure. For instance, sandy soils are better suited for recharge structures, while clayey soils are preferred for water storage structures, such as ponds (Al-Adamat *et al.* 2010; Adham *et al.* 2016). LULC informs on land surface characteristics (Ward *et al.* 2021). These latter control the quantity of runoff produced in an area, and water infiltration in soil (de Winnaar *et al.* 2007). It can be easily extracted by remote sensing data (Adham *et al.* 2016). Rainfall quantity is usually considered in WH best siting to locate areas where more runoff is generated, as it can be used to recharge the structures (Chen *et al.* 2017).

Concerning socio-economic criteria, distance from houses and cropped fields are considered because harvested water is mainly required for domestic and agricultural use, thus structures should be as close as possible to where water is needed (de Winnaar *et al.* 2007; Chen *et al.* 2017). The same occurs for roads, from which WH structures can be easily accessed. Distance from streams is sometimes considered since part of streams' discharge can be used to recharge WH structures (Al-Adamat *et al.* 2010).

Following the literature analysis, the selected parameters considered in this analysis are slope, LULC, soil type, upslope contributing area, elevation, and distance from rivers and lakes. In our case, we decided to introduce an upslope contributing area as a proxy for the presence of minimum water flow (which can feed *pokharis*) instead of estimating runoff. The upslope contributing area represents the area draining water to a downstream point, and it is measured in km^2. It is higher in riverbeds and lower in upslope zones (Chen *et al.* 2017). We did not include a rainfall layer because, due to monsoons, our study area receives enough rainwater, thus prioritizing areas with higher rainfall is unnecessary. Distance from fields and settlements was considered at a later stage of the final analysis, to include the outcomes of the field visit, as these criteria were indicated directly by local people.

Elevation and distance from rivers and lakes were both used to prioritize areas more prone to water scarcity, which are those more uphill and further away from rivers and lakes (which contribute to groundwater recharge and irrigation).

Preliminary best-siting map

The value ranges/categories of each layer were given a suitability score spanning from 1 (unsuitable values/categories) to 9 (optimal values/categories – suitability scores are referred to as x_1 in Equation (1)). Regarding the suitability score given to each value range of each criterion, we assigned them as follows: According to the literature, optimal values for slope for ponds best siting range from 0–3% to 0–5% (Ramakrishnan *et al.* 2009; Al-Adamat *et al.* 2010). On this basis, we assigned the highest suitability score to the 0–3.5% slope range. To other ranges of increasing slopes are assigned decreasing suitability values as summarized in Table 2. Regarding the upslope contributing area, we assigned the highest suitability score (9) to contributing areas of 0.09–0.18 km^2, considering that the quantity of runoff should be enough to recharge *pokharis* but not too high to prevent structural damages. This range was selected by adapting to our study area the upslope contributing area values proposed by Márquez *et al.* (2021). Consequently, upslope contributing areas higher than 0.18 km^2 were given lower suitability, while contributing areas smaller than 0.09 km^2 were given an intermediate suitability score. In the case of farm ponds, usually, the highest suitability score for land use is given to agricultural land if water is used for irrigation (Dile *et al.* 2016), while when ponds are used mainly for herding, also shrubland, bareland, and grassland areas can be marked as optimally suitable (Grum *et al.* 2016; Singhai *et al.* 2019). Another reason for which we did not assign the highest suitability score to cultivated land is that in our study area, cropped fields are terraced, thus it is not advisable to build ponds there in order to cultivate all the available surface. Moreover, in our locations of interest shrublands, barelands and grasslands are usually located near cultivations, thus the collected water can be easily accessed for supplementary irrigation as well. Furthermore, the land use map used in this work marked as 'forest' all the areas with a tree cover >10%, thus, not having information on the actual presence of open forest (which could have been very suitable for ponds) and closed forest (not suitable as there would not be enough space for a pond), we decided to assign it an intermediate suitability score. According to the literature (Al-Adamat *et al.* 2012; Adham *et al.* 2016; Ward *et al.* 2021), less-draining soils such as clay, silty loam, and loamy soils are ideal for ponds construction. We gave both silt and silty loam soil a suitability value of 9. The highest elevation and distances from water bodies were assigned maximum suitability in order to prioritize more water-scarce areas which cannot benefit from this surface water and groundwater recharge. Hence, zones closer to water bodies were given lower suitability. Table 2 summarizes suitability values assigned to each criterion category and sources used to set the suitability ranges.

Following the MCDM steps, each layer was given a weight based on the relative influence on the resulting map. Weight assignment is a complicated task that might produce misleading results (Al-Adamat *et al.* 2012). Weights of criteria for MCDM are computed mainly through the Analytical Hierarchy Process (AHP) method which allows us to assign weights by comparing the selected criteria in a pairwise matrix and inspecting the relative importance of each layer (Grum *et al.* 2016; Ward *et al.* 2021). The pairs of criteria are set according to literature and experts' opinions (Krois & Schulte 2014; Ward *et al.* 2021) or through participatory workshops with stakeholders and NGO consultants (Grum *et al.* 2016). However, AHP is highly influenced by expertise (Adham *et al.* 2016). Sometimes, weights are assigned directly on the basis of literature or opinions, without applying the pairwise matrix (Al-Adamat *et al.* 2010; Krois & Schulte 2014) or equal weights are assigned to all the criteria to avoid biases (de Winnaar *et al.* 2007; Al-Adamat *et al.* 2012). Owing to these reasons, we decided to assign equal weights to all the layers except for the distance from water bodies, which received a lower weight. Distance from rivers and lakes serves to prioritize areas that cannot likely count on groundwater recharge or surface water for agriculture. However, stream water cannot always be easily accessed in hilly areas, so this criterion has a lower influence on the final selection of best sites. We considered this weight setting since it is the most conservative and balanced.

Table 2 | Suitability scores and sources used for setting the suitability ranges used in the analysis

Suitability score	9	7	5	3	1	Range source
Slope (%)	0–3.5	3.6–7	7.1–10	10.1–15	>15.1	Literature
Upslope contributing area (km^2)	0.09–0.18	0.1809–0.269	0.27–0.359 and <0.09	0.36–0.489	>0.49	Literature and GIS analysis
LULC type and class ID number	– grassland (30) – shrubland (20) – bareland (60)	– Forest (10)	– Cropland (40)	–	– Permanent water bodies (80) – Urban/built-up land (50)	Literature
Soil type	Silty, silty loam	–	–	–	–	Literature
Elevation (m a.s.l.)	1,237–1,445	1,029–1,236	821–1,028	613–820	1–612	GIS analysis
Distance from lakes and rivers (km)	>4	3–3.9	2–2.9	1–1.9	<0.99	GIS analysis
Terraced and private areas[a]	–	–	–	–	–	GIS analysis

[a]These features have been removed using a Boolean operator which either includes or excludes a map area in the analysis according to specific criteria, thus no weight setting is required in this case. We set the operator to exclude terraced and private areas recorded on the field from the suitability map obtained through the GIS

Sensitivity analysis

We performed a sensitivity analysis to assess the variability of the results at varying weights of the best-siting criteria, as a measure of uncertainty following Dile *et al.* (2016). We first assigned equal weights to all the criteria (variation 1). We then varied the weights of each criterion using plausible strategies. In variation 2, the highest weight was given to slope, which is usually assigned a higher weight (Al-Adamat *et al.* 2010; Singhai *et al.* 2019). Usually, the soil is weighted slightly higher than land use (Mbilinyi *et al.* 2007; Krois & Schulte 2014), thus we also tested soil type influence on the result in variation 3. The fourth variation was set to give land use the highest weight, while in variation 5, we tested the influence of upslope contributing area on the result by attributing the highest weight. Table 3 summarizes the tested weight distributions.

In field bio-geophysical data checking and participatory analysis

We conducted a field visit in the study area between the 16th and the 30th of October 2021 with three specific objectives: (a) surveying additional GPS locations of existing traditional ponds beyond the 13 ones provided by Kanchan Nepal to validate the preliminary best-siting results (*pokhari* GPS survey); (b) collecting information on land ownership and terraces presence; (c) gathering information on traditional criteria for selecting suitable *pokharis*' locations.

Concerning the validation of the preliminary results, we checked in which suitability quartile the existing *pokharis* are located. Those found within the high suitability areas obtained with the preliminary GIS siting were considered successful. Field inspection of suitable areas was also carried out to check for 'false positives', i.e. areas with theoretical high suitability, but with local constraints hindering the construction of *pokharis* (e.g. presence of agricultural terraces or private land ownership).

The features resulting as 'false positives' were then removed from high suitability areas in the final map. Finally, we complemented our set of siting criteria with the indigenous knowledge (additional socio-economic criteria) to account for potentially locally relevant factors that are usually neglected in best-siting studies. Traditional criteria to locate new ponds were surveyed through two focus group discussions with two different groups of local *pokharis* building experts (i.e. four people in Hansapur and seven people in Majhkot communities). They were all elder men who accompanied the authors to visit existing *pokharis* nearby their villages together with Kanchan Nepal members. These latter had already gained local people's trust during years of collaboration in development projects carried out in the same villages. During traditional ponds inspections, the NGO members also facilitated our interviews by translating from Nepalese language to English. Villagers were asked the following simple questions.

- How do you select where to build new *pokharis*? What factors do you consider?
- Which are the features characterizing these ponds?

We then complemented our criteria set with the interviewees' socio-economic criteria to obtain the final best-siting map, including the information obtained on the field. The integration of the traditional criteria into the GIS work was done by visual inspection, owing to the lack of digital layers. We selected and displayed from the preliminary map shown in Figure 4 all the areas falling on the highest suitability quartile (Q4) and removed from them the features pointed out by local people using a Boolean operator in Qgis software.

Table 3 | Different weight variations tested in the sensitivity analysis

Criterion	Weights	Variation 1	Variation 2	Variation 3	Variation 4	Variation 5
Slope	0.19	0.16	0.35	0.15	0.15	0.15
Soil type	0.19	0.16	0.15	0.35	0.15	0.15
LULC	0.19	0.16	0.15	0.15	0.35	0.15
UCA	0.19	0.16	0.15	0.15	0.15	0.35
Elevation	0.19	0.16	0.15	0.15	0.15	0.15
Distance from water bodies	0.05	0.16	0.05	0.05	0.05	0.05

UCA, upslope contributing area.

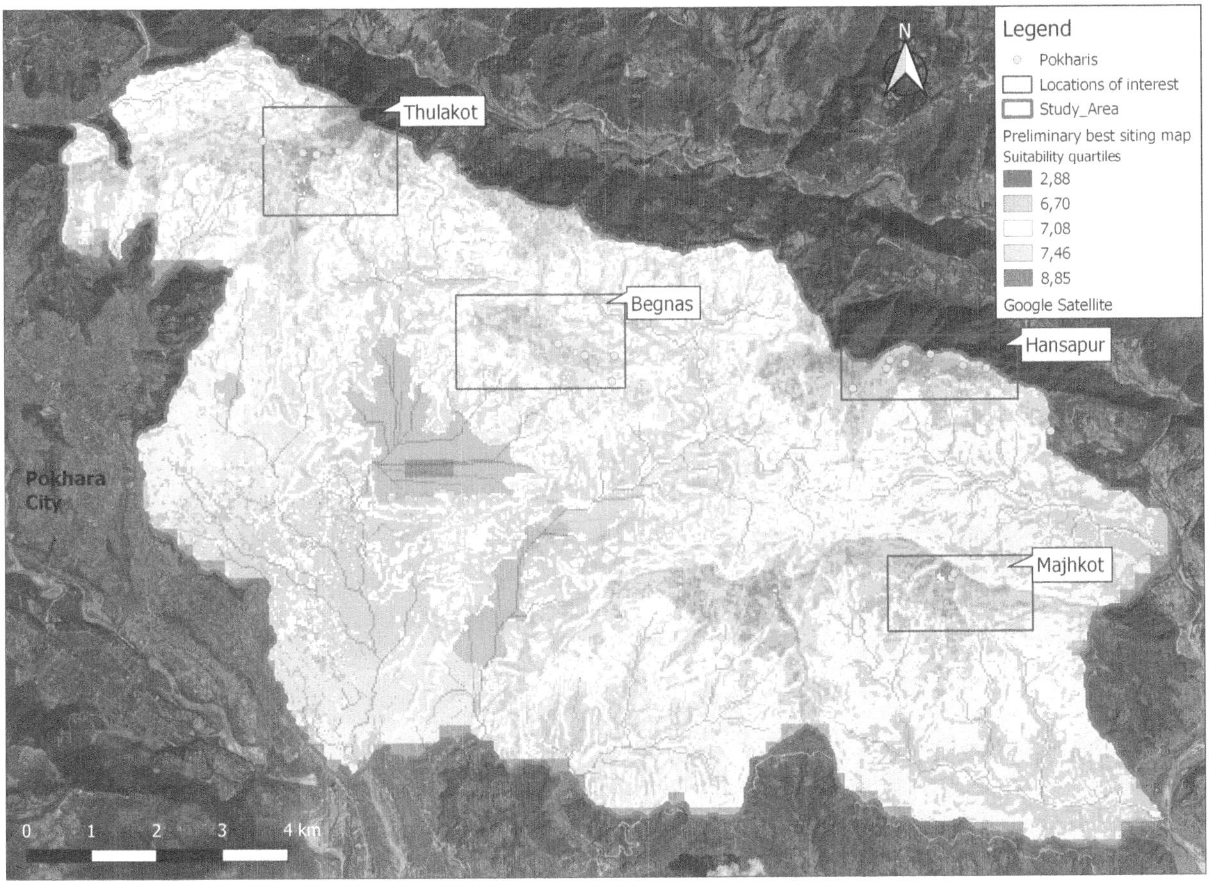

Figure 4 | Preliminary best-siting map.

RESULTS

Preliminary best-siting map

The preliminary best-siting map is shown in Figure 4. Unsurprisingly, higher suitability values lie in areas of higher elevation – where people rely on *pokharis* the most – which were given the highest priority. On the other hand, the lowest suitability values are reported in pixels corresponding to permanent water bodies followed by urban areas (Figure 4).

Sensitivity analysis

The results of the sensitivity analysis show only slight differences in the percentages of areas covered by high suitability zones (quartile 4) and unsuitable areas (quartile 1) in the six suitability maps obtained with the different weight conditions. A slightly higher variability is found in the two intermediate suitability quartiles, as shown in Figure 5. As expected, the influence of soil type on the result was negligible even in variation 3 given the homogeneity of soil types in our study area. Among the weight variations, the layer with the lowest percentage of high suitability area is variation 4, which assigned the highest relative importance to the LULC map. This is in contrast with Dile *et al.* (2016), who found rainfall to be the most restrictive criterion. Also in other works, land use is one of the most common criteria (Adham *et al.* 2016) but it is usually given a quite low weight (Mbilinyi *et al.* 2007; Singhai *et al.* 2019) while its influence on the result in our case was considerable. Furthermore, the spatial distribution of the suitability quartile is quite homogeneous among the different maps as well, as shown in Figure 6.

Preliminary best-siting map validation

After the sensitivity analysis, the Suitability score of pixels corresponding to the 20 existing *pokharis* GPS position was inspected. Figure 7 shows the suitability scores of each pond for each weight assignment.

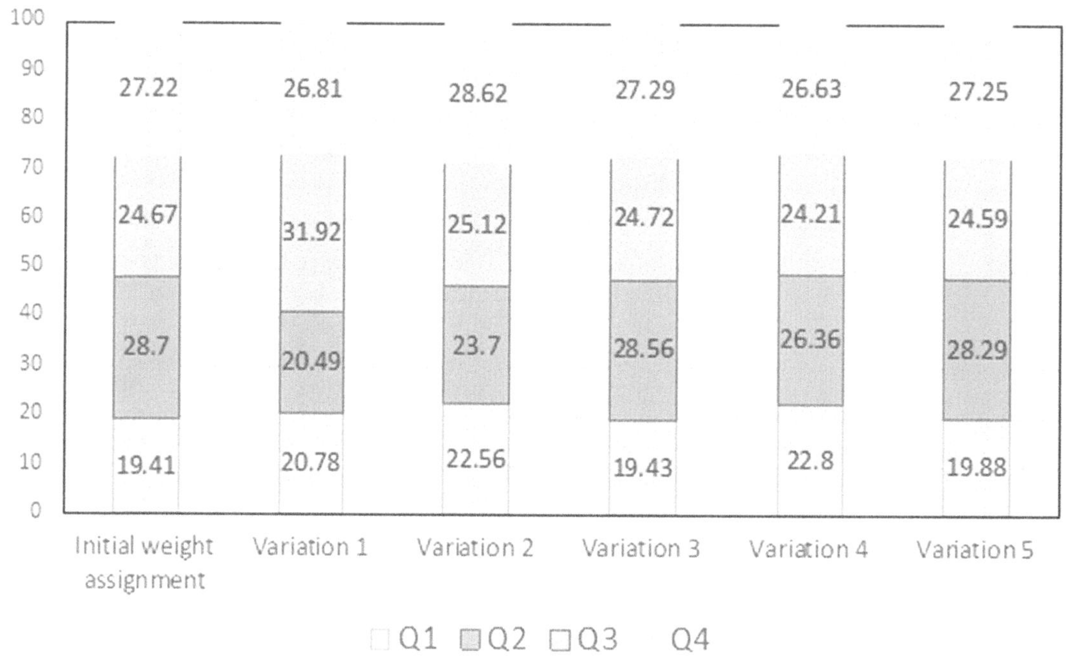

Figure 5 | Area covered (%) by each suitability quartile for each weight assignment.

In variation 1, where the distance from water bodies weight was higher, one *pokhari* laid in Q3 due to its proximity to a river, while all the other ponds were located in Q4 areas. The same occurred with another *pokhari* in the initial weight setting (the reason is addressed in the discussion).

Once we verified the robustness of the initial weight assignment, we proceeded with this preliminary map to obtain the final result. Considering the average suitability score of GPS points corresponding to existing *pokharis* and the standard deviation of all the different weight assignments (Table 4), the preliminary best-siting map can be considered successfully validated. The single suitability score of each point for each weight condition is reported in the Supplementary material.

In-field bio-geophysical data checking and participatory analysis

Kanchan Nepal members, together with local people, identified on the maps prepared for each location of interest some private lands. In the same maps, we also marked the location of terraces. These features, accounting for 3.18 km^2 (corresponding to 29% of high suitability areas within the four locations of interest in the preliminary map), were marked as unsuitable. It is important to note that the information on land ownership was not fully available for the whole study area.

Furthermore, we gathered information on both *pokharis*' characteristics and local knowledge criteria used to build these traditional ponds.

Local experts in *pokharis*' building explained that traditional criteria considered when locating new ponds are the following, ranked by importance:

- Accessibility (e.g. proximity to villages and roads or paths) – usually *pokharis* are built within 2 km from villages and as close as possible to tracks;
- Small flat areas where water stagnates over 4–5 years during monsoons.

Sometimes *pokharis* are built nearby small water springs which will feed the ponds themselves. Accessibility to *pokharis* is the most important criterion, as pond water is used mainly for cattle rearing and irrigation, which are activities carried out in areas surrounding the villages. Still for accessibility reasons, *pokharis* are located also near roads and tracks used by herders to move their cattle. *Pokharis*, especially in the past had also social functions, as people walking on mountain paths and carrying out daily activities used to rest near those ponds. Water stagnation during monsoons considered by indigenous knowledge supports our choice to include the upslope contributing area as best-siting criterion. We added these criteria to the preliminary bio-geophysical criteria set shown in Table 2. Creating our final socio-bio-geophysical criteria set. The upslope

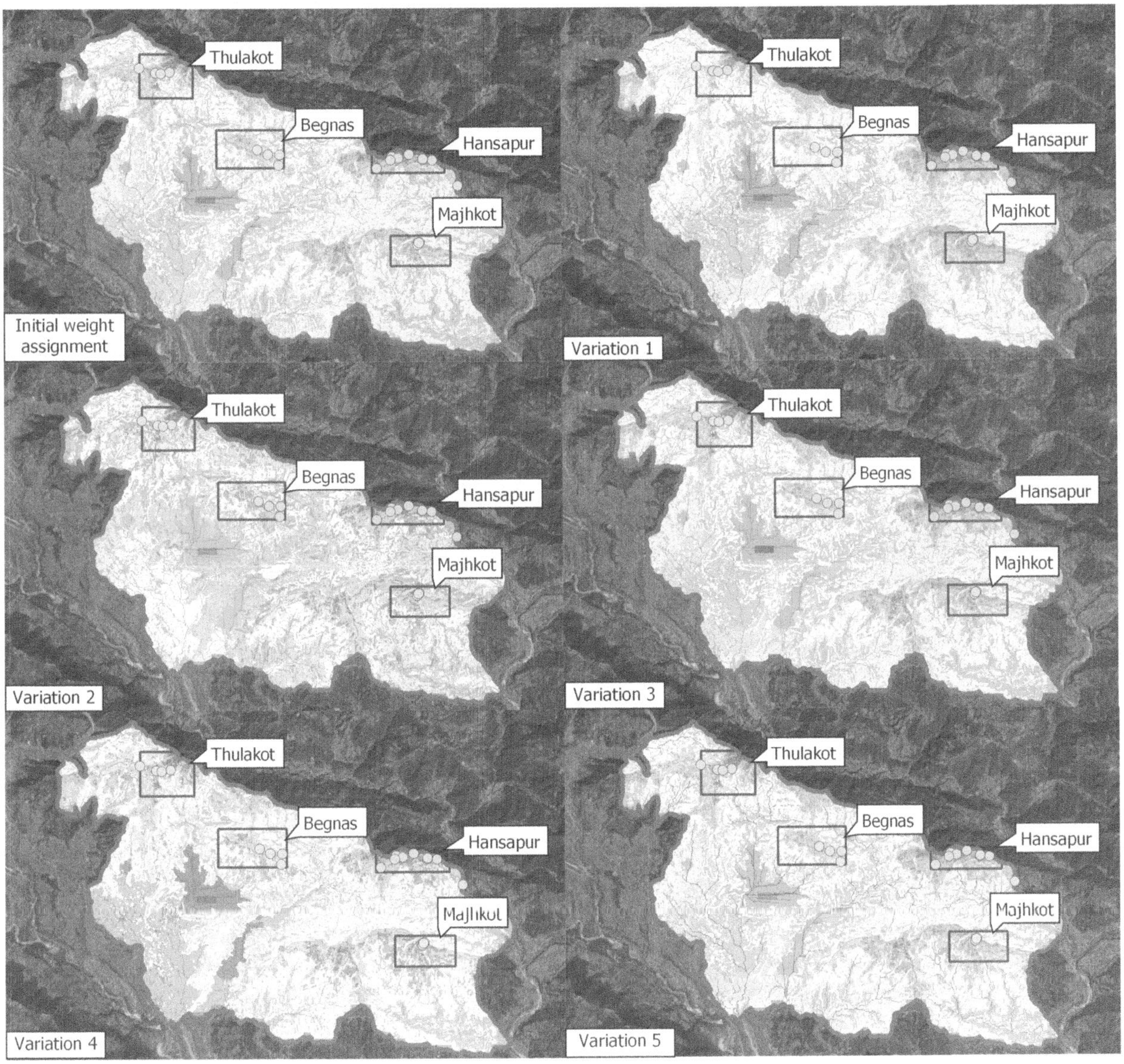

Figure 6 | Preliminary suitability maps obtained with the initial weight assignment and the five weight variations.

contributing area was already included in the MCDM analysis, we selected only Q4 suitable areas from the preliminary best-siting map and we removed those >2 km far from villages and those falling on private and terraced lands, which are unusable for community pond building, identified during the field visit.

Final best-siting map

The resulting final best sites map for *pokharis*' building is shown in Figure 8. Around 0.423 km^2 corresponding to 3.47% of the total area in the four focus zones (12 km^2) were found to be highly suitable, matching all the final socio-bio-geophysical criteria.

DISCUSSION

Pokharis' best-siting methodology

The main result of our study provides a first guidance for the identification and spread of potential sites for *pokharis* in Nepal. The sensitivity analysis confirms the robustness of our approach, which is also supported by the in-field validation. In fact,

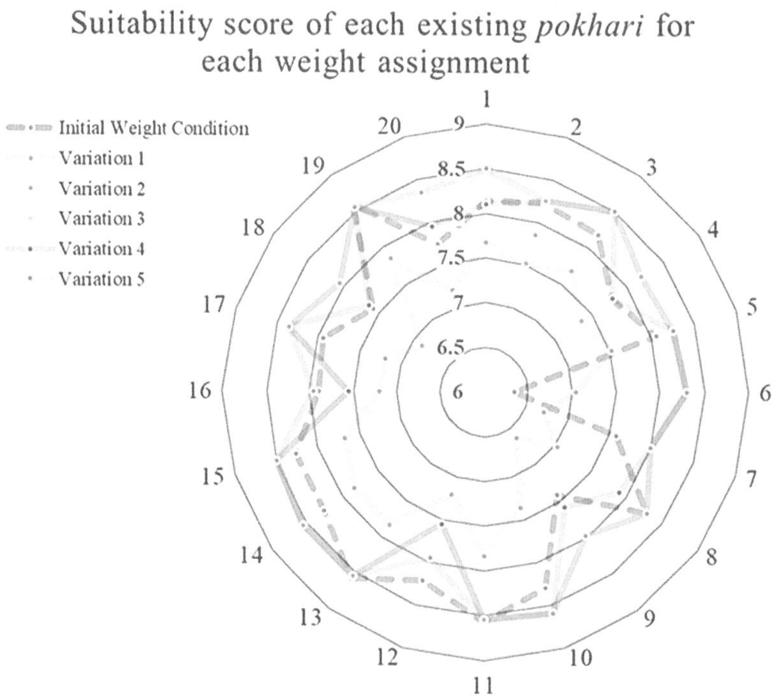

Figure 7 | Suitability score of each existing *pokhari* for each weight assignment displayed in a radar chart. Each radius of the circle represents one *pokhari*. The circles are relative to the suitability scores.

Table 4 | Average suitability score (*S*) and standard deviation of the six weight assignments

Weight assignment	Initial weight assignment	Variation 1	Variation 2	Variation 3	Variation 4	Variation 5
Mean *S*	8	7.36	8.23	8.31	8.15	8.31
Standard deviation	0.51	0.38	0.33	0.23	0.38	0.23

compared with other water harvesting best-siting analysis performed in other regions of the world, our results are among the ones with the highest correspondence between modelled sites and existing ones, with a 100% correspondence. The same percentage was obtained by Ramakrishnan *et al.* (2009) for ponds, while slightly lower accuracy was reported for other structures. Singhai *et al.* (2019) reported a suitability map accuracy of 90% with 31 surveyed farm ponds; Grum *et al.* (2016) validated check dams and percolation pond suitability, for these latter, an accuracy of 63% over 92 infiltration ponds is reported. In these studies, the number of existing structures is higher compared with our work, but the study areas are wider than ours. Moreover, the mentioned studies consider various WH structures and not just ponds as in our case. Our methodology also includes two rather uncommon criteria: upslope contributing area and elevation. This latter was introduced to prioritize higher altitude areas which are those more prone to water scarcity, while upslope contributing area was used to find minimum water accumulation potential, and it is consistent with traditional criteria reported by local experts. Moreover, the upslope contributing area can be easily computed from a DEM on GIS softwares compared with runoff estimation, which is more complicated and data-demanding.

Another key result of this study is the validated selection of siting criteria using participatory methods. We combined the benefits of an accurate validation with existing ponds with traditional knowledge to develop an efficient approach easily acceptable by local people and linked with their indigenous expertise. Other studies took into account traditional knowledge. For instance, Al-Adamat *et al.* (2012) asked local people to point out potential WH best sites on a map and to select some best-siting criteria like distance from roads, borders, villages, farmlands, and wells. Grum *et al.* (2016) also included a

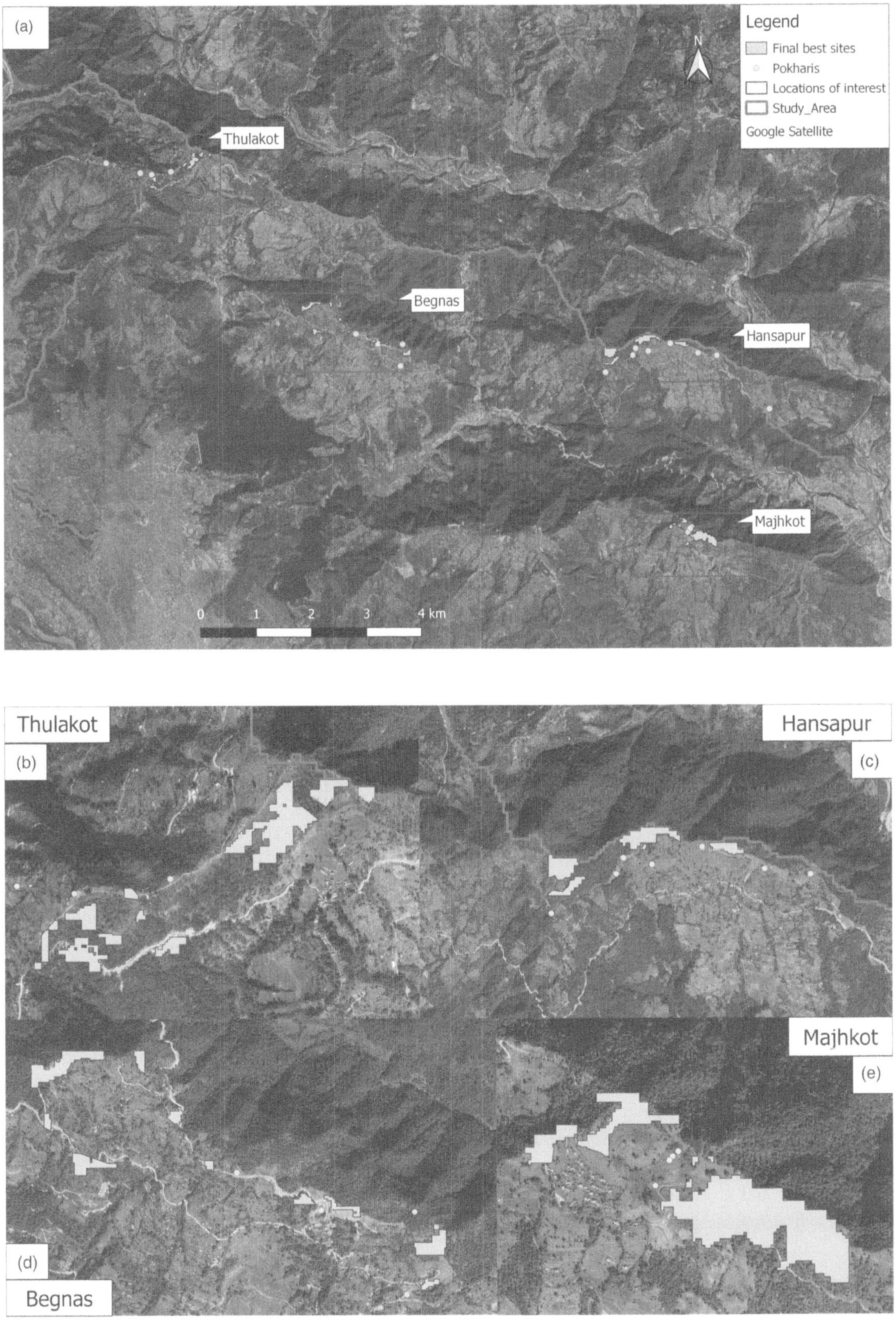

Figure 8 | Final best-siting map, including (a) the whole study area, (b) Thulakot area, (c) Hansapur area, (d) Begnas area, and (e) Majhkot area.

participatory approach to identify the most relevant criteria for local people in selecting the best WH techniques, and then they processed retrieved criteria with MCDM to sort out preferred water harvesting techniques. Finally, Mbilinyi *et al.* (2005) reported in their study carried out in Tanzania the traditional criteria used to locate new WH structures.

Concerning our comparison with traditional criteria, we added distance from villages as a socio-economic criterion pointed out by local people and upslope contributing area as a proxy for water stagnation points over years. We did not take into account springs, which was an additional indigenous criterion, as our analysis focused on rainwater storage.

Such a straightforward and validated approach for mapping best location of *pokharis* can immediately support government, NGOs and private efforts to build new distributed water points across Nepal. Increasing water available for farming, depending on the actual number of new structures and their size, can considerably improve the availability of water in the dry season, thus positively affecting both cattle health and agricultural productivity. Cascading impacts of increase water availability with *pokharis* might likely include increased food security and income stability, thus potentially reducing outmigration from rural areas, which is a pivotal cause of environmental degradation in Nepal (Pandey 2019). Other ecosystem services from *pokharis*, such as water retention and flood prevention, reduction of land erosion, cultural and social value would increase as well, improving people's wellbeing.

Main limitations

Our results report good correspondence between existing *pokharis*' locations and produced best-siting map. Thus, the presented method offers a useful guidance for WH structure best siting despite some limitations mainly due to data availability and quality. Regarding this latter issue, some considerations were made: the LULC layer was, despite being the best option available, unable to report the presence of terraces. These were reported instead as bareland or grassland (the best categories for pond building) leading to many suitable areas for pond construction. This problem was identified during the field visit and suitable areas were reduced as terraces are unsuitable for *pokhari* building. Terraced fields are usually private, and these limited lands often represent the main source of subsistence for farmers, thus it is not advisable to use those areas for community-owned ponds. Some techniques for terrace detection from remote sensing do exist (Tarolli & Sofia 2016), but the lack of high-resolution topographic data (1–2 m) hindered its estimation. Interestingly, the LULC reports the biggest *pokhari* as 'permanent water body', thanks to its good spatial resolution. Owing to this reason, the just mentioned *pokhari* is the one resulting in the third suitability quartile (see Supplementary material), because permanent water bodies were given the lowest suitability as they usually represent rivers or lakes. Actually, the water body in this case is the *pokhari* itself, thus the lower score is a 'false negative', i.e. an area resulting in a lower suitability score which, checking on the field, has instead a high suitability. This explains the 100% correspondence between retrieved best sites and existing structures despite one pond laying in the third quartile. The main limitation of the LULC map – for the scope of this work – lies in the broadness of the categories used in the classification. Particularly, pixels with a wood cover exceeding 10% are classified as tree cover even if 'other land cover classes can be present below the canopy, even with a density higher than trees' as stated in the ESA Worldcover Product User manual (Zanaga *et al.* 2021). The same occurs for grassland and shrubland. This affects our analysis as, for instance, open forests, which are very suitable, are not distinguished from closed forests, which are less suitable for WH structures having less free space. Despite being the best option available, the DEM is not ideal for best-siting analysis of relatively small structures, as in this case. A spatial resolution of 30 m cannot report small flat areas surrounded by steeper areas, which would be actually suitable for *pokharis*, especially in hilly/mountainous regions like Nepal. Also, data about land ownership were not available from institutions either before or during the field visit. However, local people provided information on some land property, which allowed us to exclude at least some private lands. Lastly, a digital map of housing was not available, so we had to remove areas too far from villages from the final best-siting map by visual inspection. The availability of a digital layer of settlements would have allowed the automatized integration of this traditional criteria into the GIS analysis, making this step faster and more accurate. Despite these limitations related to data availability and resolution, our methodology provides a first reliable guidance to large-scale planning of *pokharis*, with a good level of accuracy granted by the high validation standard and the integration with participatory methods. Particularly useful to overcome the quantitative data limitation was the integration of qualitative information from field inspection and participatory research. As in other studies (Castelli *et al.* 2018, 2021; Piemontese *et al.* 2021), the integration of indigenous knowledge was essential to (a) integrate the scarce data, (b) validate the overall approach and criteria, and (c) identify limitation and potential improvement for future work.

Further developments

With detailed soil and geological data, areas more prone to landslides (which in Nepal are common) could be estimated to remove suitable sites where water infiltration in soil could increase the risk of land collapse. This analysis could be done with the aforementioned high-resolution topography data if they will become available for Nepal too, as well as automated detection of terraces and small flat areas. Moreover, once new *pokharis* are built in at least some identified suitable spots, a thorough quantification of the benefits provided by these structures in terms of farm productivity and ecosystem services provision could be carried out. Furthermore, the proposed approach can be a guidance for similar works in other data-scarce areas with similar characteristics, to support decision-makers in sustainable land and water management and foster resilience to climate change.

CONCLUSION

We introduced a methodology for large-scale siting of traditional WH structures, which is consistent with traditional criteria. To our knowledge, it is the first best-siting analysis for Nepalese traditional ponds and one of the first best-siting works for Nepal. The best-siting analysis identified about 0.4 km^2 of highly suitable land for *pokharis*' construction, corresponding to 3.47% of the total location of interest's area covering around 12 km^2. As our analysis was successfully validated with a 100% agreement with existing *pokharis*, it can give decision-makers reliable guidance for water resources planning and management. Compared with the traditional siting methods focused on the local scale and individual structures, the combination of GIS and MCDM allows a fast and cost-effective best-siting analysis on wide areas, providing information that can support large-scale national to regional planning. Moreover, software and datasets used are freely available, so this analysis can be replicated by other institution or for other areas. Before building new structures, we recommend a field check, especially for land ownership and accessibility as this information is usually very difficult to retrieve remotely. Finally, the development of higher-resolution or better-quality datasets could improve the accuracy of our best-siting tool and improve the usefulness of our approach. The proposed methodology favours the prompt implementation of water harvesting techniques that, together with other soil and water conservation measures, can increase farming productivity through increased water availability for agriculture and livestock rearing.

ACKNOWLEDGEMENTS

This work was carried out in the framework of the project 'Rain Communities' carried out by the Swiss NGO International Rainwater Harvesting Alliance (IRHA) with its Nepalese partner NGO Kanchan Nepal. We would like to acknowledge the institutions funding the project, i.e. République et Canton de Genève, Puplinge and 1% fund. We would like to acknowledge also all the NGO members who supported us during the development of this work, especially concerning the field visit and all the villagers who collaborated with us during the participatory analysis.

The authors are thankful to editor and reviewers for evaluating and improving this manuscript.

DATA AVAILABILITY STATEMENT

All relevant data are included in the paper or its Supplementary Information.

CONFLICT OF INTEREST

The authors declare there is no conflict.

REFERENCES

Adham, A., Riksen, M., Ouessar, M. & Ritsema, C. 2016 Identification of suitable sites for rainwater harvesting structures in arid and semi-arid regions: a review. *Int. Soil Water Conserv. Res.* **4**, 108–120. https://doi.org/10.1016/j.iswcr.2016.03.001.

Adhikari, S. 2018 Drought impact and adaptation strategies in the mid-hill farming system of Western Nepal. *Environments* **5**, 101. https://doi.org/10.3390/environments5090101.

Adhikari, S. P., Timsina, K. P. & Lamichhane, J. 2018 Adoption and impact of rain water harvesting technology on rural livelihoods: the case of Makwanpur district, Nepal. *Rural Extension and Innovation Systems Journal* **14**(1), 34–40.

Al-Adamat, R., Diabat, A. & Shatnawi, G. 2010 Combining GIS with multicriteria decision making for siting water harvesting ponds in Northern Jordan. *J. Arid Environ.* **74**, 1471–1477. https://doi.org/10.1016/j.jaridenv.2010.07.001.

Al-Adamat, R., AlAyyash, S., Al-Amoush, H., Al-Meshan, O., Rawajfih, Z., Shdeifat, A., Al-Harahsheh, A. & Al-Farajat, M. 2012 The combination of indigenous knowledge and geo-informatics for water harvesting siting in the Jordanian Badia. *J. Geog. Inf. Syst.* **04**, 366–376. https://doi.org/10.4236/jgis.2012.44042.

Bastakoti, R. C., Prathapar, S. A. & Okwany, R. O. 2016 Community pond rehabilitation to deal with climate variability: a case study in Nepal Terai. *Water Resour. Rural Dev.* **7**, 20–35. https://doi.org/10.1016/j.wrr.2016.01.001.

Castelli, G., Bresci, E., Castelli, F., Hagos, E. Y. & Mehari, A. 2018 A participatory design approach for modernization of spate irrigation systems. *Agric. Water Manage.* **210**, 286–295. https://doi.org/10.1016/j.agwat.2018.08.030.

Castelli, G., Oo, W. M., di Maggio, A., Fellin, L., Re, V. & Bresci, E. 2021 Participatory analysis of sustainable land and water management practices for integrated rural development in Myanmar. *J. Water Sanit. Hyg. Dev.* **11**, 26–36. https://doi.org/10.2166/washdev.2020.166.

Chen, C., Meurk, C. D., Jia, Z., Lv, M., Wu, S. & Jia, J. 2017 Incorporating landscape connectivity into household pond configuration in a hilly agricultural landscape. *Landscapes Ecol. Eng.* **13**, 189–204. https://doi.org/10.1007/s11355-016-0317-3.

Chen, W., He, B., Nover, D., Lu, H., Liu, J., Sun, W. & Chen, W. 2019 Farm ponds in southern China: challenges and solutions for conserving a neglected wetland ecosystem. *Sci. Total Environ.* **659**, 1322–1334. https://doi.org/10.1016/j.scitotenv.2018.12.394.

de Winnaar, G., Jewitt, G. P. W. & Horan, M. 2007 A GIS-based approach for identifying potential runoff harvesting sites in the Thukela River basin, South Africa. *Phys. Chem. Earth Parts ABC* **32**, 1058–1067. https://doi.org/10.1016/j.pce.2007.07.009.

Dile, Y. T., Rockström, J. & Karlberg, L. 2016 Suitability of water harvesting in the Upper Blue Nile Basin, Ethiopia: a first step towards a mesoscale hydrological modeling framework. *Adv. Meteorol.* **2016**, 1–12. https://doi.org/10.1155/2016/5935430.

Grum, B., Hessel, R., Kessler, A., Woldearegay, K., Yazew, E., Ritsema, C. & Geissen, V. 2016 A decision support approach for the selection and implementation of water harvesting techniques in arid and semi-arid regions. *Agric. Water Manage.* **173**, 35–47. https://doi.org/10.1016/j.agwat.2016.04.018.

Japan Aerospace Exploration Agency (JAXA) 2021 *ALOS Global Digital Surface Model 'ALOS World 3D – 30 m (AW3D30)'. Version 3.2.* Available from: https://www.eorc.jaxa.jp/ALOS/en/dataset/aw3d30/aw3d30_e.htm (accessed 10 May 2021).

Krois, J. & Schulte, A. 2014 GIS-based multi-criteria evaluation to identify potential sites for soil and water conservation techniques in the Ronquillo watershed, northern Peru. *Appl. Geogr.* **51**, 131–142. https://doi.org/10.1016/j.apgeog.2014.04.006.

Lal, C. & Verma, L. R. 2008 Indigenous Technological Knowledge on Soil and Water Management From Himachal Himalaya, vol. 7, p. 9.

Márquez, J. D., Peña, L. E., Barrios, M. & Leal, J. 2021 Detection of rainwater harvesting ponds by matching terrain attributes with hydrologic response. *J. Cleaner Prod.* **296**, 126520. https://doi.org/10.1016/j.jclepro.2021.126520.

Mbilinyi, B. P., Tumbo, S. D., Mahoo, H. F., Senkondo, E. M. & Hatibu, N. 2005 Indigenous knowledge as decision support tool in rainwater harvesting. *Phys. Chem. Earth Parts ABC* **30**, 792–798. https://doi.org/10.1016/j.pce.2005.08.022.

Mbilinyi, B. P., Tumbo, S. D., Mahoo, H. F. & Mkiramwinyi, F. O. 2007 GIS-based decision support system for identifying potential sites for rainwater harvesting. *Phys. Chem. Earth Parts ABC* **32**, 1074–1081. https://doi.org/10.1016/j.pce.2007.07.014.

Mugo, G. M. & Odera, P. A. 2019 Site selection for rainwater harvesting structures in Kiambu County-Kenya, Egypt. *J. Remote Sens. Space Sci.* **22**, 155–164. https://doi.org/10.1016/j.ejrs.2018.05.003.

Nepalese National Soil Science Research Center (NARC) website. Available from: https://soil.narc.gov.np/soil/soilmap/ (accessed 21 May 2021).

Pandey, R. 2019 Farmers' perception on agro-ecological implications of climate change in the Middle-Mountains of Nepal: a case of Lumle Village, Kaski. *Environ. Dev. Sustain.* **21**, 221–247. https://doi.org/10.1007/s10668-017-0031-9.

Piemontese, L., Kamugisha, R. N., Tukahirwa, J. M. B., Tengberg, A., Pedde, S. & Jaramillo, F. 2021 Barriers to scaling sustainable land and water management in Uganda: a cross-scale archetype approach. *Ecol. Soc.* **26**, art6. https://doi.org/10.5751/ES-12531-260306.

Ramakrishnan, D., Bandyopadhyay, A. & Kusuma, K. N. 2009 SCS-CN and GIS-based approach for identifying potential water harvesting sites in the Kali Watershed, Mahi River Basin, India. *J. Earth Syst. Sci.* **118**, 355–368. https://doi.org/10.1007/s12040-009-0034-5.

Singhai, A., Das, S., Kadam, A. K., Shukla, J. P., Bundela, D. S. & Kalashetty, M. 2019 GIS-based multi-criteria approach for identification of rainwater harvesting zones in upper Betwa sub-basin of Madhya Pradesh, India. *Environ. Dev. Sustain.* **21**, 777–797. https://doi.org/10.1007/s10668-017-0060-4.

Subedi, A., Kalauni, D., Khadka, R. & Kattel, R. R. 2020 Assessment of role of water harvesting technology in vegetable-based income diversification in Palpa district, Nepal. *Cogent Food Agric.* **6**, 1758374. https://doi.org/10.1080/23311932.2020.1758374.

Tarolli, P. & Sofia, G. 2016 Human topographic signatures and derived geomorphic processes across landscapes. *Geomorphology* **255**, 140–161. https://doi.org/10.1016/j.geomorph.2015.12.007.

United Nations Development Programme (UNDP) 2020 *Human Development Report 2020.* Available from: http://hdr.undp.org/en/2020-report (accessed 20 October 2021).

Ward, M., Poleacovschi, C. & Perez, M. 2021 Using AHP and spatial analysis to determine water surface storage suitability in Cambodia. *Water* **13**, 367. https://doi.org/10.3390/w13030367.

Zanaga, D., Van De Kerchove, R., De Keersmaecker, W., Souverijns, N., Brockmann, C., Quast, R., Wevers, J., Grosu, A., Paccini, A., Vergnaud, S., Cartus, O., Santoro, M., Fritz, S., Georgieva, I., Lesiv, M., Carter, S., Herold, M., Li, L., Tsendbazar, N. E., Ramoino, F. & Arino, O. 2021 *ESA WorldCover 10m 2020 v100.* https://doi.org/10.5281/zenodo.5571936 (accessed 5 October 2021).

First received 15 February 2022; accepted in revised form 16 September 2022. Available online 5 October 2022

doi: 10.2166/nh.2023.026

Baseflow index characterization in typical temperate to dry climates: conceptual analysis and simulation experiment to assess the relative role of climate forcing features and catchment geological settings

Antonia Longobardi [a,*] and Paolo Villani [a,b]

[a] Department of Civil Engineering, University of Salerno, Via Ponte Don Melillo, Fisciano, SA 84084, Italy
[b] CUGRI Consorzio Inter-Universitario per la previsione e prevenzione dei Grandi Rischi, Via Ponte Don Melillo, Fisciano, SA 84084, Italy
*Corresponding author. E-mail: alongobardi@unisa.it

AL, 0000-0002-1575-0782

ABSTRACT

Low-flow hydrological features are crucial for efficient development and integrated water resources management. Among others, the Base-Flow Index 'BFI' is one of the most important low-flow indices. Many studies have demonstrated that it is related to several topographic parameters, climate, vegetation and soil types and to catchment geology. With the aim to enhance the knowledge about the climate and catchment properties' relative control on the 'BFI', an approach consisting of an empirical analysis, applied to a large area located in Southern Italy, characterized by a typical Mediterranean environment, is followed by a simulation experiment, considering climate settings, at the pan-European scale, typical of temperate to dry climate regimes. Main findings have revealed that (i) the correlation structure between the 'BFI' and the precipitation volume, at the annual scale, is affected by both climate variability and catchment properties; (ii) the 'BFI' variability is strongly conditioned by climate intra- and inter-annual variability; (iii) the major role is, however, assigned to the geological catchment features, with poorly and well-drained catchments behaving differently in response to similar climate forcing variability.

Key words: BFI, precipitation inter-annual variability, precipitation intra-annual variability, regional analysis

HIGHLIGHTS

- The role of climate control on the baseflow index (BFI) is explored.
- Typical temperate to dry climatic areas are investigated.
- Climate variability affects the correlation between the BFI and rainfall.
- BFI variability is strongly conditioned by climate intra and inter-annual variability.

1. INTRODUCTION

Low flow is an integral seasonal component of the hydrological regime whose quantification is crucial for efficient development and integrated water resources management and also for maintenance of quantity and quality of water for irrigation, recreation, and wildlife conservation. Considered as, in a general sense, the actual flows occurring in a river during the dry season of the year, low flows can be summarized by a number of measures or indices and among them, the BaseFlow Index, 'BFI', can be mentioned as one of the most widely used statistics. The 'BFI' is defined as the ratio between the baseflow and the total streamflow volume and many studies have demonstrated that it is related to several climatic and catchment properties (Mehaiguene *et al.* 2012; Ahiablame *et al.* 2013; Yao *et al.* 2021).

Climatic measures have been soundly and successfully used as independent variables in statistical multiple linear regression models for regional low-flow prediction. Frequently, only average climate measures, such as mean annual precipitation (MAP) and mean annual potential evapotranspiration (MAPE), have been considered (Van Dijk 2009; Beck *et al.* 2013; Lyu *et al.* 2022). But rainfall variability can directly or even indirectly affect many processes occurring within the hydrosphere. In the recent literature, scientific contributions commented on the importance of more specific climatic features to be accounted for better understanding, especially in particular climatic regions. With regard to the 'BFI', Schneider *et al.*

(2007), in a European context, argued that soil classes are in fact sufficient to predict the 'BFI' but that the variability explained through geological properties tends to decrease from Northern to Southern Europe, probably because factors such as climate and its variability, which are not obviously used to differentiate soil classes, have a greater influence, especially in specific areas. Wu *et al.* (2019) found that climate variability was the main factor altering the 'BFI' in 11 catchments located within the semi-arid Loess Plateau and the climate effect can be even larger than the human activities effect. In arid and semi-arid environments, the baseflow generation is limited by vaporization and the aridity index is the primary determinant of baseflow (Gnann *et al.* 2019). Hellwig *et al.* (2021) showed how across Germany, in a temperate climate setting, climates with higher total annual recharge, but more frequent or severe summer droughts, groundwater droughts might become less severe while the baseflow drought hazard becomes more severe.

Catchment properties and more specifically hydrogeological properties play an undeniable role in baseflow generation with specificities that are hardly recognized on a large scale. Using the hydrological regime characterization, considering the distinction between intermittent and persistent river regimes, Botter *et al.* (2013) found in a study across the United States, that erratic systems are more resilient to climate fluctuation, where perhaps catchment control is dominant. In a similar context, at the pan-European scale, Longobardi & Van loon (2018) found that systems located in the transition zone between poorly and well-drained catchments are the most vulnerable to climate fluctuation, where perhaps catchment control is dominant.

Despite a very large body of literature, there is still no general theory able to interpret the relative role of climate and catchment controls on baseflow generation (Gnann *et al.* 2019). Understanding baseflow droughts in order to achieve sustainable groundwater management, especially in light of climate change responsiveness, is crucial, therefore it is important to contribute empirical and modelling analyses to the knowledge of the scientific community in exploring 'BFI' controls.

In this research paper, an initial empirical analysis, concerning a wide region located in Southern Italy, featured by a typical Mediterranean climate, is followed by a hydrological simulation experiment concerning a broad range of climate conditions at the pan-European scale, from temperate to dry climate areas. Beyond climate properties, also a wide range of catchment geological settings were considered, from poorly to well-drained system. Both observed (empirical analysis) and synthetic (simulation experiment) streamflow time series were filtered to derive the 'BFI' and the 'BFI' variability index. Specifically, the presented research aimed at:

- Understand the role of climate properties on 'BFI' and 'BFI' variability assessment and in particular understand the specific importance of climate average and variability indices on 'BFI' assessment.
- Understand which is the relative role of climate and catchment attributes, for a wide range of hydrological system conditions.

The findings of the research will help in shading light on the identification of the hydrological system which are most impacted by climate properties as they will represent the system that will be most impacted by climate change and for which mitigation strategies are needed to preserve freshwater resources.

2. STUDY AREA AND DATA

The case study is a wide region of about 25,000 km^2, located within the Campania region, Southern Italy. Hydrologic data consist of daily streamflow time series for 28 stream gauge sites, and their locations are shown in Figure 1. For a comprehensive description of the database, refer to Longobardi & Villani (2008), where the assessment of the 'BFI', the permeability P index, assumed to represent the catchment geological settings, and the relevant empirical relationship is provided. P is computed starting from the map of hydro-geomorphological classes represented in Figure 1, which combines the information about geology, land use and catchment slope. It is defined as the ratio between the catchment area falling into categories A and B and the total catchment area. Class A corresponds to pervious and bare geological complexes whereas class B corresponds to pervious and forested geological complexes. The 'BFI' ranges from 25 to 87%, and the catchment permeability index P ranges from 3 to 97%, providing indications for a region featured by very heterogeneous geological settings.

Climate data for the studied region are provided in Longobardi & Boulariah (2022). MAP, ranges from 859 to 1,633 mm/year, and the Thornthwaite MAPE, ranges from 576 to 837 mm/year, describing a typical subhumid climate, close to the subarid limit. To account for inter-annual and intra-annual precipitation variability, respectively, the coefficient of variation of the annual precipitation (Cv(MAP)) and the Precipitation Concentration Index (PCI) were accounted for. On a climate base, the studied region falls into what is called the dry-summer subtropical climate, also known as the Mediterranean climate. Both Cv(MAP) and the PCI showed almost constant values over the studied region, providing indications for a seasonal climate

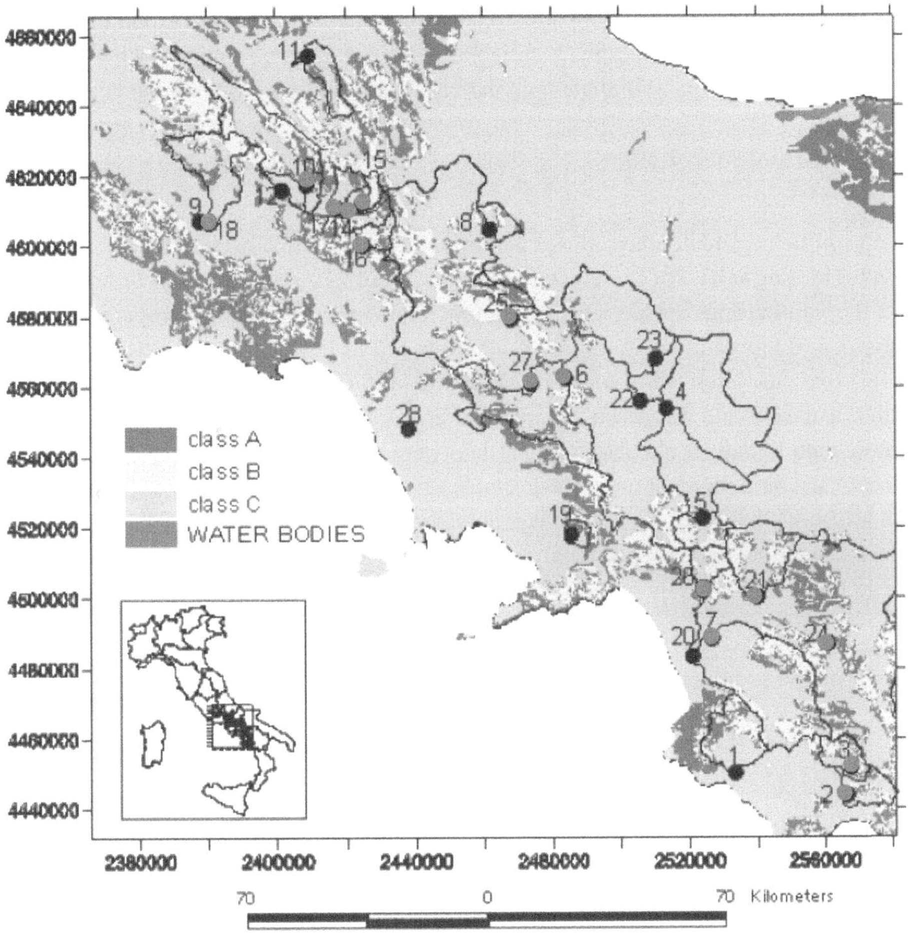

Figure 1 | Map of hydro geomorphological units for the Campania region (Class A = pervious and bare, Class B = pervious and forested; Class C = impervious, Longobardi & Villani 2008). Catchment locations are provided by circles corresponding to catchment outlets. Black circles correspond to group 1 in Figure 2. Red circles correspond to group 2 in Figure 3. Please refer to the online version of this paper to see this figure in colour: http://dx.doi.org/10.2166/nh.2023.026.

regime with moderate inter-annual variability and a substantial climate homogeneity condition (Longobardi & Boulariah 2022).

Geological influence on baseflow was found to be dominant within the studied area (Longobardi & Villani 2008; Longobardi & Villani 2013). According to the mentioned studies, for the study area, it was possible to predict the 'BFI' on the basis of the computation of a permeability index *P*, with the use of a simple univariate regression model, explaining about 70% of the variance and the mean square error of about 10%. The simple univariate model generally overestimated the 'BFI' for small *P*-values ($P < 40\%$), whereas for medium and large *P*, the model over- or under-predicted 'BFI' with no specific pattern.

3. METHODS

3.1. Empirical analysis

Provided the described importance of the geological settings on the 'BFI' prediction within the study area, the current research was aimed, furthermore, at the study of the role of climate features on the baseflow component. In the first step, this was investigated by the use of an empirical framework, based on the calibration of simple and multiple regression models, relating the 'BFI' to climate indices such as MAP, MAPE, Cv(MAP) coefficient of variation of the annual precipitation and PCI, beyond the catchment permeability index (*P*). The goodness-of-fit was measured in terms of mean squared error (mse), correlation coefficient (r^2) and jack-knife errors. To improve the interpretative performance of the regression approach, the k-means algorithm was applied to the investigated catchments in order to group them on the basis of relevant climate indices (MAP and MAPE). The weak climate variability of the study area dampened the relevant findings about climate control on the 'BFI'. For this reason, a simulation approach was set up to explore a broader range of climate conditions.

3.2. Climate forcing

A methodology for daily synthetic rainfall time series generation was identified. Gridded rainfall monthly scale data from the EU F6 WATCH project (Harding *et al.* 2011) were analyzed, for about 210 grid points, in order to cover a broad range of climate conditions, at the pan-European scale, from the temperate to the dry climate properties. In particular, data from one North–South transect and four West–East transects, as illustrated in Figure 2, were considered.

Monthly rainfall data were characterized in terms of MAP, PCI and Cv(MAP). The effectiveness of the EU F6 WATCH project data at the pan-European scale was given in Longobardi & Van Loon (2018).

To scale from monthly data to daily scale data, rainfall data were furthermore described as a sequence of instantaneous rainfall events, each carrying a random amount of water, that is a compound Poisson process (Daly & Porporato 2006). Rainfall events occurrence was modelled as a series of point events, drawn from a Poisson distribution with parameter ν, and the amount of rainfall, for each event, was drawn from an exponential distribution with parameter λ. Parameters ν and λ were described on a monthly basis. The corresponding probability density function refers to the Bessel distribution (Benjamin & Cornell 1970):

$$P[I=0]=e^{-\nu} \quad for\, I=0 \tag{1}$$

$$f_I(I)=e^{-\lambda I-\nu}\sqrt{\frac{\nu\lambda}{I}}\Im_1\left[2\sqrt{\lambda I\nu}\right] \quad for\, I>0 \tag{2}$$

where $\Im_1(x)$ is the modified order 1 Bessel function. Relations between the sample moments and the distribution parameters are:

$$\mu_I=\frac{\nu}{\lambda}=\nu\beta \quad \sigma_I^2=\frac{2\nu}{\lambda^2}=2\nu\beta^2 \quad cv_I=\frac{\sigma_I}{\mu_I}=\sqrt{\frac{2}{\nu}} \tag{3}$$

where μ, σ^2 and cv are, respectively, the precipitation mean, variance and monthly coefficient of variation and $\beta=1/\lambda$ is the average storm depth, in mm.

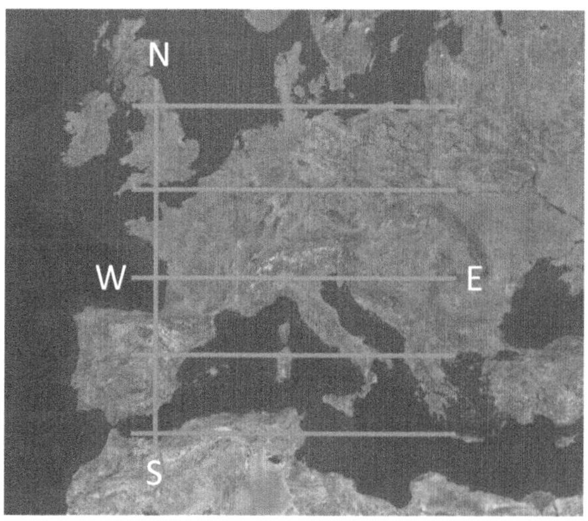

Figure 2 | Representation of the transects for which WATCH monthly precipitation data have been analyzed.

The ν and β monthly patterns and values were derived for the 210 gridded point rainfall EU F6 WATCH data. A sub-set of 30 different ν and β monthly patterns were selected, among the total estimations for the 210 gridded point data, to generate 30 synthetic daily rainfall time series over a long period (100 years), covering the pan-European temperate to dry climatic zones.

3.3. Simulation analysis

Generated synthetic daily rainfall time series were used as the input for a hydrological model in order to generate a large set of hydro-meteorological records, representative of the wide range of climate and geological properties aimed to be explored. The IHACRES (Jackeman *et al.* 1990) model has been selected in force of relevant previous applications involving the exploration of many issues, from climate and land cover change impact on the hydrological cycle to regional studies for ungauged basin prediction purposes and it was also recently applied to investigate the vulnerability of the 'BFI' to dry spell length in a pan-European scale study (Longobardi & Van Loon 2018). The IAHCRES model has six dynamic response parameters, namely '*f*' the temperature modulator factor, '*C*' the catchment storage index, 'τw' the catchment drying time constant, 'τs' the slow pathway time constant', 'τq' the quick pathway time constant, 'vs' the slow pathway proportional volume. The validation of the suitability of the IHACRES model to interpret the hydrological behaviour of the region under investigation was provided in Longobardi & Van Loon (2018) for a sub-set of the Italian catchments used in the current study. Geological properties were introduced in the simulation experiment through a conscious selection of dynamic catchment response model parameters. Since the analyses were focused on baseflow-related processes, the choice was oriented only on the assessment of the impact of two parameters, the slow pathway time constant τs and the proportional volume vs, tightly dependent on groundwater or aquifer percentage to total streamflow (Sefton & Howarth 1998).

Synthetic daily long-term total streamflow time series were processed and filtered to derive low-flow statistics. The IHACRES model provides a separation between the slow and fast pathways, but in order to consider a hydrograph filtering method that would have been independent of the hydrological catchment response simulation framework, reference was made to filtering techniques proposed in the current scientific literature. The SMT (Institute of Hydrology 1980) was selected as a simple smoothing and separation rule to separate the baseflow from the total streamflow hydrograph. Once baseflow time series were extracted from total streamflow, the following were evaluated: (i) the long-term 'BFI', as the ratio between the total baseflow volume and the total streamflow volume; (ii) an annual series of annual 'BFI', the ratio between annual baseflow and annual total streamflow, (iii) the mean and the standard deviation of the series at point (ii); (iv) the coefficient of variation of the 'BFI' Cv(BFI), as the ratio between the standard deviation and average value at point (iii).

4. RESULTS

4.1. Empirical analysis of the 'BFI' dependence on climate and geology indices

It was found that climate variables such as MAP, Cv(MAP), PCI and MAPE were only poorly related to the 'BFI'. The correlation coefficients of a simple regression model between 'BFI' and mentioned climate variables amounted respectively to 0.13, −0.01, 0.17, and −0.23. Their impact in the 'BFI' prediction was tentatively also included in a step-wise multiple regression approach. The *t*-test and *F*-test for Cv(MAP), PCI and MAPE regression coefficients showed *P*-values larger than the 5% significance level and they were excluded from the analysis. The multiple regression, based on P and MAP, had better performance indices compared to the simple regression method. The explained variance increases up to about 82%, the goodness-of-fit increase to a value of $r^2 = 0.77$ and the decreases to 7% (Table 1). Also the jack-knife errors, mean total true error (mt), mean apparent error (ma), as measures of goodness-of-fit, and mean expected excess error (me), as measures of model robustness, indicating a better performance.

To further improve the 'BFI' prediction, the k-means clustering algorithm was explored to use climate as a grouping rule rather than an independent variable in regression modelling (Laaha & Bloschl 2006). Two grouping rules were accounted for, one based on MAP values and the second based on MAP and MAPE, and a number of two clusters appeared optimal for each partition, in order to furthermore guarantee an adequate dimension of the samples to be analyzed. For each of the different resulting clustering and for each group, a simple regression model was fitted between P and the 'BFI' (Figure 2). The simple linear regression models, fitted on observed data for each group, satisfactorily described the variation of the catchment scale 'BFI' (Table 1). The explained variance was about 74–85%, depending on the group identification, however larger than the case of simple regression fitted on the whole dataset.

Both in the case of MAP and MAP + MAPE clustering (Figure 3), groups selection identified group 1 (grey circles) featured by MAP values below a threshold value of about 1,100 mm and MAPE values above a threshold value of about 700 mm

Table 1 | Regression model performances

Model	Groups	mse	r^2_{adj}	r^2 (obs-pre)	mt (%)	ma (%)	me (%)
Simple linear (P) (Longobardi & Villani 2008)	–	9.92	0.69	0.54	8.05	7.43	3.08
Multiple linear (P, MAP)	–	7.16	0.82	0.77	5.67	5.09	2.48
Simple linear (P) cluster based (MAP)	1	8.26	0.78	0.71	7.63	6.47	4.03
	2	9.22	0.74				
Simple linear (P) cluster based (MAP + MAPE)	1	9.16	0.74	0.76	7.28	5.70	4.53
	2	6.48	0.85				

Note: mse = mean squared error, r2adj = coefficient of determination adjusted for number of model parameters, r2 obs-pre = observed vs predicted scatter plot coefficient of determination, mt mean total, ma absolute and me expected jack-knife errors.

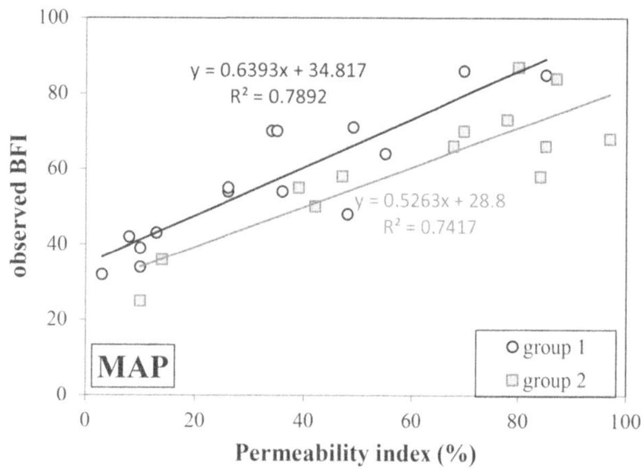

Figure 3 | A simple linear regression model between the 'BFI' and P, based on climate cluster. Please refer to the online version of this paper to see this figure in colour: http://dx.doi.org/10.2166/nh.2023.026.

(Table 2). The reverse holds in the case of group 2 (orange squares). The spatial location of group 1 and group 2 within the study region is represented in Figure 1.

In consideration of the regional annual average values of about 1,280 mm for precipitation and 690 mm for potential evapotranspiration, it could be underlined that group 1 corresponds to areas of the investigated region where the effective precipitation is low (low rainfall and high evapotranspiration). Additionally, catchments included in group 1 are characterized by a lower average permeability index P, regardless of the grouping rule (Table 2). Nevertheless, in each of the two grouping cases (Figure 2), the relative position of group 1 with respect to group 2, corresponds, for a given value of P, to a large 'BFI' for group 1.

Land cover is likely to contribute to specific effects highlighted in the case study. Forest cover is typically represented by the Mediterranean forest, which consists mainly of conifers, whose impact on the interception of rainfall may reach values up to

Table 2 | Average value for MAP, MAPE, P and FOR (percentage of forest cover), for each group identified by the two clustering partitioning methods

Methods	Groups	MAP (mm)	MAPE (mm)	P (%)	FOR (%)
MAP clustering partitioning	1	1,141	705	29	38
	2	1,447	673	58	51
MAP + MAPE clustering partitioning	1	1,119	712	33	33
	2	1,425	672	61	54

30% (Iovino *et al.* 2009). In the investigated area, the catchments belonging to group 1 (lower MAP values) have been found to be those with a moderate percentage of forest cover (Figure 4 and Table 2). This feature is also clear in Figure 1.

Instead, in group 2, characterized by the larger rainfall contribution and the smaller potential evapotranspirative loss (therefore larger effective rainfall), the fraction of wooded area is relevant and can be as high as about 90% (for a specific catchment) of the area of the basin (Figure 4). In these basins, therefore, the interception losses can be substantial, significantly reducing the effective rainfall volume and then the recharge to deep aquifers, potentially yielding smaller 'BFI' values, for an assigned level of permeability. In support of this thesis, experimental studies conducted in a dominant pine stands of a Mediterranean geographical contiguous area, have demonstrated that after thinning, a significant increase in groundwater flow has occurred, especially during spring and summer (Callegari *et al.* 2003).

4.2. Climate forcing settings characterization and rainfall generation

In order to generate daily scale rainfall data, which mimic given climate settings, the gridded monthly precipitation data from the EU F6 WATCH project (Harding *et al.* 2011) for about 210 grid point were analyzed. In particular, data from one North–South transect and four West–East transects, as illustrated in Figure 5, were considered.

At the pan-European scale, MAP, inter-annual variability Cv(MAP) (annual precipitation coefficient of variation) and intra-annual variability PCI indices appeared related to each other (Figure 5). Similar findings were also found for a specific region under investigation (Longobardi & Boulariah 2022) and at the global scale (Fatichi *et al.* 2012). A main trend exists with large values of PCI associated with large values of Cv(MAP). EU F6 WATCH data covered a large domain (Figure 6, middle plot,

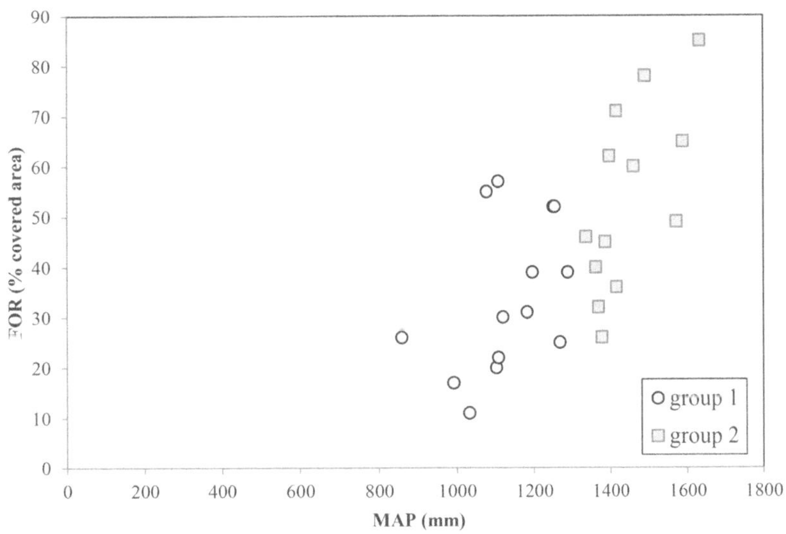

Figure 4 | MAP versus percentage of forest covered area (FOR) for the studied area. Group 1 and group 2 identification belongs to the MAP + MAPE clusters partitioning method.

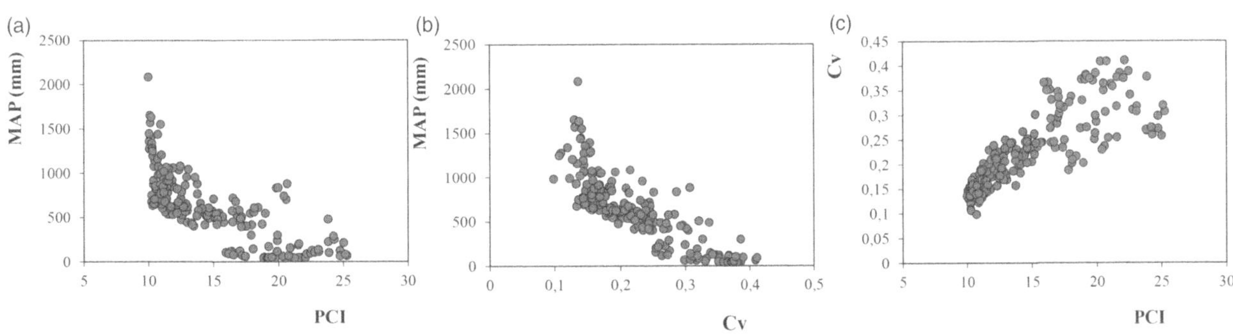

Figure 5 | Correlation between MAP, PCI, and Cv(MAP) for the analysed transects data.

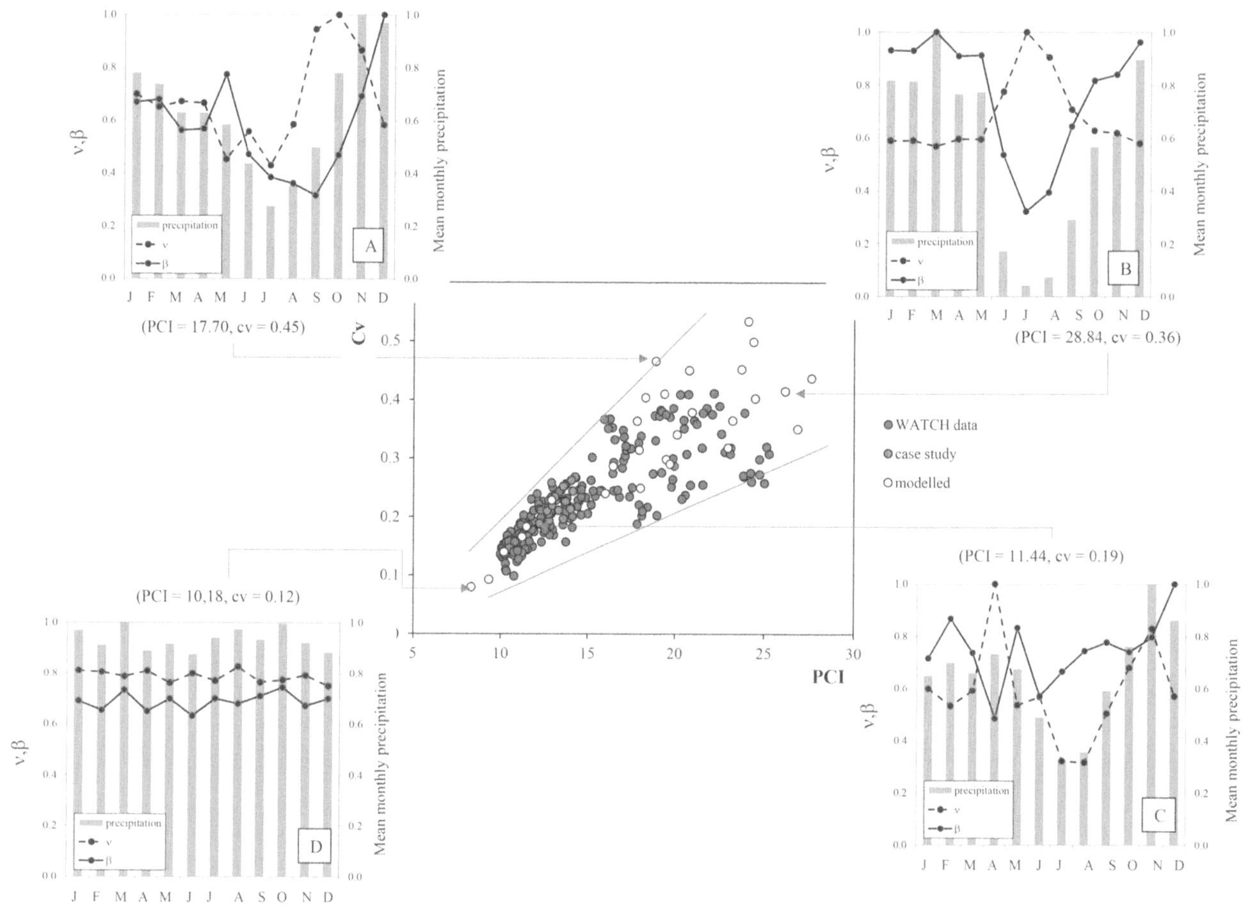

Figure 6 | PCI versus annual Cv(MAP) correlation for typical temperate to dry climate areas. In the middle, WATCH data (blue circles), case study data (red circles) and modelled (yellow circles) PCI-*cv* data. On the edges, examples of dimensionless mean monthly precipitation distribution and monthly patterns for the stochastic rainfall generator parameters ν and β, for the investigated domain. Types A and B belongs to typical dry-summer climates. Type C belongs to typical humid climate. Types D belongs to typical oceanic climates, with a uniform rainfall distribution. Please refer to the online version of this paper to see this figure in colour: http://dx.doi.org/10.2166/nh.2023.026.

blue circles), with MAP ranging between 0 and 2,000 mm (mainly increasing south to north), Cv(MAP) between 0.1 and 0.5 and PCI between 10 and 25 (Figure 5). Toseek completeness, the weak climate variability of the studied region is also represented in Figure 6 (middle plot, red circles) with a very narrow range of PCI and Cv(MAP).

The ν and β monthly patterns and values were derived for the 210 gridded point rainfall EU F6 WATCH data. They clearly affect the total amount, seasonality and variability of rainfall time series (MAP, PCI and Cv(MAP)). As an example, type A and B patterns in Figure 6, showed a seasonal pattern for the ν parameters which led to seasonal precipitation patterns and high PCI values. Analogously, the type D pattern in Figure 6 showed a rather uniform pattern for the ν parameters which led to a uniform precipitation pattern and low PCI value. A number of 30 ν and β monthly parameters were appropriately set to generate synthetic rainfall time series by an application of Equations (1)–(3). Their MAP, PCI and Cv(MAP) indices overlap the WATCH data domain, that is the whole temperate to dry pan-European climate conditions (Figure 6, middle plot, yellow circles).

4.3. Simulation analysis of the relation between the 'BFI', climate indices and catchment properties

Synthetic daily streamflow time series were derived by IAHCRES simulations. Generated synthetic daily rainfall time series were used as the input for the hydrological model. Values for τs and vs model parameters, in order to simulate a large set of hydro-meteorological records, were selected to be representative of a wide range of catchments properties, beyond climate features, aimed to be explored. Indication about the range of variability for τs and vs were drawn from Longobardi & Van Loon (2018), in an investigation, on a pan-European scale, where streamflow time series from the region under investigation in this study were also modelled. Model parameters for the sub-set of Italian catchments modelled in Longobardi & Van Loon (2018) are provided in Table 3 along with the Nash–Sutcliffe Efficiency index.

Specifically for the current study, hydrological simulations were for three different values of vs (set to 0.2, 0.5, and 0.8) to account for low to high surface soil permeability, and two values for τs (set to 30 and 300 days), to account for poorly to well-drained catchments.

The role played by catchment properties, that is by the selected model parameters, is somehow easily inferable from a conceptual perspective but the simulation approach would confirm in this sense. It is in fact quite clear that, for a given precipitation input, pervious catchments (with large vs value) would deliver larger 'BFI' than impervious ones (with small vs values). Similarly, well-drained watersheds (with large ts values) represent hydrological systems with well-sustained flows, which would imply larger 'BFI' than poorly drained watersheds (with small ts values). What is instead not inferable, from a conceptual perspective, is the role of the precipitation pattern within the year (e.g. the seasonality or intra-annual variability), here described with the PCI, and the associated inter-annual variability, here described with the annual Cv(MAP) and how they interact with the model parameters that are catchment properties. From this perspective, the simulation approach was expected to indicate basic insights into this issue.

The first result is represented in Figure 6 where, at the annual scale, the baseflow versus rainfall volume scatterplots are illustrated. Differences in precipitation seasonal patterns are underlined. Among the 30 generated rainfall time series, three daily rainfall scenarios are selected and indicated as representative of a uniform (PCI = 10.18), seasonal (PCI = 12.93), and strongly seasonal (PCI = 27.26) rainfall regime. Results are illustrated for six virtual catchments. Three different levels of proportional coefficients are considered to mimic low-permeability (vs = 0.2), intermediate-permeability (vs = 0.5) and high-permeability (vs = 0.8) catchments. Additionally, for the accounted vs values, both poorly (ts = 30 days) and well-drained (ts = 300 days) systems are modelled. As expected from a conceptual point of view, there exists a relation between the baseflow and rainfall volume, which feature clearly depends on climate settings and catchment properties. It is confirmed that, for each of the rainfall regime (PCI), and for given ts values larger 'BFI' correspond to larger vs and that for given vs values, larger 'BFI' correspond to large ts (see the location of the scatterplots in the annual rainfall-'BFI' plane).

It is more strictly gained from the simulation results that, for given ts and vs, the extent of the variability of the annual baseflow volumes increase from uniform to strongly seasonal rainfall patterns. The last observation is likely explained by the increase in rainfall annual variability when moving from uniform to seasonal climate regimes.

Besides the increase of 'BFI' variability with climate seasonality, the decrease in the annual 'BFI' is another result strictly gained from the simulation approach. This peculiarity holds in particular for well-drained catchments (ts = 300 days) and for catchments with an intermediate level of permeability (vs = 0.5). Instead, for the poorly drained systems the correlation patterns between annual 'BFI' and rainfall volume does not seem to be characterized by a definite negative sign. As the insights from Figure 7 refer only to a small number of virtual catchments, the presented results need further observation for a rational explanation. Further insights about the relationship between the 'BFI' and the rainfall properties, in particular for what concerns the rainfall variability, were gained from the simulations as below illustrated.

The left panel of Figure 8 indicates that the seasonality of the climate is relatively less important than the catchment properties for what concern the 'BFI' assessment. The 'BFI' values are almost dominated by the vs parameters and almost constant for the full range of the considered PCI, with an exception for the strongly uniform climate types (small PCI values). For a given vs, long-term 'BFI' is larger for well-drained systems. The right panel of Figure 8 instead indicates that the climate

Table 3 | IAHCRES model parameters

Catchments	f (°C^{-1})	C (mm)	τw (days)	τs (days)	τq (days)	vs	NSE
Alento	0.11	88.57	10.15	66.0	0.36	0.33	0.67
Calore	0.14	339.67	51.43	60.0	0.92	0.43	0.72
Liri	0.11	299.80	62.98	102.0	2.99	0.53	0.73
Bussento Caselle	0.07	238.72	75.90	222.7	1.35	0.81	0.67
Sele Albanella	0.08	185.05	40.07	80.0	1.17	0.55	0.72

Note: 'f' the temperature modulator factor, 'C' the catchment storage index, 'τw' the catchment drying time constant, 'τs' the slow pathway time constant', 'τq' the quick pathway time constant', 'vs' the slow pathway proportional volume

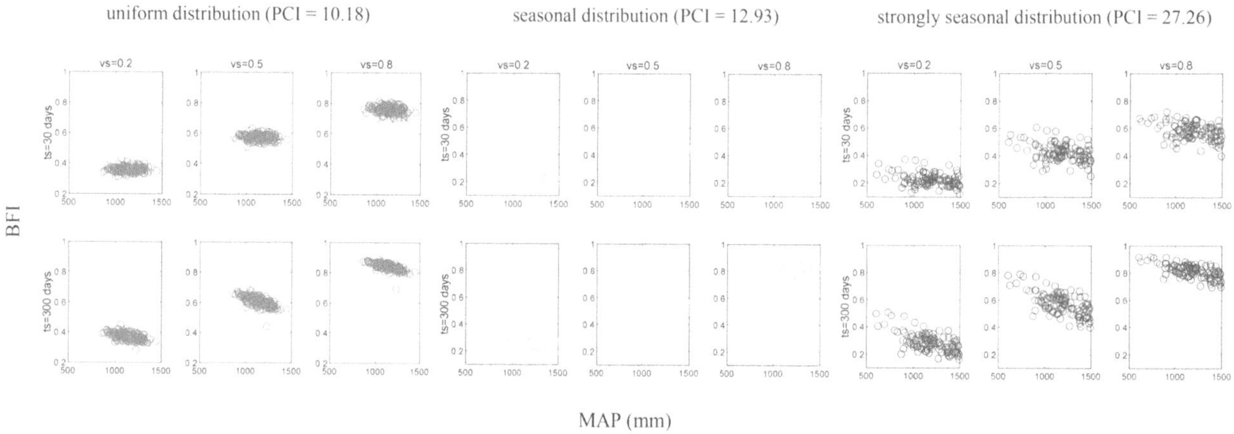

Figure 7 | Annual scale correlation between the annual 'BFI' and annual precipitation, for six virtual catchments, which differ for climate settings and model parameters.

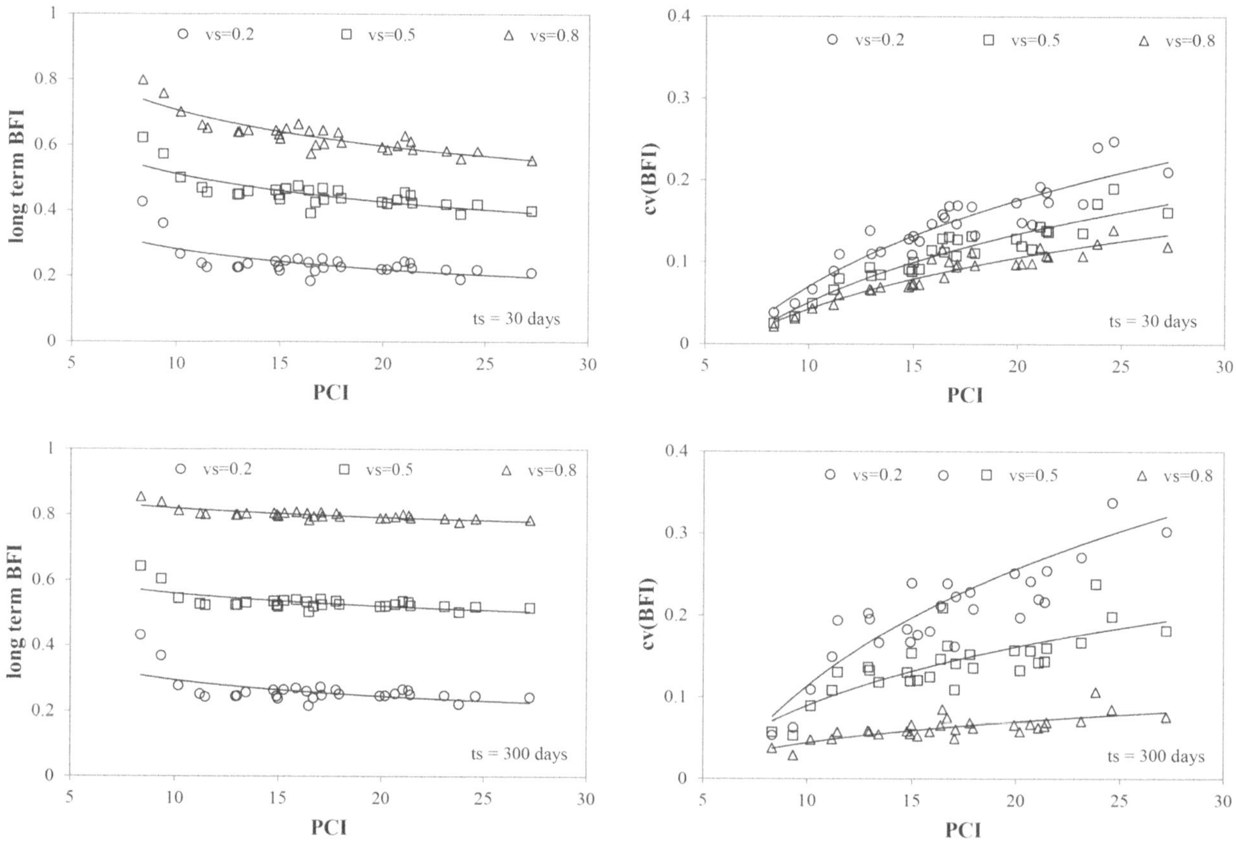

Figure 8 | The long-term 'BFI' and 'BFI' coefficient of variation dependence on the PCI, for different model parameters. (vs = 0.2, 0.5 and 0.8; ts = 30 and 300 days).

seasonality impacts the 'BFI' assessment, with large 'BFI' variability associated with large PCIs, that is with large Cv(MAP). The effect is more evident in well-drained systems.

Figure 8 provides a measure of the strength and sign of the correlation between the annual baseflow and annual rainfall, the Pearson's correlation coefficient. Pearson's correlation coefficients are plotted against a measure of rainfall variability, the annual Cv(MAP), for different catchment properties. On a general perspective, the ρ values are consistently positive and negative respectively for poorly and well-drained catchments. To explain this feature, it is important to remember that the 'BFI' is represented by the ratio between the baseflow volume Vb and the total streamflow volume Vtot and that for each hydrological system large rainfall volumes correspond to large total streamflow volumes. In the case of well-drained systems the baseflow volume is relatively constant (persistent systems) thus an increase in rainfall only generates a significant increase in Vtot, reducing thus the ratio Vb/Vtot (negative correlation). In the case of poorly drained systems, as illustrated in the upper panel of Figure 7, the 'BFI' appears larger for lower PCI values, generally associated with large rainfall amounts. Poorly and well-drained catchments differ not only in the sign of the correlation (positive and negative respectively) but also in the fact that, in the first case, larger absolute ρ values are observed for large annual *cv* whereas, in the second case, they are observed for low annual rainfall *cv*. If as illustrated (Figure 5), small annual Cv(MAP) correspond to uniform rainfall pattern and large annual Cv(MAP) corresponds to a strong seasonal rainfall pattern. The negative correlation between precipitation and baseflow for well-drained basins is stronger in uniform climate regimes: the relationship between Cv(MAP) and MAP is strongly non-linear (Figure 6) with rainfall totals that significantly increase for the lower rainfall variability, strongly reducing the 'BFI' ratio. The decrease in 'BFI' for increasing rainfall in well-drained system is also coherent with what was found by Niu *et al.* (2017), in an Amazonian catchment where the 'BFI' value decreased for annual precipitation above the 2,500 mm threshold. In the case of poorly drained systems, the positive correlation is instead stronger in the case of large Cv(MAP). Large Cv(MAP) correspond to strongly seasonal climates which, in turn, are featured by large inter-storm periods (Longobardi & Van Loon 2018). A large inter-storm period can be comparable to the modelled delay time τs: in this case, the effect of baseflow rising produced by a rainfall shot would be rapidly drained even before the occurrence of the subsequent rainfall shot. As a result, the baseflow pattern could result in a strongly intermittent process which amounts strongly depending on the amount of rainfall at the event scale. The complex nature of the baseflow pattern for poorly drained systems is responsible for the fact that although a dominant positive correlation, Figure 9 also cases for negative correlation between Cv(MAP) and the correlation coefficient can be highlighted for the whole range of considered climate variability.

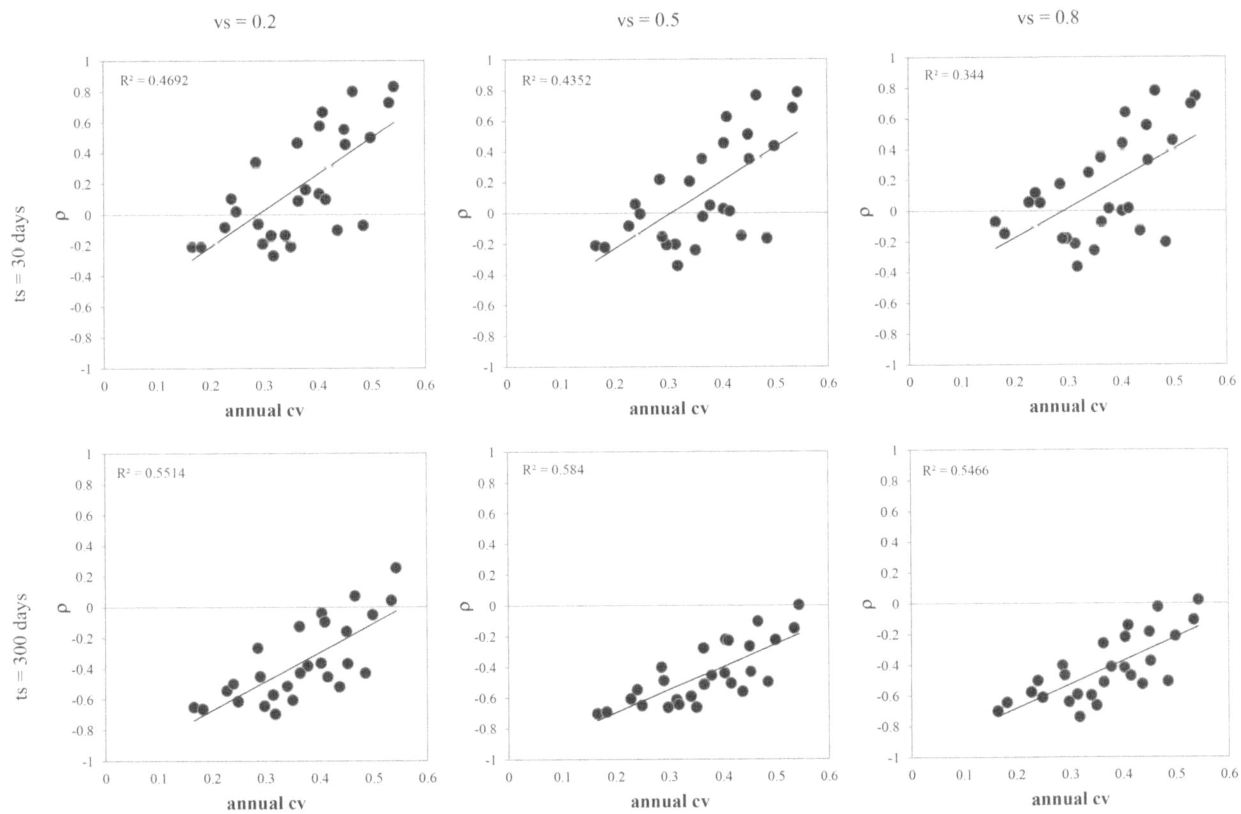

Figure 9 | Pearson's correlation coefficients between the 'BFI' and annual rainfall plotted as a function on inter-annual variability Cv(MAP), for different model parameters.

5. CONCLUSIONS AND FINAL REMARKS

This paper presented an assessment of the role of climate and catchment properties on the 'BFI'. In order to broaden the results, a coupled empirical and simulation analysis approach was adopted. In the empirical analysis, data from a Mediterranean region were analyzed to provide correlation between the 'BFI' and climate in this specific region, but the weak climate variability of the area, if considered on a very large scale, dampened the relevant findings. In the simulation analysis, rainfall scenarios were generated according to the modelling characterization given for rainfall times series provided by the WATCH data project, to describe several meteorological inputs belonging to typical temperate to dry climate domain. The IHACRES was selected to address the catchment response and synthetic total streamflow time series were processed to calculate the long-term 'BFI' and the relevant coefficient of variation Cv('BFI'). The impact of catchment properties, predictable from a conceptual perspective, was confirmed by the simulation results. The impact of other factors, such as the intra-annual and inter-annual climate variability, was only partially clear on a conceptual perspective and was inferred from the simulation approach. The following comments hold:

- Long-term 'BFI' index: the conceptual role of catchment properties was confirmed. For a given climate input large values of 'BFI' are expected for well-drained and high-permeability catchments indeed. The role played by the precipitation variability, both at the intra-annual and inter-annual scale, corresponds to a decrease in the long-term index for an increase in intra and inter-annual variability and its effect is more evident for poorly drained catchments. Poorly and well-drained catchments behave differently also with reference to the annual precipitation amount, with the 'BFI' for well-drained catchments which tends to decrease for an increase in annual precipitation.
- Cv('BFI'): as a tendency, rainfall variability propagates into 'BFI' variability, thus large Cv('BFI') are expected for large rainfall *cv* (or large PCI values). For small rainfall *cv* or PCI indices, the 'BFI' variability is comparable, independently on catchment properties. Differences arise as rainfall variability increases. Impervious watersheds, compared to previous ones, only filter a small fraction of effective rainfall through the slow pathway, the variability of total streamflow is comparable to the precipitation variability and this would entail large Cv('BFI'). The impact of the slow pathway delay time consists in large Cv('BFI') for well-drained catchments: the long recession limbs frequently intercept the effective rainfall input, increasing the 'BFI' variability contrarily to what happens for poorly drained catchments.

In a more realistic simulation experiment, it should be worth considering that the catchment response could also be affected by land use, such as considered in the empirical analysis previously reported and the nature of dominant hillslope process, which are tightly connected to climatic settings. A possible and potential solution to account for these concernments, would be the reference to a more sophisticated hydrological catchment response modelling, embedding a more detailed interaction between the soil–vegetation–atmosphere component, as well as an extension to a wider range of climatic regimes, for a more comprehensive assessment of the relative role of climate and catchment features on low-flow indices.

ACKNOWLEDGEMENTS

The authors would like to thank the anonymous referees for their encouragement and helpful comments which resulted in an improved manuscript version. The presented research results are part of the RESTORE project (PSR 2014–2020 – TIPOLOGIA 16.5.1 'Azioni congiunte per la mitigazione dei cambiamenti climatici e l'adattamento ad essi e per pratiche ambientali in corso' – Progetto RESTORE).

DATA AVAILABILITY STATEMENT

All relevant data are available from an online repository or repositories (https://www.isprambiente.gov.it/it/progetti/cartella-progetti-in-corso/acque-interne-e-marino-costiere-1/progetti-conclusi/progetto-annali).

CONFLICT OF INTEREST

The authors declare there is no conflict.

REFERENCES

Ahiablame, L., Chaubey, I., Engel, B., Cherkauer, K. & Merwadec, V. 2013 Estimation of annual baseflow at ungauged sites in Indiana USA. *Journal of Hydrology* **476**, 13–27.

Beck, H. E., van Dijk, A. I. J. M., Miralles, D. G., de Jeu, R. A. M., Bruijnzeel, L. A., McVicar, T. R. & Schellekens, J. 2013 Global patterns in base flow index and recession based on streamflow observations from 3394 catchments. *Water Resources Research* **49**, 1–21.

Benjamin, J. R. & Cornell, C. A. 1970 *Probability, Statistics and Decision for Civil Engineers*. McGraw-Hill, New York.

Botter, G., Basso, S., Rodriguez-Iturbe, I. & Rinaldo, A. 2013 Resilience of river flow regimes. *Proceedings of the National Academy of Sciences of the United States of America* **110** (32), 12925–12930.

Callegari, G., Ferrari, E., Garfì, G., Iovino, F. & Veltri, A. 2003 Impact of thinning on the water balance of a catchment in a Mediterranean environment. *The Forestry Chronicle* **79** (2), 301–306.

Daly, E. & Porporato, A. 2006 Impact of hydroclimatic fluctuations on the soil water balance. *Water Resources Research* **42**, W06401.

Fatichi, S., Ivanov, V. Y. & Caporali, E. 2012 Investigating interannual variability of precipitation at the global scale: is there a connection with seasonality? *Journal of Climate* **25** (16), 5512–5523.

Gnann, S. J., Woods, R. A. & Howden, N. J. K. 2019 Is there a baseflow budyko curve? *Water Resources Research* **55** (4), 2838–2855.

Harding, R., Best, M., Blyth, E., Hagemann, S., Kabat, P., Tallaksen, L. M., Warnaars, T., Wiberg, D., Weedon, G. P. & van Lanen, H. 2011 WATCH: Current knowledge of the terrestrial global water cycle. *Journal of Hydrometeorology* **12**, 1149–1156.

Hellwig, J., Stoelzle, M. & Stahl, K. 2021 Groundwater and baseflow drought responses to synthetic recharge stress tests. *Hydrology and Earth System Sciences* **25**, 1053–1068. https://doi.org/10.5194/hess-25-1053-2021.

Institute of Hydrology 1980 *Low Flow Studies (1–4)*. Institute of Hydrology, Wallingford, UK.

Iovino, F., Borghetti, M. & Veltri, A. 2009 Foreste e ciclo dell'acqua. *Forest@* **6**, 256–273.

Jackeman, A. J., Littlewood, I. G. & Whitehead, P. G. 1990 Computation of the instantaneous unit hydrograph and identifiable component flows with application to two small upland catchments. *Journal of Hydrology* **117**, 275–300.

Laaha, G. & Bloschl, G. 2006 A comparison of low flow regionalization methods – catchment grouping. *Journal of Hydrology.* **323**, 193–214.

Longobardi, A. & Boulariah, O. 2022 Long term regional changes in inter-annual precipitation variability in the Campania region, Southern Italy. *Theoretical and Applied Climatology* **148**, 869–879.

Longobardi, A. & Van Loon, A. F. 2018 Assessing baseflow index vulnerability to variation in dry spell length for a range of catchment and climate properties. *Hydrological Processes* **32** (16), 2496–2509.

Longobardi, A. & Villani, P. 2008 Baseflow index regionalization analysis in a Mediterranean environment and data scarcity context: role of the catchment permeability index. *Journal of Hydrology* **355**, 63–75.

Longobardi, A. & Villani, P. 2013 A statistical, parsimonious, empirical framework for regional flow duration curve shape prediction in high permeability Mediterranean region. *Journal of Hydrology* **507**, 174–185.

Lyu, S., Zhai, Y., Zhang, Y., Cheng, L., Kumar Paul, P., Song, J., Wang, Y., Huang, M., Fang, H. & Zhang, J. 2022 Baseflow signature behaviour of mountainous catchments around the North China Plain. *Journal of Hydrology* **606**, 127450.

Mehaiguene, M., Meddi, M., Longobardi, A. & Toumi, S. 2012 Low flows quantification and regionalization in North West Algeria. *Journal of Arid Environments* **87**, 67–76.

Niu, J., Shen, C., Chambers, J. Q., Melack, J. M. & Riley, W. J. 2017 Interannual variation in hydrologic budgets in an amazonian watershed with a coupled subsurface-land surface process model. *Journal of Hydrometeorology* **18** (9), 2597–2617.

Schneider, M. K., Brunner, F., Hollis, J. M. & Stamm, C. 2007 Towards a hydrological classification of European soils: preliminary test of its predictive power for the baseflow index using river discharge data. *Hydrology and Earth System Science* **11**, 1501–1513.

Sefton, C. E. M. & Howarth, S. M. 1998 Relationships between dynamic response characteristics and physical descriptors of catchments in England and Wales. *Journal of Hydrology* **211**, 1–16.

Van Dijk, A. I. J. M. 2009 Climate and terrain factors explaining streamflow response and recession in Australian catchments. *Hydrology and Earth System Science* **14**, 159–169.

Wu, J., Miao, C., Duan, Q., Lei, X., Li, X. & Li, H. 2019 Dynamics and attributions of baseflow in the semiarid loess plateau. *Journal of Geophysical Research: Atmospheres* **124** (7), 3684–3701.

Yao, L., Sankarasubramanian, A. & Wang, D. 2021 Climatic and landscape controls on long-term baseflow. *Water Resources Research* **57** (6), e2020WR029284.

First received 15 February 2022; accepted in revised form 2 December 2022. Available online 20 January 2023

© 2023 The Authors

doi: 10.2166/nh.2023.036

Land use and damaging hydrological events temporal changes in the Sarno River basin: potential for green technologies mitigation by remote sensing analysis

Mirka Mobilia [a,*], Antonia Longobardi[a], Donato Amitrano[b] and Giuseppe Ruello[c]

[a] Department of Civil Engineering, University of Salerno, Via Giovanni Paolo II, 132, Fisciano, SA 84084, Italy
[b] Italian Aerospace Research Centre, Via Maiorise snc, Capua 81043, Italy
[c] Department of Electrical Engineering and Information Technology, University of Naples 'Federico II', Via Claudio, 21, Napoli 80138, Italy
*Corresponding author. E-mail: mmobilia@unisa.it

MM, 0000-0001-7018-3592

ABSTRACT

The urban growth leads cities to adopt sustainable strategies in order to mitigate the relevant hydrological effects. In this study, the use of synthetic aperture radar SAR imagery has allowed us to demonstrate a 70% increase of the built-up area in Sarno River basin between 1995 and 2016. This increase is linked to the statistical temporal increase of the damaging hydrological events occurring during the same period. To restore the pre-development hydrological condition, a scenario analysis was undertaken where SWMM was used to simulate the hydrological effect of green roof retrofitting landscape design. SAR imagery was furthermore used to explore the potential retrofitting surfaces, leading to defining three different conversion scenarios with 5%, 30% and 100% of potential retrofitting surfaces. The study demonstrated that the pre-development hydrological condition can be never fully restored. Indeed, this scenario is partially equaled only by a 100% green conversion of the existing traditional roofs, with average runoff and peak flow reduction of 41% and 25%, respectively. Such conditions are clearly not feasible, provided the obvious retrofitting limitation for existing buildings. The use of additional nature-based techniques, beyond green roofs conversion, should be explored in the perspective of a balance for urban growth.

Key words: green roofs, MDHEs, runoff, SWMM, synthetic-aperture radar (SAR) imagery

HIGHLIGHTS

- In Sarno River basin, a correlation between the increase of built-up areas and damaging hydrological effects exists.
- Radar images help to identify the potential areas for green roof retrofit.
- Green roofs applied at a basin scale allow substantial mitigation of urban runoff.

INTRODUCTION

Nowadays over half of the world's population is living in cities and by 2050 it is estimated that roughly 6.4 billion people, almost the planet's current population, will live in a city (United-Nations 2014). Urban growth has left cities facing a number of problems and they will inevitably increase in the near future unless best management practices are adopted to mitigate those effects. As cities have grown and more development has occurred, the natural landscape has been replaced by roads, buildings, housing developments, and parking lots. Changes in land use have been followed by an increase in impervious surfaces, propagating into hydro-geological and hydraulic risk changes (Konrad 2003).

In addition to the land use changes, critical infrastructures are also becoming more sensitive to weather and climate extremes, which include a multiplicity of hazardous events falling into a wider category of the Multiple Damaging Hydrological Events (MDHEs), severe stormy periods during which floods, landslides, lightning, windstorms, hail, or storm surges can harm people (Petrucci *et al.* 2018). According to several works, the main cause of major MDHE effects in urban areas around the world is the increasing impervious surfaces. Huang *et al.* (2018) found out that in Guangzhou, China, the rapid increase in flood events was influenced by the change in the land cover. This was also affirmed by Paprotny *et al.* (2018), who argued that the upward trend of the MDHEs in Europe is linked to the rapid urbanization of the growing regions. Similar studies in different Italian regions also identified a relationship between increasing imperviousness and increasing occurrences of damaging hydrological events (Apollonio *et al.* 2016; Sofia *et al.* 2017; Bentivenga *et al.* 2020; Recanatesi & Petroselli 2020). However, contrasting messages come from the same geographical areas probably because of the urban context specificities. Indeed, Esposito *et al.* (2018) in the Pozzuoli area (Campania region) ascribed the increasing occurrence of MDHEs to the variation of the rainfall regime and not to the increasing percentage of sealed areas.

Unfortunately, conventional urban water management practices are not fully prepared to deal with this problem. Design and planning methodologies based on stationary scenarios are no longer reliable (Longobardi *et al.* 2016). Action should be taken to prepare and to adapt our cities to changing environmental conditions, in the medium-long temporal horizon.

In light of this, innovative approaches in sustainable urban development and water management, such as the SuDS (Sustainable Drainage Systems), appear as a set of multi-benefit technologies, based on the use of green and blue technologies, able to increase the urban areas' resilience to environmental changes (Sartor *et al.* 2018). SuDS are able to control runoff by reducing its quantity and improving water quality (Voyde *et al.* 2010). Among SuDS, green roofs (GRs) have proven to be a very effective measure if used in response to water quality and quantity issues at the city scale (Morgan *et al.* 2013). They allow for a reduction in runoff volume and peak flow and an increase in delay time, consequently, its application at a large scale by retrofitting the existing traditional roofs helps to significantly reduce the hydrological risk (Masseroni & Cislaghi 2016).

The effectiveness of a widespread application of GRs on stormwater generation has been tested, over time, by several authors using a number of hydrological models such as the SWMM (Storm Water Management Model), L-THIA-LID 2.1 model, Curve Number (CN) method, Visualizing Ecosystem and Land Management Assessments (VELMA v2019-07-22) model. Among these, the SWMM of the Environmental Protection Agency (EPA) has resulted in very accurate and precise reproduction of the hydrological behavior of GRs applied at a large scale (Versini *et al.* 2016). On this point, it is necessary to consider that the hydrological performances of GRs at the catchment scale strongly depend on the roof's surface suitable for GR retrofit.

The possibility to use satellite images to detect the areas with the potential for GR retrofit has been investigated during the past decades. Karteris *et al.* (2016) used high spatial resolution orthoimages from a Leica ADS40 Airborne digital sensor in the Mediterranean city of Thessaloniki, Northern Greece, and quantified the benefits of the large-scale application of GR technology in terms of energy consumption and rainwater retention. Pappalardo *et al.* (2017), in Southern Italy, derived the land use by Google satellite images, investigating the role of the SuDS in urban runoff control. Zhou *et al.* (2019) used panchromatic and multispectral images of the ZY-3 satellite to identify the building roofs suitable for greening retrofitting within the city of Beijing.

The Sarno River basin, in southern Italy, represents an emblematic case. It is a peri-urban basin highly prone to flooding events, and it is the most polluted river in Europe and one of the 10 most polluted rivers in the world (Baldantoni *et al.* 2018). In previous works, the correlation between the increasing number of MDHEs, during the last two decades, and the climate change in the Sarno River basin was investigated with negative results (Califano *et al.* 2015; Longobardi *et al.* 2016). Until now, only traditional engineering measures have been designed and realized to reduce the risk related to the occurrence of flooding events, including water detention basins and drainage channels (Di Vito *et al.* 2019). These strategies, which intended to modulate peak discharge and delay, were not significantly effective and the above-mentioned measures did not take into account the qualitative-related issues affecting the area.

With reference to the Sarno River basin, the presented research has a twofold objective. On one side, to analyze in more detail the potential cause-and-effect relationship between the increase in MDHE occurrences and the changes in the land cover. On the other side, to investigate, by a hydrological simulation approach, the impact of green infrastructure solutions as a proposal for stormwater management, conscious of the benefit they could furthermore address in terms of water quality issues in a seriously endangered environment. The link between the two objectives is represented by the use of satellite imageries which were used to both assess the changes in the land cover, over the period a raise in MDHEs was observed, and to quantify the potential for GR retrofit. Different from what was overall considered in the literature, synthetic-aperture radar (SAR) imageries, from ERS-1 and COSMO-SkyMed missions, were used for the purpose. Compared to the multispectral imagery, SAR allows for a lower computational complexity, not strictly required for the purposes of the present investigation. Multispectral images return an accurate mapping of land use and land cover features but spectral indices (normalized difference built-up index, the built-up index, normalized difference impervious surfaces index) were used to identify impervious surfaces that require higher computational effort levels (Su *et al.* 2022). The exploitation of SAR sensors allows for a quick and reliable extraction of urban agglomerates, thanks to the possibility to exploit the information carried by the phase of the radar signal. Its stability is a peculiar characteristic of the built-up feature (Franceschetti & Lanari 1999).

This procedure has not been thoroughly investigated until recently (Liang *et al.* 2022; Wu *et al.* 2022) and allows the identification of most of the impervious surfaces without the need to consider spectral information.

MATERIALS AND METHODS

The case study

The Sarno River basin is a catchment of about 450 km^2 in the Campania Region (southern Italy). It is located between the volcanic complex of Somma-Vesuvio (NW), the Sarno Mountains (NE), the Lattari Mountains (S) and the Tyrrhenian sea (W) (Figure 1(a)). Its major tributaries are the Solofrana, Cavaiola, and Nocerino channels. Based on major tributaries, the entire watershed can be divided into three main sub-basins named 'Calvagnola sub-basin', 'Solofrana sub-basin' and 'Sarno Mouth sub-basin' with areas, respectively, of about 41, 215 and 177 km^2. Sarno River rises at 30 m a.s.l. at the base of a calcareous formation of the Campanian Apennin Mountains. It is 24-km long and flows into the Tyrrhenian Sea between the towns of Torre Annunziata and Castellammare di Stabia. Its daily average flow is about 1 $m^3 s^{-1}$. Sarno Plain is almost flat with a lower longitudinal gradient. The altitude reaches almost 20 m a.s.l. at the base of the close slopes inland, 17 km from the coastline. Geologically, the watershed is dominated by marine, alluvial and volcanic deposits from the activity of the Somma–Vesuvius volcanic complex (NW). The main reliefs consist of Triassic dolomite, lower Jurassic–Cretaceous limestone, and dolomite limestone as well as Cretaceous fractured and karstified limestone (De Pippo *et al.* 2006). With regard to the soil texture characteristics of the Sarno Plain, loamy sand can be found in surface layers and silty loam in deep horizons (Adamo *et al.* 2014).

From the climate point of view, the Sarno River basin is characterized by a warm-summer Mediterranean climate with warm, dry summers, and cool mild winters and with most of the annual precipitation falling in the autumn-winter seasons.

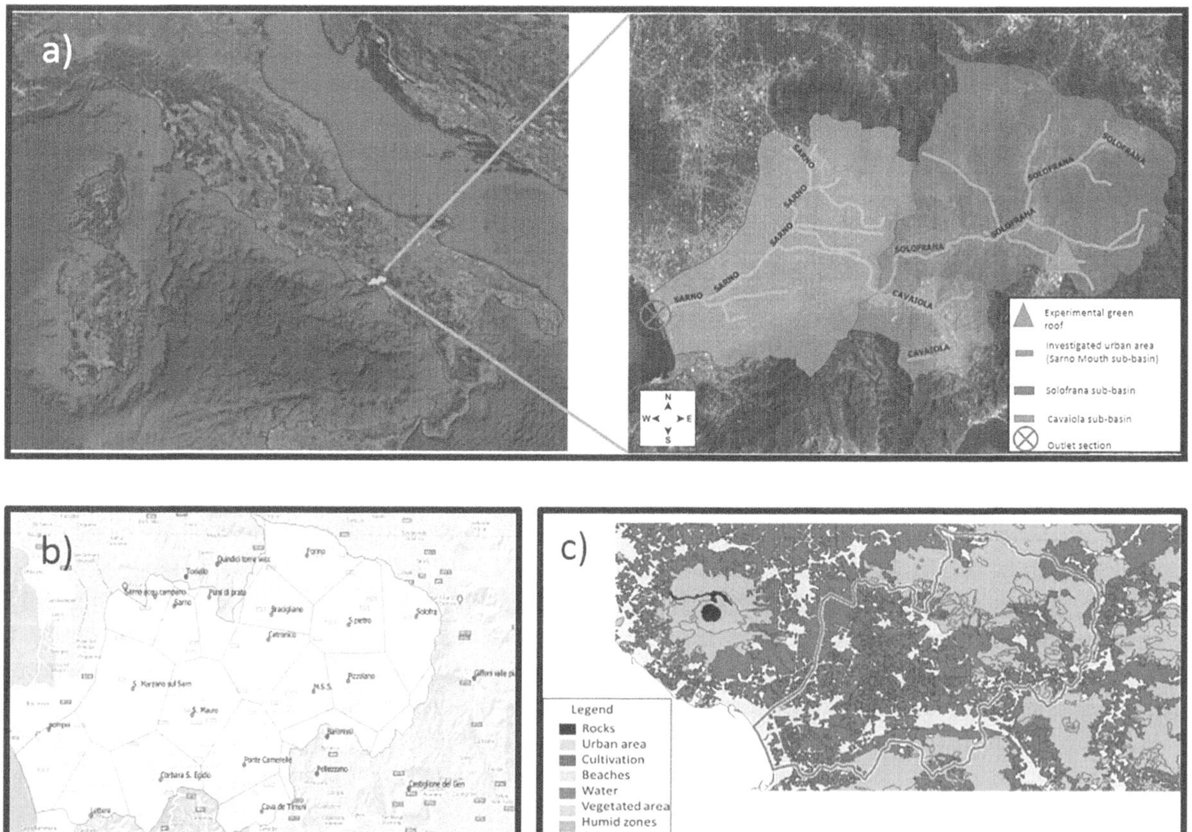

Figure 1 | (a) Location of the study basin; (b) the network of rain gauges of the Sarno River basin; and (c) land use of the Sarno River basin. Please refer to the online version of this paper to see this figure in colour: http://dx.doi.org/10.14273/unisa-1301.

The area presents an average annual air temperature of about 17 °C and average annual rainfall of approximately 1,200 mm. The rain gauge monitoring network includes 25 stations for which rainfall data observations are available from 1990 to the present at the sub-daily scale. In Figure 1(b), the available rain gauges (red circles) and their area of influence resulting from the Thiessen polygons method have been shown.

The basin is a flood-prone area which, in time, has been the cause of several hazardous hydrological events including flash floods and landslides (Califano *et al.* 2015; Longobardi *et al.* 2016). The vulnerability associated with urban flooding is considerably high since the basin is populated by more than 1 million people with an average population density of about 1,800 inhabitants per km^2 and a peak density of up to 2,200 inhabitants per km^2, which make it among the most densely populated and urbanized areas in Italy. The peak density is reached along the coastal belt and the mouth of the Sarno River (Figure 1(c)) and it is lower in the rest of the plain, which is used for intensive agriculture of vineyards, chestnuts, etc.

MDHEs and temporal analysis within the Sarno River basin

The Sarno River basin is a flood-prone area that has been affected over time by several damaging hydrological events. Longobardi *et al.* (2016) investigated the occurrence of MDHEs in a relevant peri-urban sub-catchment, the Solofrana watershed (Figure 1(a)), showing a temporal increase in their frequencies. For the period between 1951 and 2014, a total of 45 events causing seriously damaged urban infrastructures and loss of human lives were recorded. Temporal changes in MDHE occurrences over this period of time were investigated in the light of a climate change perspective with no significant results (Longobardi *et al.* 2016). With the aim to further investigate and address the possible motivation besides the increase in MDHE occurrences in this geographical region, in the current research, the collection of the events was extended to the whole catchment. The collected information includes:

- affected areas;
- the date of occurrence; and
- the source of information.

Only MDHEs that caused damage to structures such as bridges, levees, and roads and/or economic and social impact on properties and/or damage to land and loss of natural resources and in extreme circumstances, losses of life and injury to people, were taken into consideration. The date of occurrence of these events is available with daily accuracy. Papers, books, articles from local newspapers, event databases from state civil protection agencies, and scientific and technical reports were used as the source of information about the MHDEs. Rainfall data were collected from the regional rain gauge network (Figure 1(b)) in order to characterize the properties of the triggering rainfall events. Unfortunately, streamflow data for the Sarno River basin are only available starting from 2001, thus it was not possible to have a further characterization of flow rates.

The temporal distribution of the collected MDHEs and the possible existence of any temporal trends or patterns in the occurrence of these events were examined.

In particular, the existence of linear trends in the number of MDHE records was inspected using statistical parametric (Pearson test) and non-parametric tests (Mann Kendall test and Sen's test).

Pearson's parametric test considers the linear regression of the random variable Y on time X. The regression coefficient or the Pearson correlation coefficient is computed as:

$$\rho = \frac{co\text{-}deviance(X, Y)}{\sqrt{dev(X)dev(Y)}} \tag{1}$$

High values of the coefficient indicate that the number of occurrences varies jointly to the time variable, both in ascending or descending tendencies. The test statistic follows Student's t distribution:

$$t = \frac{\rho\sqrt{n-2}}{(1-\rho^2)^{0,5}} \tag{2}$$

where n is the sample size.

The null hypothesis (no trend) H_0: $\rho = 0$ is tested against the alternative hypothesis H_1: $\rho \neq 0$ at a level of significance α. The null hypothesis is rejected when the t value is greater in absolute value than the critical value $t_{\alpha/2}$.

One of the most common non-parametric approaches for trend analysis is the Mann Kendall test, which is widely used for detecting trends in hydrological time series. In this method, the test statistic S is calculated as:

$$S = \sum_{k=1}^{n-1} \sum_{j=k+1}^{n} sgn(X_j - X_k) \tag{3}$$

where n is the number of observations, X_j and X_k are the jth and kth data of the sample while 'sgn' is the sign function.

Finally, the standardized Z statistic is given as:

$$Z = \begin{cases} \frac{S-1}{\sigma} & if \quad S > 0 \\ 0 & if \quad S = 0 \\ \frac{S+1}{\sigma} & if \quad S < 0 \end{cases} \tag{4}$$

where σ is the variance. Negative and positive Z values show downward and upward trends, respectively. The hypothesis that there is no trend (null hypothesis) is rejected when Z is greater, in an absolute value than the critical value Z_α, for the considered level of significance α.

The previous tests show the trend existence and direction but, using Sen's test, it is possible to estimate the trend magnitude. Sen's slope estimator 'S' is calculated as:

$$S = Median \frac{x_j - x_k}{j - k} \; for \; i = 1 \ldots, N' \tag{5}$$

where N' indicates the pairs of data, x_j and x_k are the data values at times j and k ($j > k$).

$S > 0$ indicates an upward trend in the time series and vice versa.

The confidence interval for S with the level of significance α can be computed as follows:

$$C_\alpha = Z_{1-\alpha/2}\sqrt{Var\ (S)} \tag{6}$$

where $Z_{1-\alpha/2}$ is obtained from the standard normal distribution table and $Var(S)$ is the variance of S.

Next, $M_1 = (N' - C_\alpha)/2$ and $M_2 = (N' + C_\alpha)/2$ are computed. The lower and upper limits of C_α, Q_{min} and Q_{max} are the M_1th and the $(M_2 + 1)$th of the N' ordered slope estimates. The slope S is statistically different from zero if the two limits Q_{max} and Q_{min} have the same sign.

Land cover change analysis by SAR images elaboration

The processing and elaboration of SAR images acquired over the Sarno River basin were performed in order to track the urban development of the area and investigate the changes in the extension of the impervious areas during the last decades. Due to the availability of radar imagery, the period considered for the present investigation covers about 20 years from 1995 to 2016. Indeed, the first SAR images covering the study area have been provided by the ERS-1 sensor that was launched in 1995 by the European Space Agency (ESA) but ended its operations in 2000, so, in order to integrate these data, the satellite images provided by the COSMO-SkyMed mission and released by the Italian Space Agency (ASI) were used. The mission launched four identical satellites between 2007 and 2010 and all are still operating. Provided the spatial resolution of the remote sensing data, once the percentage of the total impervious area within the basin is identified by the SAR detection analysis, the use of building construction indexes will allow the areal extension of buildings over the whole sealed area to be quantified, which is essential to investigate the hydrological benefits associated with a widespread GR retrofitting of the existing traditional roofs. In this context, the land use scenario detected in 1995 will be referred to as the pre-development scenario, whereas the 2016 land use scenario will be referred to as the post-development scenario.

The process here proposed for the analysis of SAR images can be organized into three major blocks: data acquisition, pre-processing, and feature extraction (Figure 2).

The first step was the data download. Two sets of archive satellite images were used. The first one consisted of six images acquired between March and December 1995 by the ERS-1 satellite. The second dataset was acquired by the COSMO-SkyMed mission between May and December 2016, and consisted of seven images. The main characteristics of the two aforementioned sensors are summarized in Table 1. Details about the processed datasets are provided in Table 2. These two datasets provided, respectively, the two Sarno River basin impervious area scenarios under analysis.

The pre-processing block includes data coregistration and the estimation of the interferometric coherence. Coregistration is the process of alignment of images with respect to one of them selected as the reference (Franceschetti & Lanari 1999). In our

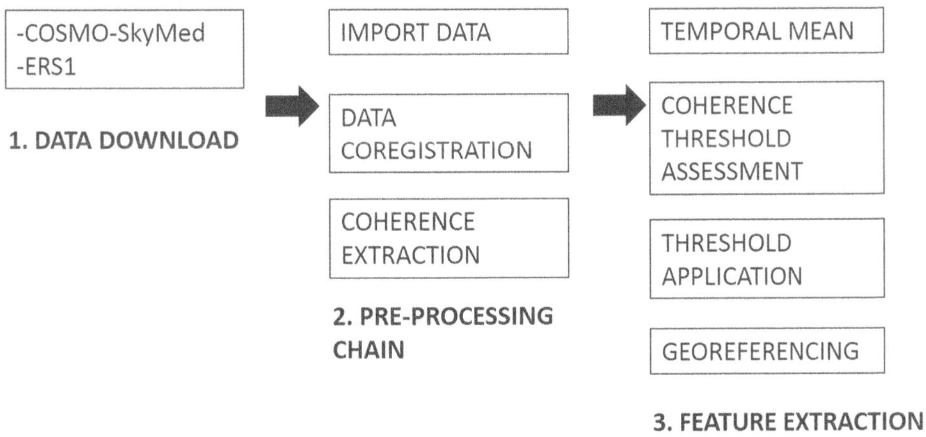

Figure 2 | Process for the elaboration of SAR images.

Table 1 | Characteristics of the two sensors

Characteristics	COSMO-SkyMed	ERS1
Pixel size	3 m × 3 m	8 m (range) × 4 m(azimuth)
Spectral band	X	C
Frequency	9.6 GHz	5.3 GHz
Wavelength	3.1 cm	5.8 cm
Orbit period	97.2 min	100 min
Nominal repeat cycle	16 days	35 days

Table 2 | Two sets of SAR images

ERS1	COSMO-SkyMed
24/03/1995	03/05/2016
08/07/1995	20/06/2016
12/08/1995	06/07/2016
21/10/1995	08/09/2016
25/11/1995	24/09/2016
30/12/1995	27/11/2016
–	13/12/2016

case, reference (or master) images for the ERS1 and COSMO-SkyMed time series were acquired, respectively, on March 1995 and May 2016.

The second step of the pre-processing block was the interferometric coherence estimation. It is a measure of the stability of a target with respect to the phase of the complex signal (Franceschetti & Lanari 1999). Considering two co-registered images, it is computed through the following relation:

$$\gamma = \frac{|\Sigma S_1(x) \cdot S_2^*(x)|}{\sqrt{\Sigma |S_1(x)|^2 \cdot |S_2(x)|^2}} \tag{7}$$

where S_1 and S_2 represent the complex image values for the first and the second image, respectively, and * stands for complex conjugation operation.

The coherence image for all interferometric pairs for the years 1995 and 2016 was produced using as the reference image, the master image of the previous step. The coherence ranges between 0 and 1. The higher the coherence, the higher the phase stability of the target. Built-up areas are typically stable with respect to the phase, so they are expected to exhibit high values of the interferometric coherence (Franceschetti & Lanari 1999), even when images are acquired with a long-time baseline. On the other side, natural land cover is expected to exhibit low coherence values, even when images are acquired at short time scale. This is due to the intrinsic variability of features like crops or bare soils. The feature extraction block includes the temporal mean of the coherence maps grouped per year. In other words, all the coherence maps relevant to images acquired in the same year are stacked together and processed for temporal average. This is useful to enhance the presence of man-made stable targets at the expense of natural scene features.

The obtained temporal coherence maps were then treated with threshold segmentation to isolate urban pixels. The threshold process is implemented using Otsu's method (Otsu 1979). The assumption of this procedure is that the image has bimodal distribution and contains two classes of pixels, which in the present study represent the urban and non-urban areas. It iteratively calculates, among all possible threshold values, the optimum threshold of intensity level '*t*' separating the two classes that minimize the weighted within-class variance of the background and foreground pixels according to the relation as follows:

$$\min(\sigma_w^2(t)) = \min(W_b(t) \cdot \sigma_b^2(t) + W_f(t) \cdot \sigma_f^2(t)) \rightarrow \mathit{for} \quad t = \min \ldots \max \tag{8}$$

where 'min' and 'max' are, respectively, the minimum and maximum intensity, $\sigma^2{}_w$ is the intra-class variance, σ_{2f} and σ_{2b} are the variances of the foreground and background classes of pixels, W_b and W_f are the weights of the two classes.

This procedure returned two cluster-based threshold images relating to the years 1995 and 2016 from which it was possible to quantify, for each year, the amount of pervious and impervious areas.

SWMM and initial setting

The SWMM model was used to assess the hydrological response of the Sarno River basin, both in the pre-development (1995) and post-development (2016) scenarios, showing the hydrological changes related to the changes in land cover. Furthermore, the SWMM model was used to assess the hydrological performance of a GR retrofitting stormwater management plan, which, conscious of the importance of SuDS areal extension and investment costs, takes into account different scenarios of GRs retrofitting, considering different scenarios, corresponding to 5, 30, and 100% of traditional roofs conversion to GRs.

The SWMM is a free and open-source model implemented by the United States Environmental Protection Agency (USEPA) with the aim to simulate runoff quantity and quality in urban/suburban areas. The SWMM includes a number of 'blocks' which are computational modules tasked with the hydrologic, pollutant generation and transport, and hydraulic calculations (Figure 3).

The RUNOFF block is used to produce the runoff hydrograph from sub-catchments. Indeed, the algorithm allows us to generate surface runoff in response to precipitation. The method employs the surface water budget approach as shown in Figure 3 with a single inflow represented by the precipitation and several discharges including the runoff '*Q*', the infiltration '*f*', and the evapotranspiration 'ET'.

ET rates can be stated as a single constant value, a set of monthly average values, a user-defined time series of values, values computed from the daily temperatures contained in an external file, and daily values read from an external file. In the present

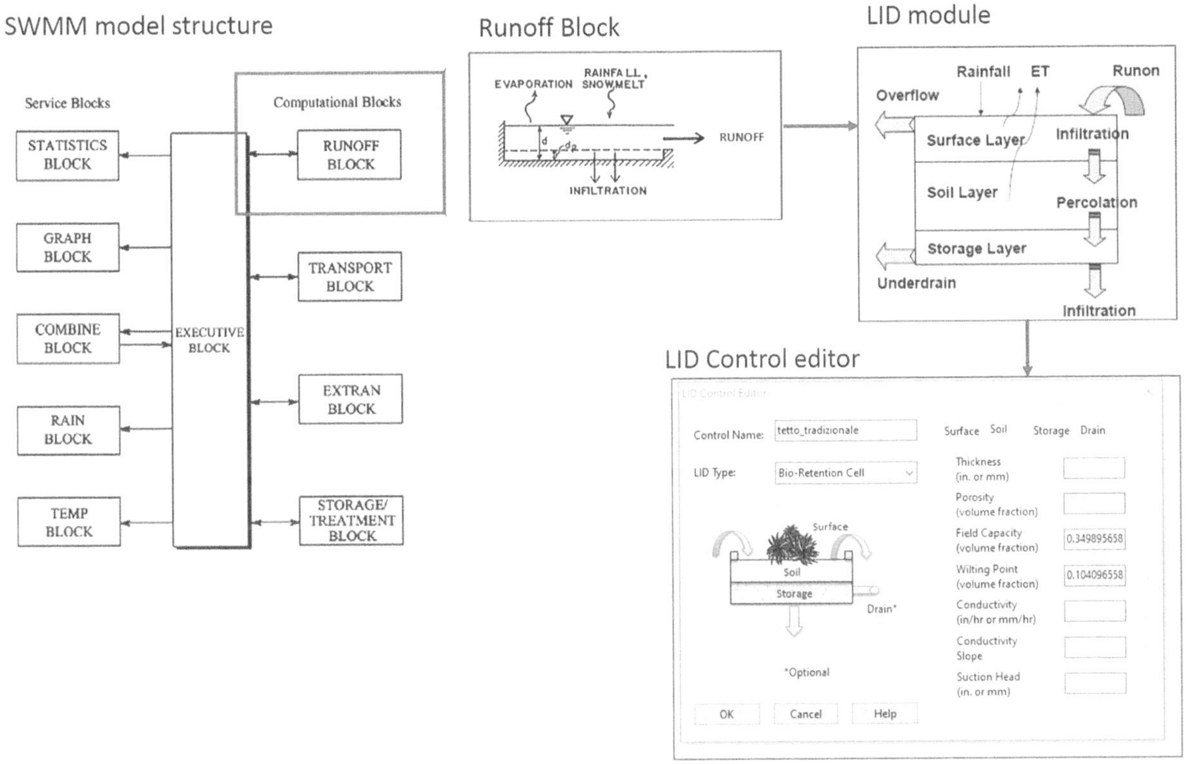

Figure 3 | Left side panels: Processing blocks of the SWMM and the surface water balance employed by the RUNOFF module. Right side panels: the water balance occurring within the LID module and an overview of the main parameters required by the LID control editor to run the model.

study, it has been neglected during the calibration procedure carried out at an event scale while it has been set as a constant value during the simulations of the GR conversion scenarios performed considering a longer time scale. It has been estimated according to the Blaney–Criddle model (Blaney & Criddle 1950).

The rainwater infiltration '*f*' can be calculated using one of three available methods, namely the Green–Ampt infiltration model Equation (9), the Horton model Equation (10), and the CN method Equation (11) which can be selected by the user:

$$f_{Green-Ampt} = k_{sat}\left[1 + \frac{(\phi - \theta)\psi}{F}\right] \quad (9)$$

$$f_{Horton} = f_c + (f_0 - f_c)e^{-kt} \quad (10)$$

$$f_{CN} = \frac{P_e}{P - I_a}S \quad (11)$$

where f_c is the minimum rate on the Horton infiltration curve, f_0 is the maximum rate on the Horton infiltration curve, k is a decay constant for the Horton infiltration curve, t is the infiltration time, k_{sat} represents the saturated hydraulic conductivity, ϕ is the soil porosity, θ equals the water content, ψ corresponds to the suction head, F is the cumulative amount of infiltrated water, I_a is an initial abstraction, P is the rainfall, P_e is the rainfall excess, and S is the potential maximum soil moisture retention after runoff begins. The CN method was selected for the current study.

Surface runoff occurs, according to Manning's equation (12), when the depth of water over the sub-catchment '*d*' exceeds the maximum depression storage 'd_p' which is the maximum surface storage provided by ponding, surface wetting, and an interception. The water in storage can also be depleted by infiltration and evaporation.

$$Q = W \cdot \frac{1}{n} \cdot (d - d_p)^{\frac{5}{3}} S_0^{1/2} \tag{12}$$

where Q is the surface overflow rate (m^3/s), S_0 represents the surface slope (m/m), n is Manning's roughness coefficient (-), and W is the sub-catchment width (m).

Provided the urban development spatial features, while the land cover change analysis and the temporal investigation of MDHE occurrences refer to the whole Sarno River basin, the hydrological modeling is only applied to the downstream area of the Sarno River basin (Figure 1(a)), hereinafter called Sarno Mouth sub-basin or investigated urban area.

The inflows from the peri-urban area located in the upstream part of the Sarno River basin (Solofrana and Cavaiola sub-catchments in Figure 1(a)) were not considered because a previous study demonstrated that the hydrological connection between the peri-urban and the urban system (Sarno Mouth sub-catchment in Figure 1(a)) can be neglected (Mobilia 2018).

The Sarno Mouth sub-basin was divided into six catchments numbered from 10 to 15 and then sketched in the SWMM as shown in Figure 4. The six catchments were delineated based on the characteristics of the relevant drainage network.

The average surface slope of the catchments was obtained using the slope analysis of the Digital Elevation Model in QGIS while the width of the overland flow path was given, according to the SWMM user's manual, by the catchment area divided by the average maximum overland flow length. After the GR placement, the sub-catchment width has required adjustment to compensate the amount of the initial watershed area replaced by the green technology. The original sub-catchment width has been modified taking into account the width of the installed GRs intended as the total length along the edge of the roof where runoff is collected. More details about the experimental GR dimensions are provided in Table 3.

The percentage of imperviousness derived by SAR images elaboration

These data were used to define the CN. The CN is tabulated in the literature and, for open spaces with grass cover on 75% or more of the area and hydrological soil group (HSG) B, it assumes the value of 61, whereas for grass cover on 50–75% of the area and B-HSG, it equals 69. The B group includes soils with a moderate infiltration rate attributable to the texture of silt loam or loam which is the dominant soil texture of the Sarno Plain (Adamo *et al.* 2014). In Table 4, more details about the input parameters of the sub-catchments, conduits and nodes are available.

The SWMM is equipped with an LID tool. It has been specifically designed for modeling the storm water management performance of various types of LID practices, including GRs, bioretention cells, permeable pavements, etc.

According to several studies (Burszta-Adamiak & Mrowiec 2013; Cipolla *et al.* 2016; Limos *et al.* 2018), the bioretention cell LID module has been selected in this study to mimic the GR hydrological behavior. In detail, this LID type consists of a number of horizontal layers including the surface layer, the soil layer, and the storage substrate layer with only vertical movement of water within and between them. Beyond the literature studies, this particular model fits the building practices of the experimental GRs used to calibrate the model, as later explained.

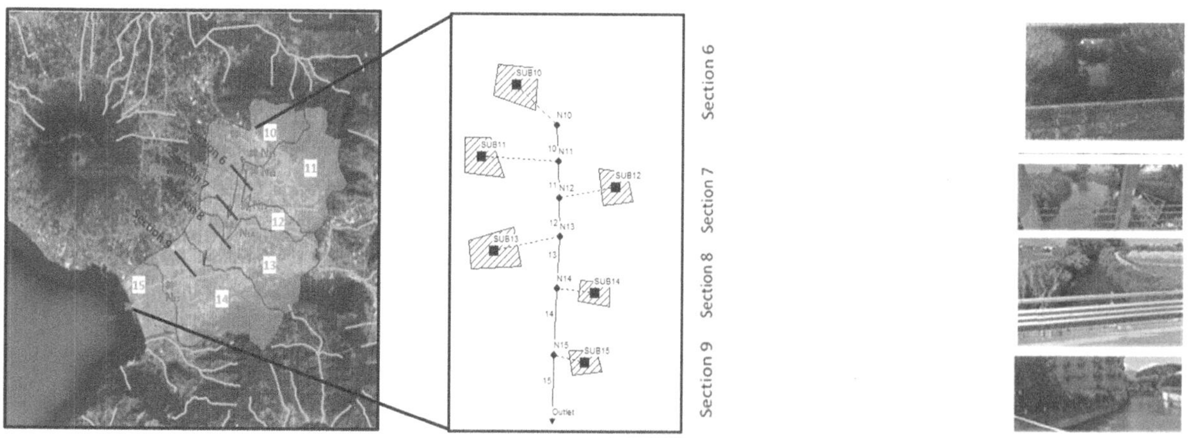

Figure 4 | An urban drainage network sketched in the SWMM.

Table 3 | The properties of the sub-catchments, of the conduits, and of the nodes

	Sub-catchments				
ID	**Area (km^2)**	**Width (m)**	**Slope (%)**	**Imperviousness 1995/2016**	**CN (–)**
SUB10	30.48	14.14	30.89	4.71/8.2	61
SUB11	38.24	32.97	7.73	5.75/10.1	61
SUB12	12.75	3.75	0.72	8.39/14.7	61
SUB13	38.05	19.88	2.29	12.53/21.9	61
SUB14	44.14	7.03	9.35	15.02/26.3	69
SUB15	13.99	3.96	22.92	18.63/32.6	69
	Trunks				
ID [Length (km)]		10 [2.16]; 11[1.16]; 12[3.39]; 13[1.91]; 14[6.28]; 15[3.53]			
	Nodes				
ID [Altitude (m.a.s.l.m.)]		N10 [24]; N11 [22]; N12 [20]; N13 [16], N14 [13]; N15 [5]; Outlet[0]			

Table 4 | Input parameters for LID module

Parameter	**U.M.**	**Initial value**	**Data sources**
Surface layer			
Surface roughness	(–)	0.24	Literature data
Surface slope	(m/m)	0.01	System geometrical characteristic
Soil (Loam)			
Thickness	(mm)	100	System geometrical characteristics
Porosity	(–)	0.85	Material supplier
Field capacity	(–)	0.2	Literature data
Wilting point	(–)	0.1	Literature data
Conductivity	(mm/h)	3.302	Literature data
Conductivity slope	(–)	35	Literature data
Suction head	Ψ (mm)	88.9	Literature data
Storage layer (Expanded clay)			
Thickness	(mm)	50	System geometrical characteristics
Void ratio	(–)	0.75	Literature data
Drain			
Drain coefficient	(–)	1.88	Literature data
Flow exponent	(–)	0.5	Literature data
Offset	(mm)	0	System geometrical characteristics

Figure 3 illustrates the conceptual model of the processes that the SWMM accounts for. Rainfall and evapotranspiration for LIDs are the same as used in the SWMM runoff module, while infiltration is calculated with the Green–Ampt equation (Equation (9)), which is also one of the options in the groundwater module. The wetted front created by Green–Ampt infiltration moves from the surface layer downward into the soil layer and finally percolates into the storage layer with a velocity determined by Darcy's law (Wang *et al.* 2019). Drain flow in SWMM LIDs is calculated according to the following equation:

$$q = Ch^n \tag{13}$$

where q is the outflow; C and n are, respectively, the drain coefficient and exponent; h is the height of the drain above the bottom of the unit's storage layer.

The input parameters required by the bioretention module and used in the present study refer to an experimental GR and are listed in Table 4.

In detail, the values of the thickness and the slope of the layers are geometrical characteristics of the system while the roughness of the surface layer and the void ratio of the storage layer were sourced from the user's manual of the SWMM which draws on the literature data. Concerning the soil layer, the value of the porosity was provided by the material supplier while the values of the other parameters were derived by Rawls *et al.* (1983) who suggested average values of the infiltration parameters for different kinds of soil. With reference to the drain parameters, the drain offset is a geometrical characteristic of the system, while the user's manual suggests values for the flow exponent when the drain acts like an orifice, and for the drain coefficient estimated to be 60,000 times the ratio of the total slot area to the LID area.

The experimental roof is located within the campus of the University of Salerno, in Fisciano, Southern Italy. It was selected as a test bed since it falls within the study area (Figure 1(a)). It is of extensive type and has a surface of 2.5 m^2 (2.5 m × 1 m; width × length) and a depth of 15 cm. The support layer is made up of a mix of peat and zeolite while the storage layer is made up of expanded clay. The vegetation layer hosts succulent plants called Mesembryanthemum. Further details about the GR test bed are available at Longobardi *et al.* (2019).

The SWMM hydrological modeling required a spin-up period of 72 h in order to give the model the time to reach a state of equilibrium after being started from initial conditions. The soil moisture content of the LID module before each simulated event was fixed as the initial condition. As suggested by Cipolla *et al.* 2016, with the aim of reducing the dependency of the model on the degree of saturation of the LID, the value of this parameter of the day before the rainfall event was set to 100% so that it could be estimated by the model as a function of the weather data.

Calibration procedure

For accurate simulations in SWMM, model parameters need to be calibrated. A double calibration was performed in the presented research. It involved both the RUNOFF module and the GRs LID module, by a comparison of modeled and observed streamflow (RUNOFF module) and runoff (LID module). The optimal parameters were selected corresponding to the maximization of the NSE (Nash–Sutcliffe Efficiency Index) (Nash & Sutcliffe 1971) given by the following equation:

$$NSE\,(-) = 1 - \frac{\sum_{i=1}^{N}(\mathrm{OBS_i} - \mathrm{SIM_i})^2}{\sum_{i=1}^{n}(\mathrm{OBS_i} - \overline{OBS})^2} \tag{14}$$

where N represents the length of the sample, i is the time index in an hour, SIM and OBS, respectively, represent the predicted and measured values, and $\overline{OBS}$ is the mean of the measured values. The calibration of the RUNOFF module was performed by comparing modeled and observed streamflow at the event hourly scale, at the outlet of the Sarno Mouth sub-catchment (Figure 1(a)). The selection of the calibration parameters was made in accordance with Behrouz *et al.* (2020). Based on the review of several previous studies, they suggested that parameters exhibiting the largest sensitivity, or that were most frequently reported as sensitive in the literature, were the catchment imperviousness, the width, the impervious depression storage coefficient and the channel Manning's roughness coefficient. In the present research, the percentage of imperviousness is a known parameter deriving from SAR image elaboration while the width has been obtained by dividing the catchment area by the average maximum overland flow length as suggested by the SWMM user's manual. Consequently, the impervious depression storage coefficient d_p and channel Manning's roughness coefficient n were considered as the only parameters to be calibrated. These parameters are supposed to vary uniformly across the basin and so they assume the same value for each sub-catchment and model link. The range of expected values for each parameter, which considers the physically plausible values, has been shown in Table 5, according to the review of previous studies (Barco *et al.* 2008; Rossman 2010; Sadeghi *et al.* 2022).

Four rainfall–runoff events were selected for the calibration of the RUNOFF module. They belong to the historical MDHEs database previously described. Being all hazardous events, they were chosen according to the following selection criteria:

Table 5 | Range of values for d_p and n

Parameter	Minimum value	Maximum value
dp (mm)	0	1.27
n (–)	0.013	0.027

- The number of affected areas;
- The rainfall return period; and
- The rainfall intensity (5-min interval)

A single event can affect more than one municipality, and in this case, it probably has a major impact, so, in the present study, the rainfall events which triggered urban MDHEs in several municipalities were considered.

The 10-year return period is generally selected as the design standard for most urban drainage systems and when it is exceeded, the systems are overloaded and damaging hydrological events can occur. In light of this, only the rainfall events with a return period higher than 10 years were selected for the analysis.

Many authors have asserted that rainfall intensities are of prime importance for flood generation (e.g. Pitlick 1994); indeed, for higher rainfall intensities, higher hydrological responses are expected (Habonimana 2014). In this research, the triggering rainfall events which occurred in at least three locations of the basin with an intensity higher than 15 mm/h (heavy events) were chosen.

For the reasons related to the modeling of the LID tool, the selected rainfall events were divided into convective and stratiform categories based on their rainfall shape profile (Terranova & Iaquinta 2011). They were separately modeled since it was shown by Mobilia & Longobardi (2020) that SWMM LID tool is affected by the behavioral effects of events with different shapes of rainfall profiles. Two different calibration phases were performed for the two classes of events. Their relevant properties are illustrated in Table 6.

The calibration of the LID module was carried out using rainfall/runoff events collected at an experimental roof located within the investigated river basin, at the university campus of Salerno. The experimental extensive GR is fully equipped

Table 6 | Rainfall characteristics of the selected events

		Rain gauges						
Type of event	**Criteria**	**Corbara**	**Lettere**	**Pompei**	**San Marzano**	**San Mauro**	**Sarno**	**Sarno acq.**
Stratiform	*31.10.2012*							
	Rainfall Intensity (5 min) (mm/h)	44.4	32.4	12	12	21.6	4.8	10.8
	Return Period (years)	>100	16.18	2.17	2.33	4.75	1.68	1.82
	Affected areas				7			
	21.01.2014							
	Rainfall Intensity (5 min) (mm/h)	24	19.2	26.4	19.2	16.8	13.2	16.8
	Return Period (years)	10.38	7.61	4.51	4.94	6.78	3.88	4.25
	Affected areas				6			
Convective	*13.09.2012*							
	Rainfall Intensity (5 min) (mm/h)	115.20	84.00	85.20	48.00	48.00	74.40	55.20
	Return Period (years)	>100	60.70	35.95	9.32	>100	12.21	6.70
	Affected areas				14			
	20.11.2013							
	Rainfall Intensity (5 min) (mm/h)	42.00	68.40	10.80	25.20	52.80	25.20	20.40
	Return Period (years)	5.71	10.20	1.75	2.57	5.27	2.64	2.14
	Affected areas				8			

with a rain gauge for recording the precipitation and a digital scale for weighing the runoff volume since 2017. The rain gauge is part of a weather station, Watchdog 2000 Series (Model 2550) while the digital scale is a LAUMAS AC60 Kg and above it, a tank is placed where the rainwater is collected. The measurements are taken at 5 min time step and aggregated at the desired work scale (Longobardi *et al.* 2019).

According to a number of sensitivity analyses from several experimental studies in this area of research, the suction head Ψ is the most sensitive parameter (Baek *et al.* 2020; Paithankar & Taji 2020; Lisenbee *et al.* 2022). For example, in the case of the RUNOFF module, stratiform and convective events were clustered. A complete list of 29 rainfall events recorded during the period 2017–2019, used for the LID calibration phase, and their characteristics such as rainfall intensity, duration, cumulative, and retention coefficient (RC) are shown in Table 7.

Table 7 | Characteristics of the selected rainfall–runoff events

Date	Duration [min]	Cumulative rainfall [mm]	Peak Intensity [mm/h]	Retention coefficient [%]
		Convective events		
25/07/2017 07:20	420	2.8	2.032	82
07/09/2017 12:10	540	4.6	2.286	69
07/11/2017 10:00	360	15.2	9.398	79
10/01/2018 00:15	540	30.2	10.160	45
11/01/2018 17:00	960	20.1	9.398	22
17/01/2018 08:15	180	1.3	1.016	77
17/01/2018 19:45	60	3.6	3.556	78
01/02/2018 21:15	300	3.3	1.524	80
13/02/2018 17:35	60	0.8	0.762	81
02/03/2018 09:55	240	3.3	1.778	38
09/04/2018 06:00	180	6.1	3.556	80
17/04/2018 16:50	360	5.8	5.334	83
23/05/2018 10:00	300	13.0	4.826	11
07/11/2018 04:00	360	16.0	6.350	25
08/12/2018 06:15	60	2.79	2.794	80
08/12/2018 10:00	120	5.84	4.572	43
10/12/2018 19:00	120	1.78	1.524	82
10/01/2019 01:00	360	6.60	2.286	79
22/04/2019 18:15	120	3.56	1.778	91
Average value	*297*	*7.71*	*3.944*	*61*
		Stratiform events		
03/02/2018 12:50	1,200	12.4	3.810	37
07/02/2018 14:00	840	11.2	4.826	7
20/02/2018 02:01	1,080	11.4	2.286	19
03/03/2018 06:57	720	11.4	2.286	28
03/05/2018 16:00	180	7.1	4.826	78
20/12/2018 15:00	240	3.81	2.032	76
22/12/2018 02:00	240	6.35	2.286	37
09/01/2019 12:00	180	5.33	3.302	70
11/02/2019 14:40	180	4.32	2.54	79
12/04/2019 21:35	180	5.33	3.556	26
Average value	*504*	*7.87*	*3.175*	*46*

The calibration procedures returned the best parameters for LID and RUNOFF modules to be used to run SWMM for subsequent analysis. In detail, the average value of the suction heads obtained by the calibration of the convective events was used for the simulation of events with this shape of rainfall profile and the same applies to the stratiform events. On the other side, the calibrated values of d_p and n for each of the four selected events were used for the reproduction of the same events in the modeled scenarios.

Evaluation of the hydrological effect of different land use scenarios

The SWMM model was used to simulate the hydrological behavior of the Sarno Mouth sub-catchment, under the two land use scenarios (1995 and 2016) and in the case of GR retrofitting scenarios. The comparison among the different scenarios was performed in terms of the percentage of reduction in runoff volume (ΔR_V), in peak flow (ΔP_F) and in the number of flooded section (ΔF_S) and increase in the delay time (ΔD_T). The indices were estimated as follows:

$$\Delta R_V = \frac{R_{V,0} - R_{V,1}}{R_{V,0}} \cdot 100 \tag{15}$$

$$\Delta P_F = \frac{P_{F,0} - P_{F,1}}{P_{F,0}} \cdot 100 \tag{16}$$

$$\Delta F_S = \frac{F_{S,0} - F_{S,1}}{F_{S,0}} \cdot 100 \tag{17}$$

$$\Delta D_T = D_{T,0} - D_{T,1} \tag{18}$$

The land use cover corresponding to 2016 (post development) was set as the baseline scenario and differences with pre-development (1995) and greening technologies were considered. Thus, $R_{V,0}$, $P_{F,0}$, $F_{S,0}$, $D_{T,0}$ are, respectively, the runoff volume (m^3), the peak flow (m), the number of flooded sections (–) and the delay time (hours) referred to the post-development scenario, while $R_{V,1}$, $P_{F,1}$, $F_{S,0}$, $D_{T,1}$ are the same parameters but referred both to the different greening scenarios and to the pre-development scenario.

RESULTS AND DISCUSSION

MDHEs and temporal analysis within the Sarno river basin

The database of the MDHEs occurred between 1951 and 2016 within the Sarno River basin, and the related information such as the areas hit by the events, the date of occurrence and the documentary resources is accessible in Table S1 (see Supplementary Materials).

The data set includes 102 MDHE records. About 55% of the events affect more than one municipality. The temporal distribution of MDHEs was investigated. The monthly distribution of the events suggests that May is the month with the lowest occurrence while most events occurred between September and January with a peak in October (25 events) which are the wettest months of the year as appears from the values of the mean maximum daily precipitation (Figure 5). Indeed, the monthly distribution of the mean maximum daily precipitation recorded at the rain gauge of Sarno (Figure 1(b)) shows values higher than 25 mm during the winter and autumn months and lower values during the warmer months.

Figure 6 shows the temporal evolution of MDHEs between 1951 and 2016. At a first visual inspection, an increasing trend over the years could be detected. It seems that the temporal growth in the number of events increases faster than linear and approaches almost an exponential function. A significant increase is mostly appreciable from 2011 to 2016. Before that year, the number of events is never higher than 5.

As SAR images and consequently land uses change is only available from 1995 to 2016, a statistical analysis was performed for trend detection in this specific period in order to assess the significance of the temporal change in MDHE occurrences (Figure 6). The Mann–Kendall test, Pearson test and Sen's Slope estimator test were applied for detecting the existence of a possible monotonic trend direction and the magnitude of change at 5% significance level. Each of the three tests rejects the null hypothesis, that is the presence of a statistically significant trend is confirmed (Table 8). In addition, the positive value of the test-statistics suggests an upward trend direction, as evident in Figure 6.

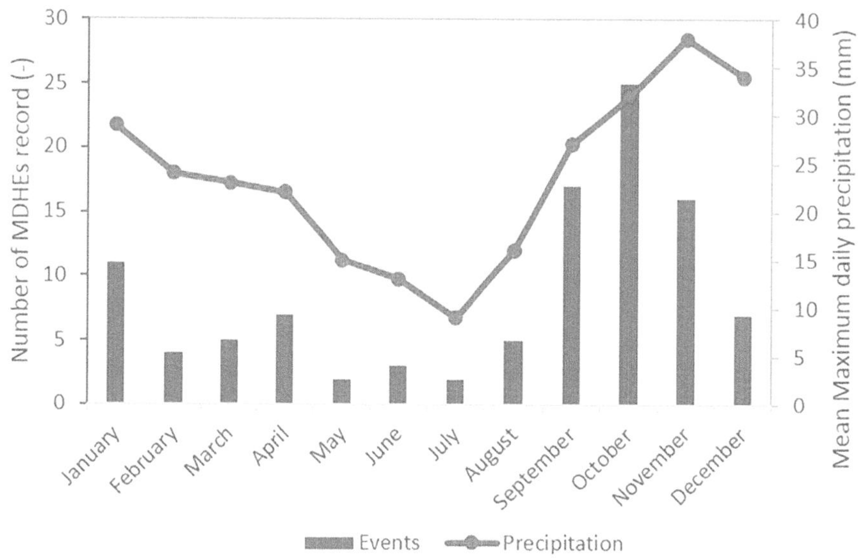

Figure 5 | Number of recorded MDHEs and mean maximum daily precipitation monthly distribution.

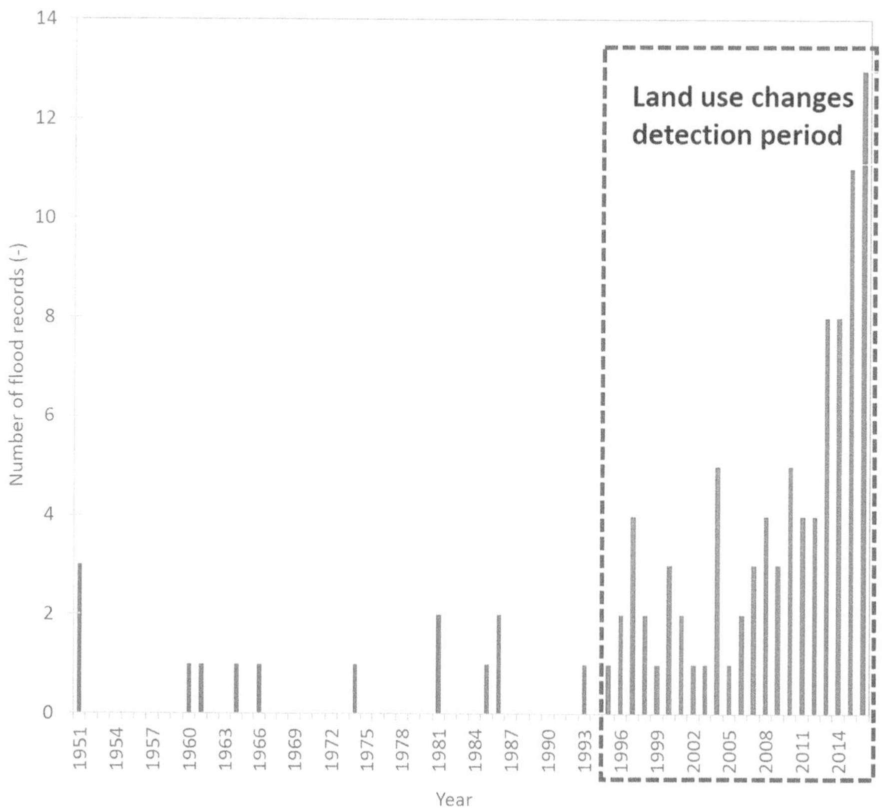

Figure 6 | Annual distribution of the number of records of floods from 1951 to 2016.

Various sources of uncertainty might affect the objectivity and the outcome of the statistical analysis. The urban growth is inevitably associated with an increase in the urban infrastructure exposed to risk such as the use of an increasing number of media in recent years may have led to a physiological increase in the records of flood event occurrences compared to the past.

Table 8 | Results of the statistical tests

Test	Test statistic	Null hypothesis	Direction of trend	*p*-value
Mann–Kendall	$Z = 3.7$	Rejected	Upward	0.000174315
Pearson	$t = 5.2$	Rejected	Upward	0.000021153
Sen	$Q_{min} = 0.17$; $Q_{max} = 0.5$; $S = 0.33$	Rejected	Upward	–

It is however extremely difficult to identify them, but in a general context in which society's perception is an effective increase in the risk of urban flooding, these limitations in objectivity seem altogether acceptable.

Results are furthermore supported and confirmed by previous research which investigated the variation of the frequency of flash floods in urban areas, during time periods similar to the one considered in this study. In detail, within the same region, Esposito *et al.* (2018) observed an increase in flooding events between 2000 and 2014 in the Pozzuoli area. They ascribed this increasing trend to the variation in the rainfall regime since urbanization has increased only for 0.7% of the watershed area. In a geographically close region, Bentivenga *et al.* (2020) showed a recent increase in the frequency of flooding events in the Ionian Belt of Basilicata Region confirming the role played by both climate changes and human activities. Apollonio *et al.* (2016), found a good correlation between flooding areas and land use changes from 1984 to 2011 in the Cervaro river basin in Apulia Region. In the Lazio region, Recanatesi & Petroselli (2020) investigated the relationship between land cover change and flood risk during the period from 1954 to 2018, showing how the flood risk changes due to an increase in urbanization. In the Veneto Region, Sofia *et al.* (2017) coupled the increase in floods from 1900 to 2010 to both climate and land use change.

Other works, worldwide, confirmed the upward trend of urban flood events. For instance, according to Huang *et al.* (2018), Guangzhou, China experienced a rapid increase in flood events during the period of 2009–2015 influenced by both change in precipitation and impervious surface. In Europe, the amount flooding in the affected areas has seen a rise since 1870 and among the main causes, the changes in the distribution of land cover/use can be included (Paprotny *et al.* 2018).

SAR images elaboration

The Otsu thresholding algorithm was applied to distinguish between urban areas and natural land cover. The resulting classifications for the years 1995 and 2016 are shown in Figure 7.

As shown in Table 1, images acquired by the two sensors have slightly different pixel size. This is not expected to influence significantly the segmentation, especially over dense urban areas. Misclassification of small agglomerates and/or isolated structures can occur using ERS-1 images. However, the impact of these areas on the calculation of the percentage of soil sealing, which is the parameter of interest for hydrological modeling, is expected to be negligible (see Table 9). This analysis was performed at the sub-catchment scale, in order to identify the areas of the basin which, in time, were most affected by the land use change.

The percentage of soil imperviousness heavily increases from 1995 to 2016 for each sub-basin. The built-up area in the whole basin changes from 7.4 to 12.38% in 2016, doubling its original value in 20 years. The basin with the lowest increment of impervious area is the Calvagnola with an increase of about 30% between 1995 and 2016, while the catchment which experiences the highest variation of the urbanized surfaces, around 75%, is the Sarno Mouth basin. These findings, on one side, reinforce the choice to focus the hydrological analysis on the downstream part of the Sarno River basin, where the increase of built-up area is considerable. On the other side, they also support the idea of the increase in flooding risk due to land cover changes. As reported in the introduction section, for what concerns the Sarno river basin, in previous research, the temporal variability of the precipitation over the same time-horizon was studied (Califano *et al.* 2015). The findings excluded both the existence of a significant variability and the increase in storm events severity. Despite the discussed limitations of the statistical empirical analysis and considering the presented results in terms of land use changes, it is reasonable to assume that the cause of the recent rise of MDHEs is most likely related to the increase in the impervious surface. This result would suggest that a valid solution to reduce the frequency of occurrence of damaging events could be found in sustainable landscape planning, by green technologies implementation.

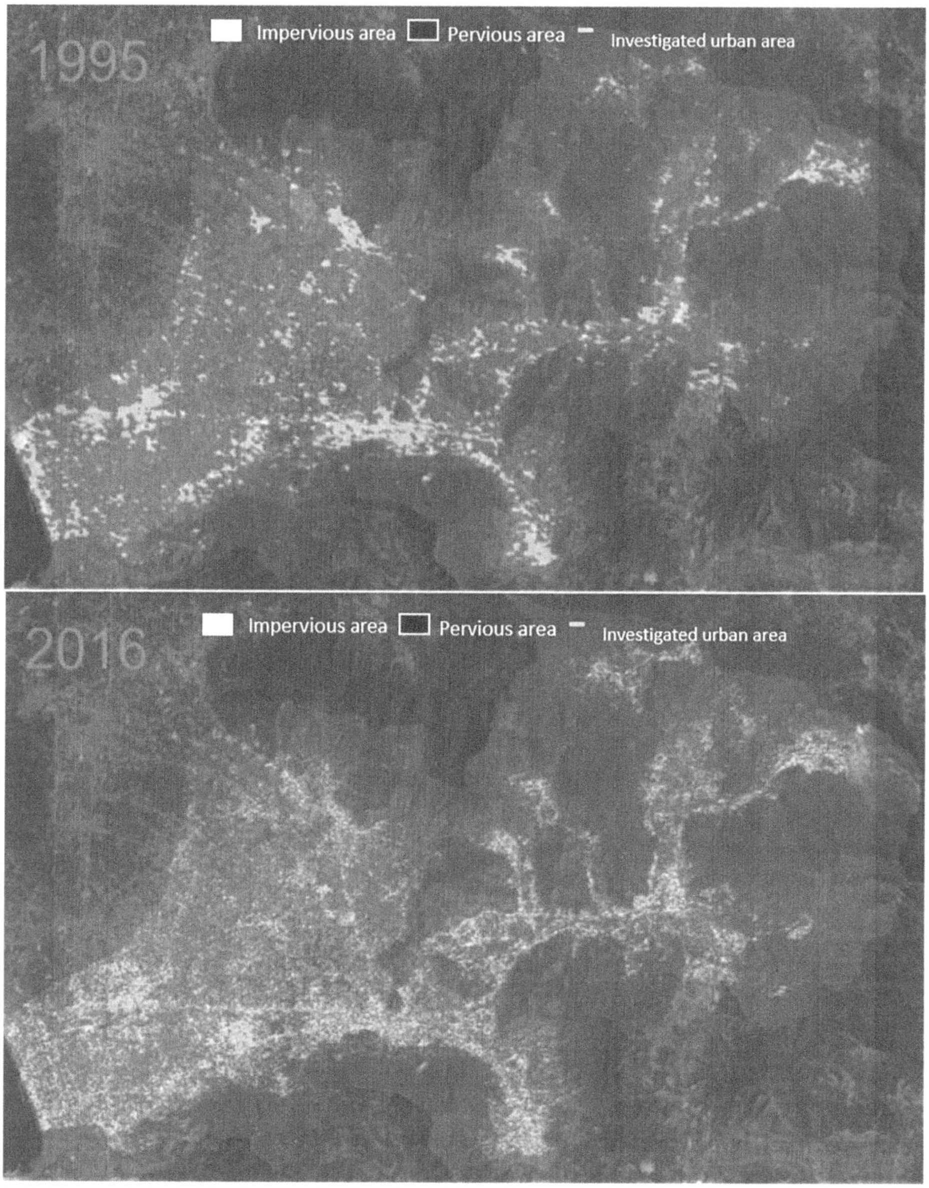

Figure 7 | The cluster-based threshold images of the Sarno river basin: (a) year 1995 and (b) year 2016.

Table 9 | Percentage of built-up area for each sub-basin in 1995 and 2016

Basin/sub-basin Name	Total area Pixel	1995		2016	
		Impervious area (Pixel)	Built-up area (%)	Impervious area (Pixel)	Built-up area (%)
Sarno	736,237	54,494	7.40	91,123	12.38
Calvagnola	71,087	7105	9.99	9203	12.95
Sarno Mouth	300,314	31,638	10.53	55,320	18.42
Solofrana	364,836	15,751	4.32	26,600	7.29

Calibration results

In order to return accurate simulations of the GR hydrological response at the basin scale, the LID and RUNOFF modules of the SWMM were calibrated.

For what concerns the event scale calibration of the RUNOFF module, hourly rainfall and streamflow data occurring on 31.10.2012, 21.01.2014, 20.11.2013, and 13.09.2012 were used (see Table 6). The calibration parameters are the impervious depression storage coefficient d_p and channel Manning's roughness coefficient n and their input values were iteratively changed in order to reach the maximum value of the NSE index for each event. The optimal values of the calibration parameters and the corresponding NSE are shown in Table 10.

The NSE is larger overall than 0.7 which means a satisfactory model's ability to simulate observed flows. In detail, the 20.11.2013 event, which is the one with the lowest return periods, never exceeded 10 years and with some of the highest rainfall intensities (about 68 mm/h), produced the smallest relative error in terms of total flow volume (NSE = 0.84). On the other side, the event occurring on 31.10.2012 with return periods up to 100 years and rainfall intensities among the lowest ones (4.8 mm/h) (Table 6), returned the highest error (NSE = 0.71). The Manning's roughness coefficient assumes almost the same value for all of the simulated events. The impervious depression storage coefficient is higher during the events of 2012 reaching the upper constraint of 1.27 mm, and lower during the remaining two events with values of 0.15 and 0 mm, respectively, for the event that occurred in 2013 and 2014. Graphically, the fitting between observed and modeled flow rates appears in Figure 8. It confirms that the model is able to capture the shape, total volume, and peak flow of the measured outlet hydrograph.

For the calibration of the LID module, in general, the model appears to be suitable to reproduce the experimental GR behavior. The values of NSE for each modeled rainfall/runoff event are shown in Figure 9(a), the average NSE ($NSE_{average}$) is greater than 0.5 for both the types of events which confirms an acceptable model performance. The model seems to better perform with stratiform events for which the average NSE reaches a value of 0.81 while the accuracy is lower for the convective events with an $NSE_{average}$ of 0.74.

With the aim of interpreting the findings related to the GR effectiveness in mitigating the runoff at the basin scale, it is useful to observe that the retention capacities of the two groups of events substantially differ in quantitative terms: the convective events present an average RC of about 60% which is higher than that related to the stratiform events which approaches 45% (Figure 9(c) and Table 7). With the average rainfall intensities being similar for the two groups, the lower retention capacity of the stratiform group is probably due to the fact that the average duration of the rainfall events belonging to this category (504 min) is higher than in the case of a convective group (297 min), confirming the importance of more rainfall properties beyond the relevant severity (Longobardi *et al.* 2019; Mobilia & Longobardi 2020).

For illustrative purposes, two modeled/observed rainfall–runoff events are shown in Figure 10 in terms of hourly depths, the one corresponding to a convective precipitation input (10.01.2018) and the other to a stratiform rainfall event (20.02.2018).

Besides the NSE index, which describes the model goodness-of-fit at the event scale, it is possible to observe how well the simulations match the measured data for both types of events. Indeed, the model, besides the runoff volume, is able to well reproduce the temporal pattern of the events too.

The calibration process returns values of Ψ ranging from 3 to 300 mm (the last is an outlier in the series of estimated parameters) for convective events and between 6 and 100 mm for the stratiform ones with average values, respectively, of 49 and 39.8 mm. The boxplots in Figure 9(b) show that the empirical distribution of Ψ values is actually wider for convective than for the stratiform events which means higher uncertainty in the selection of the most appropriate value of the calibration parameter for the first type of event.

Table 10 | Values of the calibrated parameters

Type of event	Date	*n* (–)	dp (mm)	NSE (–)
Stratiform	31.10.2012	0.0155	1.27	0.71
	21.01.2014	0.0155	0	0.72
Convective	13.09.2012	0.027	1.27	0.73
	20.11.2013	0.0155	0.15	0.84

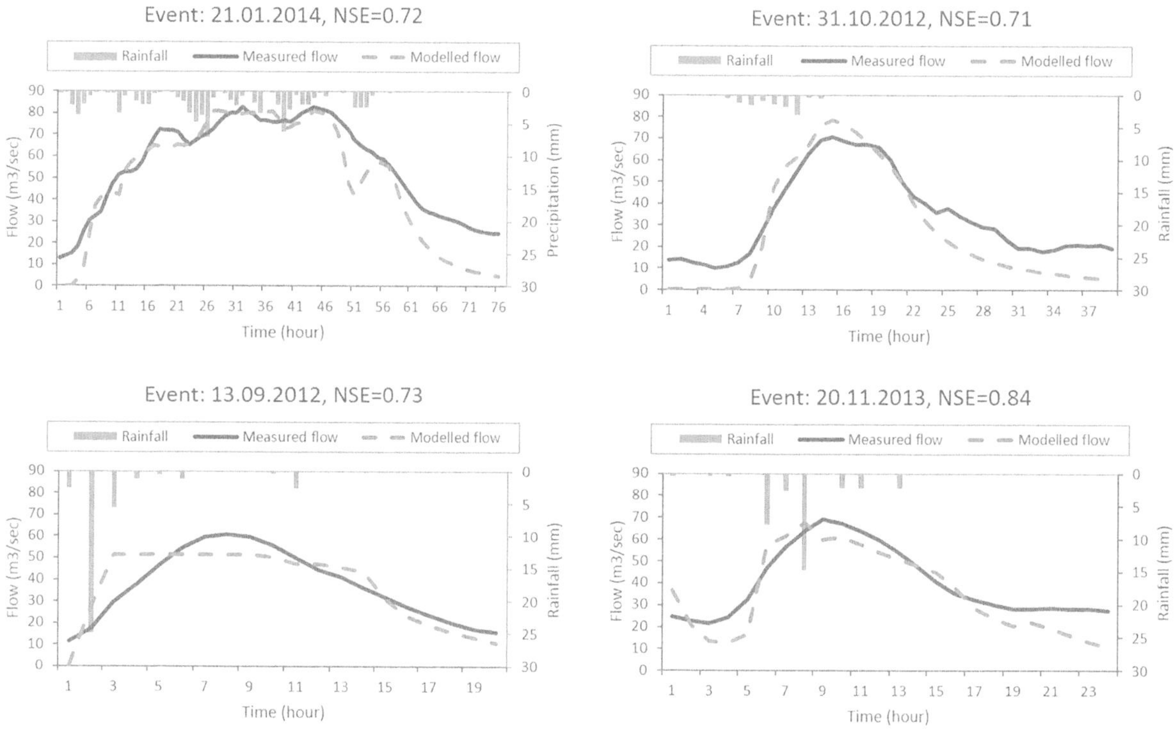

Figure 8 | Graphical fitting between observed and modeled flow rates.

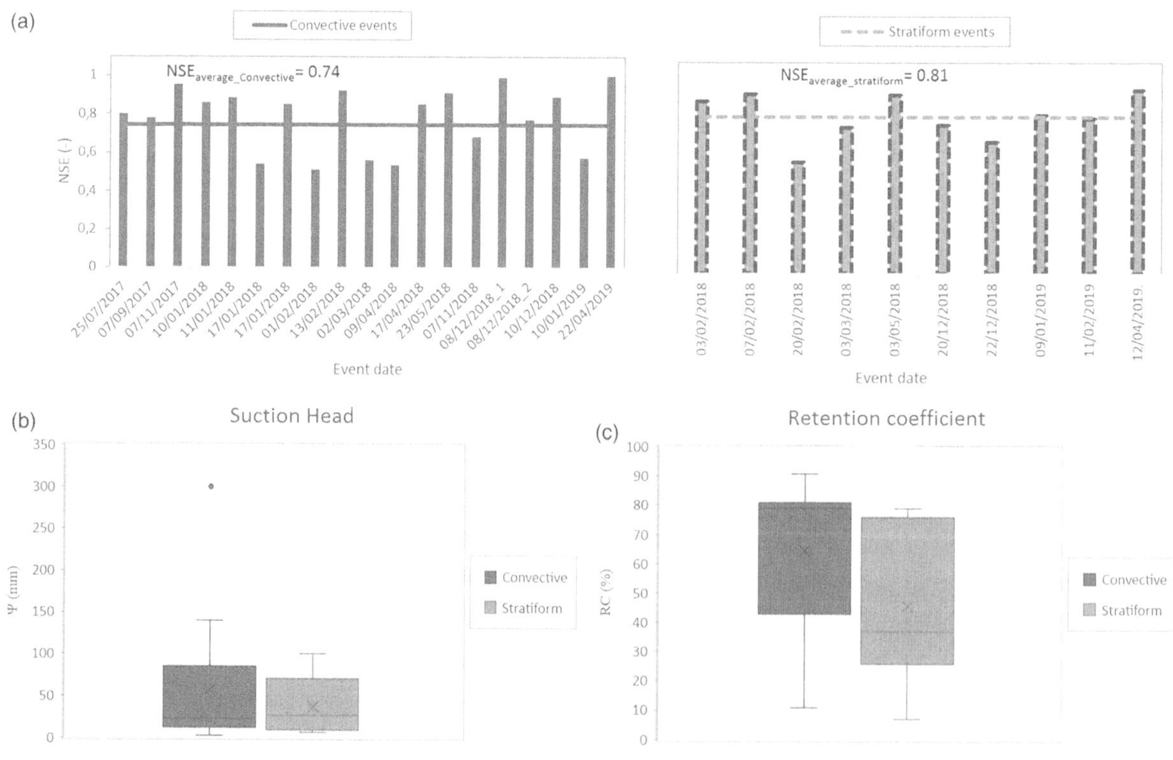

Figure 9 | Model performances for convective and stratiform events. (a) Values of NSE for each modeled rainfall/runoff event; (b) empirical distribution of Ψ values; (c) distribution of the values of the retention coefficient RC.

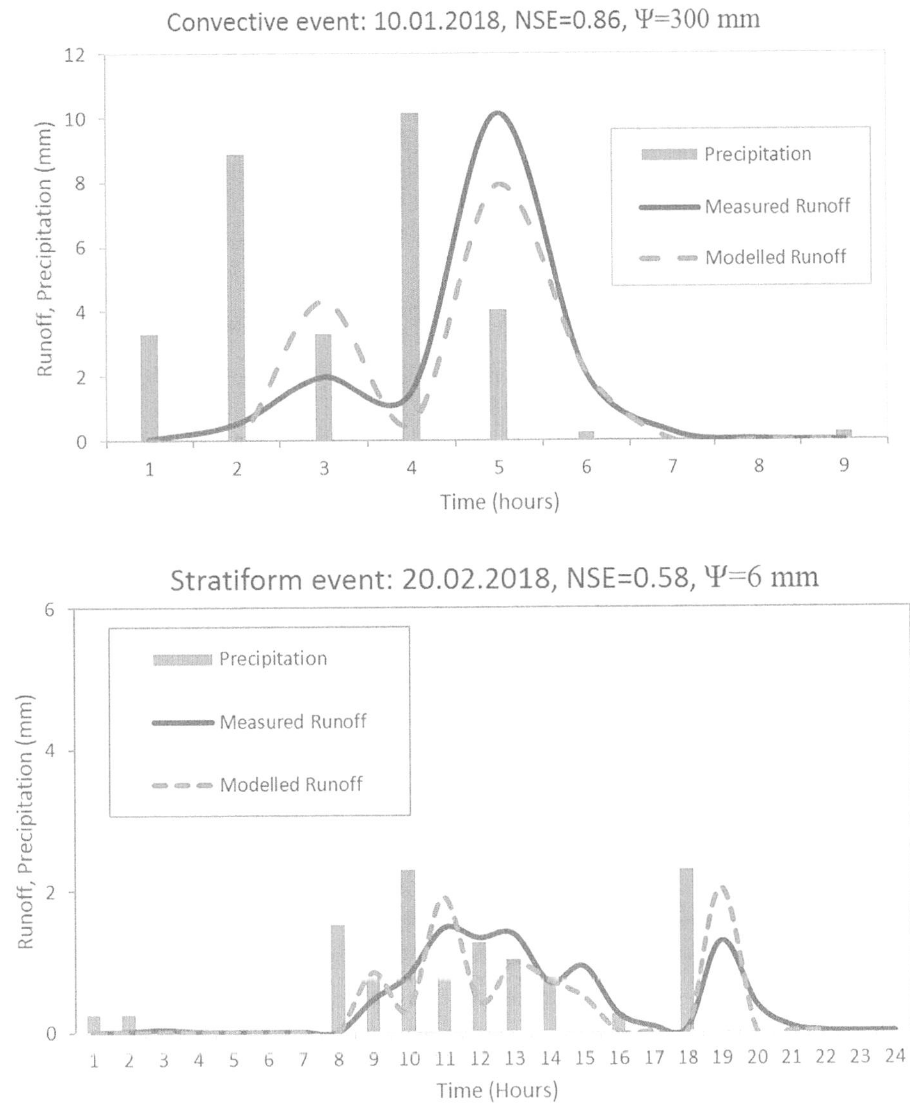

Figure 10 | Example of modeled stratiform and convective rainfall/runoff events.

Indeed, the value of the suction head for stratiform and convective events is never higher than 100 and 140 mm, respectively, with an outlier represented by the event with $\Psi = 300$ mm. These values are in line with the maximum value suggested by Rossman & Huber (2016) for the LID module which was set at 4 inches. The large variability of the values assumed by Ψ can be explained by the choice of using only this single variable as the calibration parameter. If from one side, the system has been forced to be governed by only one parameter, from the other side, this choice avoids an overparameterization and reduces the complexity of the retention process schematization that would otherwise have to take into account multiple parameters.

Modeling and comparing different land use scenarios in the SWMM

The results of SAR images elaboration suggest that the increase in the number of MDHEs within the studied basin is potentially related to the rapid expansion of the built-up area, therefore the GR retrofitting of existing roofs, with the consequent reduction of impervious surfaces and runoff production, could represent a valid solution for the flood risk management in the investigated catchment. In light of this, the hydrological behavior of large-scale GRs retrofitting scenarios were furthermore compared to the pre-development scenarios of 1995 and post-development scenarios of 2016. The specific modeling was considered for the four events used for the calibration of the RUNOFF module.

The comparison allows us, on one hand, to investigate how the variation of impervious areas between 1995 and 2016 affects the runoff production and, on the other hand, if the implementation of GRs at a wide scale could effectively help to restore the hydrological behavior of the system in the pre-development scenario. For the greening scenarios, a reduction of the impervious area taken by traditional roofs in 2016 is planned. In detail, it is assumed to replace 5, 30, and 100% of the traditional rooftops with GRs. Since the SAR technology allows us to identify the total paved areas without distinguishing among concrete roofs, walkways, sidewalks, and streets, the percentage of surface hosting traditional covers, suggested by SNPA (2019), was used in this study. According to this source, in Southern Italy, building area corresponds to 35% of the total impervious surface after 2005 and to 26% in the previous years.

In Figure 11, details about the area occupied by the building over the total area of the basin and the area considered for GR retrofit for each scenario are sketched.

At a first visual inspection of the modeled runoff hydrographs (Figure 12), the runoff volume and the peak flow appear higher for the post-development scenario (2016) than for the pre-development (1995) scenario and for the GR retrofit scenarios, regardless of the type of event.

For comparison in quantitative terms, the evaluation of the four assessment indices (Equations (15)–(18)) was carried out (Table 11). The reduction of the runoff volume detected after the GR conversion of the existing traditional roofs ranges from a maximum value of 49.7%, for the scenario with a conversion of 100%, to a minimum value of 16.8% for the scenario with a retrofit of 5%. In general, ΔR_V is higher for the scenarios with a higher green requalification than for the ones with a lower implementation of green technologies; indeed, the average runoff reduction reaches average values across all rainfall events of about 41, 27, and 22%, respectively, for 100, 30, and 5% GR retrofit scenarios. The 1995-scenario, where the percentage of imperviousness is nearly half the 2016-scenario (10.53 against 18.42%), is the best performing scenario with an average runoff mitigation of about 47%. The 100%-scenario can only partially achieve this behavior. This finding shows the negative impact of sealing within the basin in terms of runoff production, which cannot even be compensated by a complete green conversion of the existing traditional roofs. With reference to the shape of the rainfall profile of the events, the largest reduction of the runoff volume occurs for the stratiform events characterized by lower average intensity (about 19 mm/h) and longer duration (between 16 and 58 h). For these events (21.01.2014 and 31.10.2012), the average ΔR_V across all the scenarios is, also in this case, around 40% against the 29% of the convective events (20.11.2013 and 13.09.2012) characterized by higher average intensities (respectively, of 34 and 72 mm/h) and shorter duration (between 12 and 14 h). Similar considerations can be extended to the reduction of the peak flow. ΔP_F ranges between 40.8 and 0%. In detail, for the stratiform events, its average

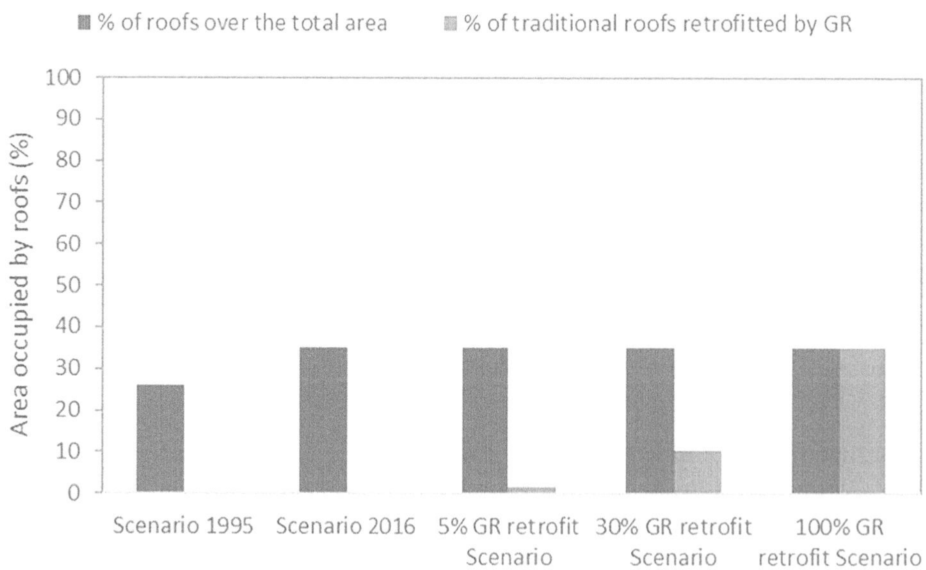

Figure 11 | Percentage of the rooftop and GR retrofitted for each scenario.

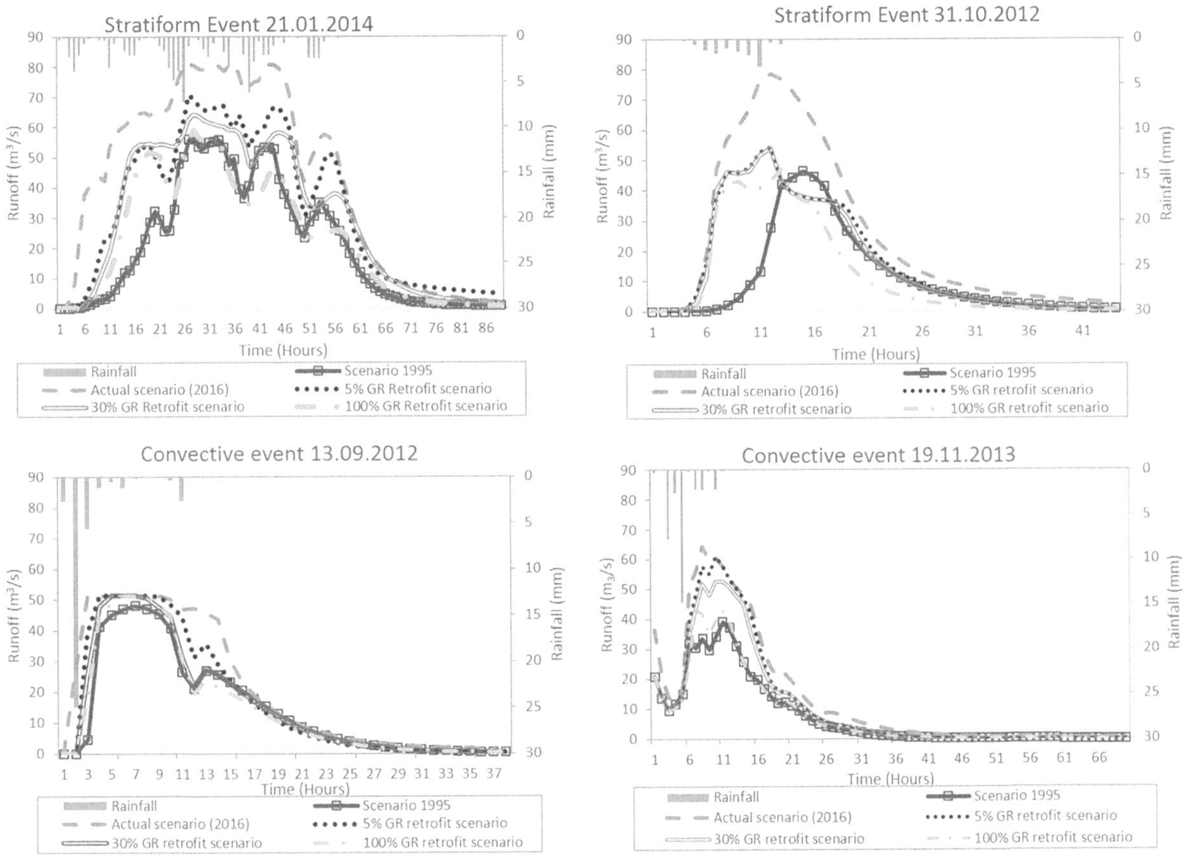

Figure 12 | Runoff hydrographs for the actual and GR retrofit scenarios.

Table 11 | Values of the indices of performance

Event type	Date	GR scenario	ΔR_V (%)	ΔP_F (%)	ΔF_S (%)	ΔD_T (h)
Stratiform	21.01.2014	100%	45.4	27.3	50.0	1.0
		30%	28.5	20.7	50.0	1.0
		5%	23.1	12.7	25.0	0.0
		1995	51.1	30.5	50.0	1.0
	31.10.2012	100%	49.7	39.9	66.7	1.0
		30%	35.0	31.2	66.7	0.0
		5%	31.9	30.6	66.7	0.0
		1995	55.5	40.8	100.0	3.0
Convective	13.09.2012	100%	27.7	0.0	75.0	3.0
		30%	22.6	0.0	75.0	1.0
		5%	16.8	0.0	75.0	1.0
		1995	37.1	6.4	100.0	3.0
	19.11.2013	100%	41.9	33.0	100.0	3.0
		30%	23.9	18.0	100.0	2.0
		5%	17.5	5.2	100.0	2.0
		1995	44.8	39.3	100.0	3.0

value is around 29% while for the convective ones, it approaches 13%. It is confirmed that the reduction is higher for the 1995-scenario and 100% GR retrofit scenario with average values, respectively, of 29 and 25% and values lower than 15% on average in the other cases. The reduction of the flooded sections is significant both for stratiform and convective events and reaches maximum and minimum values of 100 and of 25%, respectively. Here too, on average, the best performing scenarios are the ones corresponding to the 1995 scenario and to the 100% GR conversion scenario. Finally, the increase in delay time is more evident for the events occurring on 13.09.2012 and 20.11.2013. For these events, the delay time ranges between 1 and 3 h. For the other two events, the time of peak discharge is delayed at most for 1 h.

A due consideration is that the suitability for GR retrofit of traditional roofs depends on a number of criteria, beyond the potential retrofitting surface, including the slope and the orientation of the roof, the number of stories of the building and the number of site boundaries (Wilkinson & Reed 2009). The percentage of buildings that meet all the required attributes and are physically suitable for GR adaptation is never higher than 20% (Wilkinson & Reed 2009; Karteris *et al.* 2016; Mobilia & Longobardi 2021). In light of this, the really feasible scenarios are the ones corresponding to 5 and 30% of GR conversion. The 100% GR conversion scenario was analyzed in order to perform an exhaustive analysis, also taking into account ideal conditions.

Based on these considerations, the condition occurring in 1995 can never be fully restored. Indeed, this scenario is partially equaled only by a complete (100%) green conversion of the existing traditional roofs but this condition is clearly not feasible. This finding confirms the negative effect of imperviousness on the stormwater production. An attempt to restore the hydrological behavior of the system before the massive increase of paved areas in 2016 could be made using GR infrastructures in combination with other types of LID practices (bioretention cells, permeable pavements).

The results of this study are consistent with previously published research which confirms that the diffuse installation of GRs allows a reduction of both runoff volume and flow peak, which are variables with the rainfall intensity and the duration of the event. In detail, the performances are better for events with a smaller magnitude than for the ones with larger magnitude. The scientific literature also confirms that the reduction of runoff volume is higher for the 100% GR conversion scenarios and smaller for a lower percentage of green requalification. Consistent is also the data around the volume reduction which ranges from 35 to 50% for conversion of 100% and goes from 30 to 51% for the peak flow reduction (Palla *et al.* 2008; Qin *et al.* 2013; Masseroni & Cislaghi 2016; Ercolani *et al.* 2018).

CONCLUSION

With the support of an archive collection of damaging events that occurred within the study area between 1951 and 2016, the present study has shown that in the Sarno basin, the MDHE occurrences have been characterized by an overall upward trend. Contextually, an increase in pervious surfaces has been detected by the use and analysis of SAR images by the Cosmo-SkyMed constellation, comparing SAR images referring to 1995 (pre-development scenario) and 2016 (post-development scenario). These results have brought to light a potential correlation between the two phenomena (MDHE occurrence and soil sealing), hence the possibility to act in a sustainable landscape planning context, with implementation of GRs at the basin scale to mitigate the effects related to excessive runoff production. Three GRs retrofitting scenarios (corresponding to retrofitting 5, 30, 100% of roofs areal extension as detected by SAR images elaboration) have been analyzed using the SWMM and compared to the pre- and post-development scenarios using performance indices. The findings suggest that soil sealing has a negative impact on catchment hydrology, indeed, in the post-development scenario, the catchment is expected to see increased surface runoff. On the other side, simulation results point out that the extensive use of GRs, even on only 5% of all roofs of the catchment, allows us to mitigate the negative effects of the stormwater excess. In this case, the reduction of runoff is recorded at most as 32%, whereas 31% reduction of the peak flow can be reached. The hydrological performances of the 100% greening scenario, with a maximum percentage of runoff and peak flow reduction, respectively, of 50 and 40%, appears excellent but, however good, it is not able to totally restore the hydrological pre-development conditions of the watershed. In addition, it should be considered that the spatial potential for GR retrofitting can hardly reach a percentage of 100% due to the building attributes required for the conversion.

The present work confirms that GRs are an effective stormwater mitigation strategy for the selected basin and suggests that a massive green restoration planning for the existing roofs of the area can improve the environmental benefits from a hydrologic point of view. The possibility to combine GR technologies with other SuDS practices within the study basin, with the aim to mimic a pre-development hydrological condition, can be explored in future works. The present investigation can be considered paradigmatic for any degraded and vulnerable urban context interested in a high frequency of MDHEs and where an extreme land use change occurred rapidly with visible results in just a few decades.

ACKNOWLEDGEMENTS

The authors would like to thank anonymous reviewers for their helpful comments and suggestions which resulted in an improved manuscript version. This study was carried out using CSK Products ASI (Italian Space Agency), delivered under an ASI licence to be used in the framework of 'COSMO-SkyMed Open Call for Science Initiative' (Proposal ID 452 – 'Use of high-resolution SAR images for monitoring urban dynamics'-Principal investigator: Antonia Longobardi).

DATA AVAILABILITY STATEMENT

All relevant data are included in the paper or its Supplementary Information.

CONFLICT OF INTEREST

The authors declare there is no conflict.

REFERENCES

Adamo, P., Iavazzo, P., Albanese, S., Agrelli, D., De Vivo, B. & Lima, A. 2014 Bioavailability and soil-to-plant transfer factors as indicators of potentially toxic element contamination in agricultural soils. *Sci. Total Environ.* **500**, 11–22.

Apollonio, C., Balacco, G., Novelli, A., Tarantino, E. & Piccinni, A. F. 2016 Land use change impact on flooding areas: the case study of Cervaro Basin (Italy). *Sustainability* **8** (10), 996.

Baek, S. S., Ligaray, M., Pyo, J., Park, J. P., Kang, J. H., Pachepsky, Y. & Cho, K. H. 2020 A novel water quality module of the SWMM model for assessing low impact development (LID) in urban watersheds. *J. Hydrol.* **586**, 124886.

Baldantoni, D., Bellino, A., Lofrano, G., Libralato, G., Pucci, L. & Carotenuto, M. 2018 Biomonitoring of nutrient and toxic element concentrations in the Sarno River through aquatic plants. *Ecotoxicol. Environ. Saf.* **148**, 520–527.

Barco, J., Wong, K. M. & Stenstrom, M. K. 2008 Automatic calibration of the US EPA SWMM model for a large urban catchment. *J. Hydraul. Eng.* **134** (4), 466–474.

Behrouz, M. S., Zhu, Z., Matott, L. S. & Rabideau, A. J. 2020 A new tool for automatic calibration of the Storm Water Management Model (SWMM). *J. Hydrol.* **581**, 124436.

Bentivenga, M., Giano, S. I. & Piccarreta, M. 2020 Recent increase of flood frequency in the Ionian Belt of Basilicata Region, Southern Italy: human or climatic changes? *Water* **12** (7), 2062.

Blaney, H. F. & Criddle, W. D. 1950 Determining water requirements in irrigated areas from climatological and irrigation data. *Soil Conserv. Serv.* **96**, 32–33.

Burszta-Adamiak, E. A. & Mrowiec, M. 2013 Modelling of green roofs' hydrologic performance using EPA's SWMM. *Water Sci. Technol.* **68** (1), 36–42.

Califano, F., Mobilia, M. & Longobardi, A. 2015 Heavy rainfall temporal characterization in the Peri-Urban Solofrana River Basin, Southern Italy. *Procedia Eng.* **119**, 1129–1138.

Cipolla, S. S., Maglionico, M. & Stojkov, I. 2016 A long-term hydrological modelling of an extensive green roof by means of SWMM. *Ecol. Eng.* **95**, 876–887.

De Pippo, T., Donadio, C., Guida, M. & Petrosino, C. 2006 The Case of Sarno River (Southern Italy). effects of geomorphology on the environmental impacts. *Environ. Sci. Pollut. Res.* **13** (3), 184–191.

Di Vito, M. A., Calcaterra, D., Petrosino, P., Zanchetta, G., de Vita, S., Marotta, E., Cesarano, M., De Simone, A., Sansivero, F. & Rucco, I. 2019 Landslides, volcanism and volcano-tectonics: the fragility of the Neapolitan territory. *Geolog. F. Trips Maps* **11** (1), 53.

Ercolani, G., Chiaradia, E. A., Gandolfi, C., Castelli, F. & Masseroni, D. 2018 Evaluating performances of green roofs for stormwater runoff mitigation in a high flood risk urban catchment. *J. Hydrol.* **566**, 830–845.

Esposito, G., Matano, F. & Scepi, G. 2018 Analysis of increasing flash flood frequency in the densely urbanized coastline of the Campi Flegrei volcanic area, Italy. *Front. Earth Sci.* **6**, 63.

Franceschetti, G. & Lanari, R. 1999 *Synthetic Aperture Radar Processing*. CRC Press, Boca Raton.

Habonimana, H. V. 2014 *Integrated Flood Modeling in Lubigi Catchment Kampala*. Master's Thesis, University of Twente, Enschede, Netherlands.

Huang, H., Chen, X., Zhu, Z., Xie, Y., Liu, L., Wang, X. & Liu, K. 2018 The changing pattern of urban flooding in Guangzhou, China. *Sci. Total Environ.* **622**, 394–401.

Karteris, M., Theodoridou, I., Mallinis, G., Tsiros, E. & Karteris, A. 2016 Towards a green sustainable strategy for Mediterranean cities: assessing the benefits of large-scale green roofs implementation in Thessaloniki, Northern Greece, using environmental modelling, GIS and very high spatial resolution remote sensing data. *Renewable Sustainable Energy Rev.* **58**, 510–525.

Konrad, C. P. 2003 *Effects of Urban Development on Floods*. United States Geological Survey Fact Sheet, Tacoma, Washington.

Liang, X., Lin, Y. & Zhang, H. 2022 Mapping urban impervious surface with an unsupervised approach using interferometric coherence of SAR images. *IEEE J. Sel. Top. Appl. Earth Obs. Remote Sens.* **15**, 2734–2744.

Limos, A. G., Mallari, K. J. B., Baek, J., Kim, H., Hong, S. & Yoon, J. 2018 Assessing the significance of evapotranspiration in green roof modeling by SWMM. *J. Hydroinf.* **20** (3), 588–596.

Lisenbee, W. A., Hathaway, J. M. & Winston, R. J. 2022 Modeling bioretention hydrology: quantifying the performance of DRAINMOD-Urban and the SWMM LID module. *J. Hydrol.* **612**, 128179.

Longobardi, A., Diodato, N. & Mobilia, M. 2016 Historical storminess and hydro-geological hazard temporal evolution in the solofrana river basin – Southern Italy. *Water* **8** (9), 398.

Longobardi, A., D'Ambrosio, R. & Mobilia, M. 2019 Predicting stormwater retention capacity of green roofs: an experimental study of the roles of climate, substrate soil moisture, and drainage layer properties. *Sustainability* **11** (24), 6956.

Masseroni, D. & Cislaghi, A. 2016 Green roof benefits for reducing flood risk at the catchment scale. *Environ. Earth Sci.* **75** (7), 579.

Mobilia, M. 2018 *Sustainable Managment of Stormwater in A Changing Environment Under Mediterranean Climate Conditions*. Master's Thesis, University of Salerno, Salerno, Italy.

Mobilia, M. & Longobardi, A. 2020 Impact of rainfall properties on the performance of hydrological models for green roofs simulation. *Water Sci. Technol.* **81** (7), 1375–1387.

Mobilia, M. & Longobardi, A. 2021 Using SAR satellite imagery for potential green roof retrofitting for flood mitigation in urban environment. In: *Proc. of GISTAM*, pp. 59–66.

Morgan, S., Celik, S. & Retzlaff, W. 2013 Green roof storm-water runoff quantity and quality. *J. Environ. Eng.* **139** (4), 471–478.

Nash, J. E. & Sutcliffe, J. V. 1971 River flow forecasting through conceptual models. Part 1: a discussion of principles. *J. Hydrol.* **10** (3), 282–290.

Otsu, N. 1979 A threshold selection method from gray-level histograms. *IEEE Trans. Syst. man Cybern.* **9** (1), 62–66.

Paithankar, D. N. & Taji, S. G. 2020 Investigating the hydrological performance of green roofs using storm water management model. *Mater. Today: Proc.* **32**, 943–950.

Palla, A., Berretta, C., Lanza, L. G. & La Barbera, P. 2008 Modelling storm water control operated by green roofs at the urban catchment scale. In: *Proc. of 11th International Conference on Urban Drainage*, Edinburgh, Scotland, UK.

Pappalardo, V., La Rosa, D., Campisano, A. & La Greca, P. 2017 The potential of green infrastructure application in urban runoff control for land use planning: a preliminary evaluation from a southern Italy case study. *Ecosyst. Serv.* **26**, 345–354.

Paprotny, D., Sebastian, A., Morales-Nápoles, O. & Jonkman, S. N. 2018 Trends in flood losses in Europe over the past 150 years. *Nat. Commun.* **9** (1), 1–12.

Petrucci, O., Salvati, P., Aceto, L., Bianchi, C., Pasqua, A. A., Rossi, M. & Guzzetti, F. 2018 The vulnerability of people to damaging hydrogeological events in the Calabria Region (Southern Italy). *Int. J. Environ. Res. Public Health* **15** (1), 48.

Pitlick, J. 1994 Relation between peak flows, precipitation, and physiography for five mountainous regions in the western USA. *J. Hydrol.* **158** (3–4), 219–240.

Qin, H. P., Li, Z. X. & Fu, G. 2013 The effects of low impact development on urban flooding under different rainfall characteristics. *J. Environ. Manage.* **129**, 577–585.

Rawls, W. J., Asce, M., Brakensiek, D. L. & Miller, N. 1983 Green-ampt infiltration parameters from soils data. *J. Hydraul. Eng.* **109** (1), 1316.

Recanatesi, F. & Petroselli, A. 2020 Land cover change and flood risk in a peri-urban environment of the metropolitan area of Rome (Italy). *Water Resour. Manage.* **34** (14), 4399–4413.

Rossman, L. A. 2010 Modeling low impact development alternatives with SWMM. *J. Water Manage. Model.* **11**, 167–182.

Rossman, L. A. & Huber, W. C. 2016 *Storm Water Management Model Reference Manual Volume III – Water Quality*. US EPA National Risk Management Research Laboratory, Cincinnati, OH, USA.

Sadeghi, F., Rubinato, M., Goerke, M. & Hart, J. 2022 Assessing the performance of LISFLOOD-FP and SWMM for a small watershed with scarce data availability. *Water* **14** (5), 748.

Sartor, J., Mobilia, M. & Longobardi, A. 2018 Results and findings from 15 years of sustainable urban storm water management. *Int. J. Saf. Secur. Eng.* **8** (4), 505–514.

SNPA 2019 *Consumo di suolo, dinamiche territoriali e servizi ecosistemici, Report n. 8/2019.*

Sofia, G., Roder, G., Dalla Fontana, G. & Tarolli, P. 2017 Flood dynamics in urbanised landscapes: 100 years of climate and humans' interaction. *Sci. Rep.* **7** (1), 1–12.

Su, S., Tian, J., Dong, X., Tian, Q., Wang, N. & Xi, Y. 2022 An impervious surface spectral index on multispectral imagery using visible and near-infrared bands. *Remote Sens.* **14** (14), 3391.

Terranova, O. G. & Iaquinta, P. 2011 Temporal properties of rainfall events in Calabria (southern Italy). *Nat. Hazards Earth Syst. Sci.* **11**, 751–757.

United-Nations 2014 *World Urbanization Prospects, Technical Report*. Department of Economic and Social Affairs, New York, USA.

Versini, P. A., Jouve, P., Ramier, D., Berthier, E. & De Gouvello, B. 2016 Use of green roofs to solve storm water issues at the basin scale–Study in the Hauts-de-Seine County (France). *Urban Water J.* **13** (4), 372–381.

Voyde, E., Fassman, E. & Simcock, R. 2010 Hydrology of an extensive living roof under sub-tropical climate conditions in Auckland, New Zealand. *J. Hydrol.* **394** (3–4), 384–395.

Wang, M., Zhang, D., Cheng, Y. & Tan, S. K. 2019 Assessing performance of porous pavements and bioretention cells for stormwater management in response to probable climatic changes. *J. Environ. Manage.* **243**, 157–167.

Wilkinson, S. J. & Reed, R. 2009 Green roof retrofit potential in the central business district. *Prop. Manage.* **27** (5), 284–301.

Wu, W., Shao, Z., Teng, J., Huang, X., Zhao, X. & Guo, S. 2022 Urban impervious surface extraction using seasonal time series SAR images. *IEEE J. Sel. Top. Appl. Earth Obs. Remote Sens.* **15**, 6033–6045.

Zhou, D., Liu, Y., Hu, S., Hu, D., Neto, S. & Zhang, Y. 2019 Assessing the hydrological behaviour of large-scale potential green roofs retrofitting scenarios in Beijing. *Urban For. Urban Greening* **40**, 105–113.

First received 18 March 2022; accepted in revised form 25 January 2023. Available online 14 February 2023

doi: 10.2166/nh.2023.081

Application of Benfratello's method to estimate the spatio-temporal variability of the irrigation deficit in a Mediterranean semiarid climate

Marco Peli *, Cesare Rapuzzi, Stefano Barontini and Roberto Ranzi

Dipartimento di Ingegneria Civile, Architettura, Territorio, Ambiente e di Matematica (DICATAM), Università degli Studi di Brescia, via Branze 43, Brescia 25123, Italy

*Corresponding author. E-mail: marco.peli@unibs.it

MP, 0000-0003-3481-2296

ABSTRACT

This work presents a novel, spatially distributed, GIS-based application of Benfratello's conceptual method (developed in the 1960s) to estimate the climatic water deficit and the irrigation deficit at the field and basin scales. Explicit analytical relationships are obtained to define the deficit uncertainty on the basis of the interannual variability of temperature and precipitation. With this model, we aim at proposing a rather simple and effective tool to deal with the complicated issues of assessing the soil water balance, determining the irrigation deficit and managing the water resources in semiarid agricultural environments, in the context of climatic, land-use and anthropogenic changes. In order to test this new application, the model was applied to estimate the irrigation deficit of the Bonifica della Capitanata consortium in the Apulia region, one of the most important agricultural districts in Southern Italy and in the whole Mediterranean area, in four different historical land-use scenarios. The first results of the application seem encouraging, as by using a limited amount of parameters we estimated an irrigation demand which is in agreement with the irrigation volumes supplied by the consortium. The different land-use cases are discussed in the light of an application of the Budyko curve.

Key words: Capitanata agricultural district, climate change adaptation, climatic water deficit, crop water requirement, wadi hydrology, water balance

HIGHLIGHTS

- We present a GIS-based implementation of Benfratello's method to assess the climatic water deficit and the irrigation water deficit.
- Explicit relationships are proposed to initialize the method and to determine the variability of the annual deficit.
- Starting from the Budyko curve, we evaluate the degree of optimization of water resource consumption.
- We apply the method to the Capitanata agricultural district (Southern Italy).

1. INTRODUCTION

In an era of fast climatic evolution as it is the present one, where the hydrological cycle is expected to experience meaningful alterations due to the increased global average air temperature, agriculture, which is still the main driver of freshwater consumption worldwide (UN-Water 2022) and, at the same time, the main driving sector in poverty reduction and sustainable development (FAO 2021), may face meaningful water shortages. In fact, focusing on the areas surrounding the Mediterranean basin, which are mainly characterized by semiarid and arid conditions, there is high confidence that in future climatic scenarios, the warming will be greater than the world average, and that the increased atmospheric energy and the increased evaporative capacity of the atmosphere will emphasize the hydrological cycle by intensifying the wet season, by increasing the length and severity of droughts and by exposing the soil to more severe dry conditions (Arias *et al.* 2021). Climate scenarios used by Braca *et al.* (2019, 2021) to force a water balance model designed to describe the Italian territory at both national and regional levels depict a meaningful decrease in the groundwater recharge due both to a decrease in precipitation and an increase in air temperature. As an example, they report that the resulting expected decrease in groundwater recharge for the mid-21st century is estimated at about 10 and 17% for the whole Italian territory under scenarios RCP2.6 and RCP4.5, respectively. The expected decrease is greater in Southern regions, e.g. in the Campania region it is predicted at about 13 and

22% under the same scenarios. Analogous results were found by Brussolo *et al.* (2022) for the Piedmont Alpine area considering the present temperature-increasing trend, even if in absence of meaningful precipitation trends.

Irrigation practices often heavily impact both surface and groundwater resources. Therefore the assessment of the consequences on food production and crop yield, of a reduction of water availability, may benefit from simplified water balance models that provide information on the expected irrigation deficit in different climatic and land-use scenarios. With this work, we aim at providing a robust and easy-to-use tool that allows us to make such estimates at the local and district scale. The model is based on a theoretical improvement and on a Geographic Information System (GIS) application of Benfratello's water balance method (Benfratello 1961), originally developed for semiarid and arid agricultural districts in Sicily (Southern Italy), characterized by a Mediterranean climate.

We believe that due to its simplicity, to its elegance and to the small number of needed parameters, Benfratello's method might be regarded as an effective tool to assess the effects of climatic, land-use and anthropogenic change scenarios on the soil water balance and on the irrigation deficit of agricultural crops. In fact, despite the presence, today, of more complex and complete, distributed and semi-distributed hydrological models which well perform at the considered scales (see Xu & Singh 1998; Pereira *et al.* 2020, for a review on this topic), the presented model meets the key need of numerical approaches to find a compromise between the solidity and accuracy characteristics of the solver and the high computational cost associated with the vast spatial and temporal scales typical of environmental problems (Farthing & Ogden 2017). Moreover, as far as in the last decade online water management and irrigation advisory tools have become a widely used resource (Chen *et al.* 2010; Verma *et al.* 2012), the model could fit an application within a webGIS environment, such as a geospatial decision support system to be run in real-time.

From the second half of the 20th century, focusing in particular on the Italian hydrological community, the soil–water balance has been studied both by means of statistical approaches (Viparelli 1956a, 1956b; Quignones 1974; Mantica 1975; Viparelli & Versace 1979) and with conceptually based (Benfratello 1961; Melisenda 1964; Santoro 1991; Casadei *et al.* 1993; Bartolini *et al.* 2011; Braca *et al.* 2019) as well as physically based approaches (Santini 1979, 1992; D'Urso *et al.* 1999; Scalenghe & Ferraris 2009; Romano 2014). In this context, and with a particular focus on agricultural soil, a *classical formulation* (Santoro 1991) was proposed by Benfratello (1961). Benfratello's model is based on a dimensionless form, introduced by De Varennes e Mendonça (1958), of Thornthwaite and Mather's water balance model (Thornthwaite & Mather 1955a, 1955b, 1957). The original Thornthwaite and Mather's model estimates the actual evapotranspiration as a function of Thornthwaite's *potential evapotranspiration* (Thornthwaite 1948), as maximum allowable evapotranspiration – defined as a climatic property – and uses a soil desiccation law that is an exponential function of the potential soil–water loss. The model is still used both for climate characterization (Dourado-Neto *et al.* 2010; Braca *et al.* 2019) and for irrigation deficit assessment (da Silva Tavares *et al.* 2018).

The generalization proposed by Benfratello allows to modulate the soil desiccation law as a power law of the potential soil–water loss, that accounts for the different crop reactions to the water stresses. At its first appearance in 1961, the model was used to estimate the irrigation deficit in Palermo (as representative of a semiarid climate where the field capacity is refilled during the wet season) and in Trapani (as representative of a semiarid climate where the field capacity is not completely refilled during the wet season), both in Sicily. After that, Benfratello (1964) reported about one of the first automatized applications of the model performed by means of the electronic calculator of the Remington Rand center for electronic calculus of the University of Milan, in view of extending the water balance to the whole of Sicily. The method was then used by Melisenda (1965) to explore the variability of the evapotranspiration rate in some representative basins of Sicilian climatology. At the same time Melisenda (1964) proposed a synthetic formulation to immediately calculate the annual irrigation deficit – for the specific case in which Benfratello's method is reduced to Thornthwaite and Mather's one –, based on the definition of a climatic index given by the ratio between the wet-season potential soil–water gain and the dry-season potential soil–water loss. This new formulation was applied to the cases of Catania and Agrigento, as a matching pair of the cases of Palermo and Trapani presented in the 1961 paper.

This methodology was then used by Melisenda (1967b) to define regional climatic maps of Sicily, and it was later made suitable for assessing the monthly irrigation deficit by Santoro (1970) who restored the generality of Benfratello's approach. It is noteworthy that until that time the method was typically applied in view of climatic analyses and by performing the calculations at a monthly scale. In this sense, it is coherent with the purposes of the use of Thornthwaite's method to assess the maximum evapotranspiration as a climatic factor that is not directly related to the crop. In the same direction, in another work by Benfratello *et al.* (1979), the method was used as an indirect procedure to estimate the groundwater recharge for the Sicilian aquifers. On the other hand, Melisenda (1967a) applied the method, still in Thornthwaite and Mather's simplified form, to design the irrigation schedule at a weekly scale and applied it to the case of Lentini agricultural lands. Later, the model was applied to assess the water balance also in other climatic conditions, both in Italy (Casadei *et al.* 1993) and abroad (Dumitru *et al.* 2009).

In this work, we come back to the original Benfratello's model and we propose a theoretical development that allows (*a*) to explicitly initialize the model for two cases of Benfratello's desiccation power law and (*b*) to explicitly determine the deficit uncertainty on the basis of the knowledge of the interannual variability of temperature and precipitation.

We propose a GIS-based implementation of the model (with QGIS 3.10 and custom Python scripts) and we present an application to the study of the case of the Capitanata district (Apulia, Southern Italy), which is one of the main agricultural districts in the Mediterranean basin (see, e.g. Lamaddalena *et al.* 2004, 2008; Portoghese *et al.* 2005; Guyennon *et al.* 2016, 2017; de Vito *et al.* 2019; Dragonetti *et al.* 2020). The model was applied to determine both the climatic water deficit, i.e. by estimating the maximum required evapotranspiration as Thornthwaite's potential evapotranspiration, and the agricultural water deficit, i.e. by estimating the maximum required evapotranspiration by means of Hargreaves' method modulated with the FAO procedure (Allen *et al.* 1998), in four different land-use conditions provided by the Apulia administrative authority.

In the following, after recalling the original structure of Benfratello's water balance method and presenting our theoretical development (Section 2), we introduce the Capitanata case study (Section 3) and the GIS-based application of the model (Section 4). In the Discussion (Section 5), we present the potentialities of this application also on the basis of the classical Budyko (1974) curve and of an irrigation-deficit curve obtained by transforming the Budyko one.

2. BENFRATELLO'S ESTIMATE OF IRRIGATION DEFICIT

2.1. Original estimate of the expected deficit

2.1.1. Problem statement

In the following, we will present Benfratello's method as it was first introduced in 1961 (see Figure 1 for a sketch of the control volume). According to the original hypotheses, there is no water capillary rise from the groundwater table to the soil, and runoff and percolation take place only after the soil water reached the field capacity (U). The simplified soil water balance of the ith month and of the jth year is therefore given by:

$$\Delta A_{i,j} = P_{i,j} - ET_{i,j} - S_{i,j} \tag{1}$$

where $A_{i,j} \leq U[L]$ is the equivalent depth of the available stored water at the end of the month, $\Delta A_{i,j} = A_{i,j} - A_{i-1,j}$ is the variation of the available water, $P_{i,j}[L]$ is the monthly precipitation, $ET_{i,j}[L]$ is the monthly (actual) evapotranspiration and $S_{i,j}[L]$ is the cumulated water exceedance, i.e. the summation of runoff and percolation. The usage of the letters D and S, standing for the Latin words *deficit* (*it lacks*) and *superavit* (*it exceeded*), was early introduced by De Varennes e Mendonça (1958) and later maintained by Benfratello (1961).

The monthly deficit $D_{i,j}$ is calculated as the amount of water needed to fill the gap between the maximum required evapotranspiration $ET_{\max,i,j}$ and the actual one $ET_{i,j}$. It is worth recalling that the definition of Thornthwaite's potential evapotranspiration, and the values calculated by means of Thornthwaite's formula, have a climatic meaning and they are not considered by the FAO procedure in order to estimate the maximum required evapotranspiration for irrigation purposes (Allen *et al.* 1998). According to this procedure, in fact, the maximum required evapotranspiration $ET_{\max,i,j}$ is estimated by modulating the reference crop evapotranspiration with coefficients that account for the considered crop and for its phenological development. The reference crop evapotranspiration is in turn estimated by means of Penman–Monteith's formula (as in, e.g., Licciardello *et al.*, 2011), or in case of scarce data, by means of Hargreaves' formula. In this paper, $ET_{\max}$ was calculated both with Thornthwaite's formula and with FAO procedure, in order to explore the possibility of using the method to assess both the climatic water deficit and the irrigation water demand.

The calculations are performed on a climatic basis, which means that taking the expectation of the water balance, the method provides the expectation of the monthly deficit. Therefore, once defined the expectations of $A_{i,j}$, $P_{i,j}$, $ET_{i,j}$, $S_{i,j}$ and

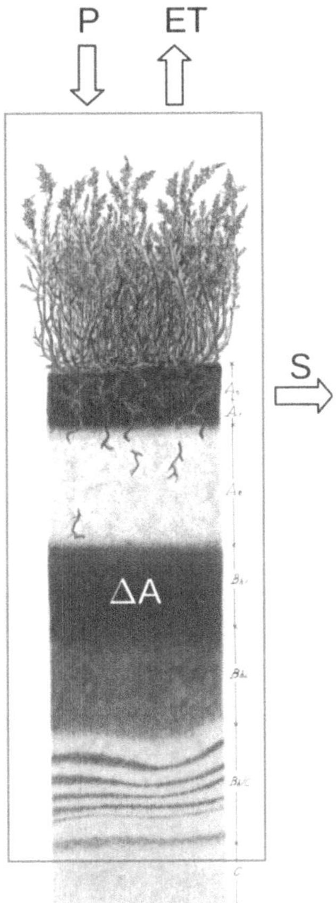

Figure 1 | Control volume of Benfratello's water balance sketched on the soil n.1 of Plate 22 by Kubiena (Humus podsol). Symbols meaning is explained in the text.

$D_{i,j}$, as A_i, P_i, ET_i, S_i, and D_i, respectively, the water balance and the irrigation deficit are rewritten as:

$$\Delta A_i = P_i - ET_i - S_i \tag{2}$$

$$D_i = ET_{\max,i} - ET_i \tag{3}$$

In the original method $ET_{\max,i}$ was a climatic factor, estimated as Thornthwaite's potential evapotranspiration (Thornthwaite 1948). The method is anyway flexible and the maximum required evapotranspiration may alternatively be estimated as the maximum evapotranspiration of a specific crop according, e.g., to the consolidated FAO procedure.

In order to perform the calculations, Benfratello divided the year into a dry season and into a wet season, on the basis of a comparison between the monthly precipitation P_i and the monthly maximum required evapotranspiration $ET_{\max,i}$. This is possible for climates having one major rainfall season, as the Mediterranean one is. The dry season starts when:

$$P_i < ET_{\max,i} \tag{4}$$

and ends when:

$$P_i \geq ET_{\max,i} \tag{5}$$

It is assumed that during the wet season:

$$ET_i = ET_{\max,i} \tag{6}$$

and all the precipitation excess firstly refills the deficit with respect to field capacity and only when the latter is reached, runoff and groundwater recharge are generated. Therefore, during the wet season, the water balance is fully defined as:

$$\Delta A_i + S_i = P_i - ET_{\max,i} \tag{7}$$

$$S_i > 0 \quad \text{when } A_i = U \tag{8}$$

During the dry season instead, the water exceedance S_i is null, because the evapotranspiration demand is greater than the precipitation, and the soil water partially compensates for it. It is therefore not possible that the field capacity is reached during the dry season, a part that at its beginning, and the soil water balance is rewritten as:

$$\Delta A_i = P_i - ET_i \tag{9}$$

2.1.2. Derivation of the desiccation law

Let us fix the origin of the time axis $t = 0$ at the beginning of the dry season. The dry-season water balance is then fully defined once it is possible to determine one between the actual evapotranspiration $ET(t)$ and the desiccation law of the soil $A(t)$. In order to do so, Benfratello rewrote the water balance in terms of cumulated values from the beginning of the dry season $t = 0$:

$$A(t) - A_{\max} = P(t) - ET(t) \tag{10}$$

where $A_{\max}$ is the maximum soil water content, reached at $t = 0$. It may be $A_{\max} \leq U$ depending on the local climate, as it is defined later. In Equation (10), $P(t)$ and $ET(t)$ are the cumulated precipitation and evapotranspiration at a certain instant t. With the symbols of the 1961 paper, the potential soil water loss $-L(t)$ (with $L(t)$ intrinsically negative) and the actual soil water loss $B(t)$ are introduced as:

$$-L(t) = ET_{\max}(t) - P(t) \tag{11}$$

$$B(t) = ET(t) - P(t) \tag{12}$$

The fundamental hypothesis of the method, needed to link the soil desiccation law to the actual evapotranspiration, is now introduced:

$$\frac{\mathrm{d}B/\mathrm{d}t}{-(\mathrm{d}L/\mathrm{d}t)} = \left(\frac{A}{U}\right)^m \tag{13}$$

This law expresses the concept that the ratio of the actual to the potential water losses is proportional to the ratio of soil water content to field capacity. According to Equation (13) the ratio between the actual and the potential rates of soil water loss is a power function of the available soil water normalized with the field capacity. According to the water balance of the dry season (Equation (10)) $\mathrm{d}B = -\mathrm{d}A$, so that Equation (13) is simplified as:

$$\frac{\mathrm{d}A}{\mathrm{d}L} = \left(\frac{A}{U}\right)^m \tag{14}$$

The power $m \geq 0$ accounts for the different attitudes of the crop to react to water stress and it generalizes the Thornthwaite and Mather's approach in which $m = 1$.

Following De Varennes e Mendonça (1958), the dimensionless variables $\alpha(t)$ and $\lambda(t)$ were introduced:

$$\alpha(t) = \frac{A(t)}{U} \tag{15}$$

$$\lambda(t) = \frac{L(t)}{U} \tag{16}$$

Equation (14) is rewritten as:

$$\frac{d\alpha}{d\lambda} = \alpha^m \tag{17}$$

with the initial condition:

$$\alpha(0) = \alpha_{\max} \tag{18}$$

where $\alpha_{\max} = A_{\max}/U(\leq 1)$. In Figure 2 (left), Equation (17) is plotted for different values of m and for $\alpha_{\max} = 1$. In case $m > 1$ the crop reacts to water stress by strongly reducing its transpirative capacity, on the contrary, if $m < 1$ the crop reacts by only weakly reducing it. The limit case $m = 0$ means that the crop does not react to the water shortage and always tries to satisfy the maximum transpirative demand.

The solution of Equation (17) with the condition (18) provides the (normalized) desiccation law of the soil $\alpha(\lambda)$ as a function of the cumulated and normalized potential soil–water loss, which is a known function of the climate and of the crop. Four cases, as reported in Table 1, are usually given for the solution, depending on $\alpha_{\max} = 1$ or $\alpha_{\max} < 1$ and for $m = 1$ or $m \neq 1$. In Figure 2 (right) some solutions of Equation (17) are plotted for different values of m and for $\alpha_{\max} = 1$. It is worth noting that for $m < 1$ the soil completely dries out for finite values of the potential soil–water loss $-\lambda$, while for $m = 1$ the soil only asymptotically dries out.

2.1.3. Model initialization

In order to apply the method a procedure is now needed to determine whether $\alpha_{\max}$ is ≤ 1. Let us define the soil water gain during the wet season Σ_w, and its dimensionless form σ_w, as it was previously defined L:

$$\Sigma_{w,i} = P_i - ET_{\max,i} > 0 \tag{19}$$

$$\sigma_{w,i} = \frac{\Sigma_{w,i}}{U} \tag{20}$$

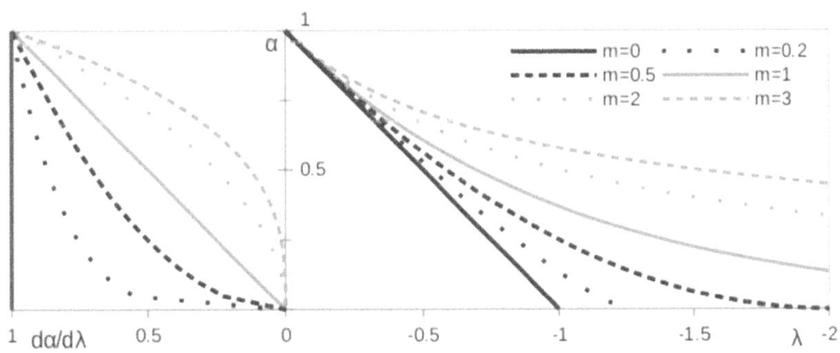

Figure 2 | Fundamental laws of Benfratello's method. Left: evapotranspiration law $d\alpha/d\lambda$ as given by Equation (17). Right: desiccation law $\alpha(\lambda)$ as given in Table 1.

Table 1 | Solutions of Equation (17) with the condition (18), for different cases of $\alpha_{\max}$ and m. The meaning of the symbols is reported in the text

	$m = 1$	$m \neq 1$
$\alpha_{\max} = 1$	$\alpha = e^{\lambda}$	$\alpha = [1 + (1 - m)\lambda]^{1/(1-m)}$
$\alpha_{\max} < 1$	$\frac{\alpha}{\alpha_{\max}} = e^{\lambda}$	$\frac{\alpha}{\alpha_{\max}} = \left[1 + (1 - m)\frac{\lambda}{\alpha_{\max}^{1-m}}\right]^{1/(1-m)}$

Let us define $-\lambda_{\min}$, $\alpha_{\min}$ and $\sigma_{w,\max}$ as the dimensionless forms of the maximum potential loss during the dry season, of the minimum available soil water at the end of the dry season, and of the maximum soil water gain at the end of the wet season, respectively. In case $\alpha_{\max} < 1$, i.e. the precipitation surplus during the wet season is not enough to refill the soil field capacity, the annual water balance is written as:

$$\alpha_{\min} = \alpha(\lambda_{\min}) \tag{21}$$

$$\alpha_{\max} = \alpha_{\min} + \sigma_{w,\max}, \tag{22}$$

where Equation (21) stands for the dry season and Equation (22) stands for the wet season. From Equations (21) and (22), $\alpha_{\max}$ and $\alpha_{\min}$ are determined. In case the computed value of $\alpha_{\max} > 1$, it is a signal that $\alpha_{\max}$ should be posed equal to 1, since the precipitation surplus is effective at refilling the field capacity, and the water exceedance will accordingly emerge from the water balance of Equations (7) and (8).

The system given by Equations (21) and (22) is analytically solvable in some cases as e.g. those for $m = 0.5$, $m = 1$ and $m = 2$. The solutions are provided in Table 2.

2.1.4. Deficit estimate

The expectation of the annual deficit D is given by the summation of the monthly deficit D_i given by Equation (3), over the whole dry season. Having defined t_w the time at which the wet season begins, with the same convention introduced for Equation (10), the cumulated irrigation deficit at the end of the dry season is given by:

$$D(t_w) = ET_{\max}(t_w) - ET(t_w) \tag{23}$$

According to Equations (11) and (12), and accounting for the simplified water balance of Equation (10), we can write:

$$D(t_w) = -L(t_w) - B(t_w) \tag{24}$$

$$D(t_w) = -L(t_w) + A_{\min} - A_{\max} \tag{25}$$

where $A_{\min}$ and $A_{\max}$ are the available soil water at the end and at the beginning of the dry season, respectively. In the dimensionless form we obtain:

$$\delta(t_w) = -\lambda_{\min} + \alpha_{\min} - \alpha_{\max}, \tag{26}$$

where $\delta = D/U$ is the dimensionless deficit.

2.1.5. Method application

The method is traditionally applied as follows:

1. The field capacity U and the power m are chosen.
2. The value of $\alpha_{\max}$ is calculated on the basis of the system of Equations (21) and (22), or with the solutions of Table 2.
3. Starting from the beginning of the dry season, the cumulated value of $L(t)$ given by Equation (11) and of the corresponding dimensionless form $\lambda(t)$ is calculated at the end of each month of the dry season.

Table 2 | Values of $\alpha_{\max}$ for different values of m

$m = 0.5$	$m = 1$	$m = 2$
$\alpha_{\max} = \left(\frac{\sigma_{\max}}{-\lambda_{\min}} - \frac{\lambda_{\min}}{4}\right)$	$\alpha_{\max} = \frac{\sigma_{\max}}{1 - e^{\lambda_{\min}}}$	$\alpha_{\max} = \frac{\sigma_{\max}}{2}\left(1 + \sqrt{1 - \frac{4}{\sigma_{\max}\lambda_{\min}}}\right)$

4. The values of $\alpha(\lambda)$ are determined at the same times, with the solutions reported in Table 1, and consequently the values of the available water A_i at the end of each ith month of the dry season are calculated.
5. The actual evapotranspiration ET_i is determined via the water balance of Equation (9).
6. From the end of the dry season, i.e. for the whole wet season, the water balance of Equations (7) and (8) hold and allow us to determine the soil water refill ΔA_i and the water exceedance S_i for each month.
7. At the end of the wet season, the initial state of the soil A_{max} is restored.
8. The monthly irrigation deficit D_i is calculated during the dry season at the end of each month by means of Equation (3).
9. The yearly irrigation deficit D is finally calculated by means of Equations (24) and (25).

As an example of Benfratello's method application at the field scale, Figure 3 shows the water balance components in Foggia, determined with $U = 300\text{mm}$, $A_{max} = 64\text{mm}$, $m = 0.5,\ 1,\ 2$, paired with the local climograph for the period 1950–2007 to show the precipitation and temperature regimes.

2.2. Proposal of an estimate of the deficit uncertainty

The traditional application of Benfratello's method is meant to provide an estimate of the expectations of the monthly and annual components of the water balance in a semiarid climate. Though the climate shows interannual variability, so that, in order to use the obtained results as design input for irrigation purposes, it is relevant to determine also the uncertainty of the irrigation deficit, as a consequence of the interannual variability of the climatic variables, viz temperature and precipitation. In the following lines, we will present an improvement of Benfratello's method to do so.

As a first step, we need to define the interannual variability of L and Σ_w. According to the method, in fact, all the climatic variability affects the monthly values of L_i and $\Sigma_{w,i}$. For each ith month during the dry or the wet season, respectively, the variabilities $\sigma(L_i)$ and $\sigma(\Sigma_{w,i})$ are given as a function of the interannual standard deviations of the maximum required evapotranspiration $\sigma(ET_{max,i})$ and of the precipitation $\sigma(P_i)$. Assuming the independence of the maximum evapotranspiration

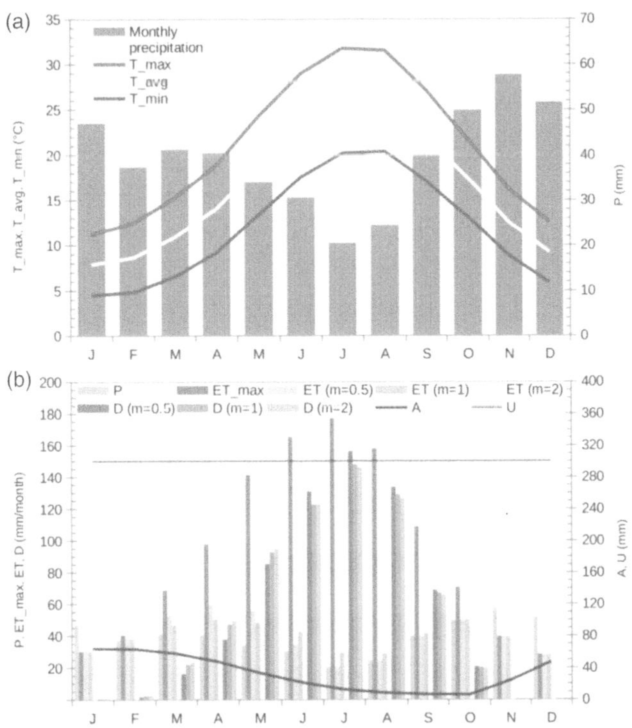

Figure 3 | (a) Climograph of Foggia (1950–2007) and (b) water balance components in Foggia, according to Benfratello's method with $U = 300$ mm, $A_{max} = 64$ mm, $m = 0.5, 1, 2$.

$ET_{\max,i}$ and of the precipitation P_i, the variabilities $\sigma(L_i)$ and $\sigma(\Sigma_{w,i})$ are given by:

$$\begin{aligned}\sigma(L_i) &= \sqrt{\sigma^2(ET_{\max,i}) + \sigma^2(P_i)}, \quad i \in \text{dry season},\\ \sigma(\Sigma_{w,i}) &= \sqrt{\sigma^2(ET_{\max,i}) + \sigma^2(P_i)}, \quad i \in \text{wet season}.\end{aligned}$$

If $ET_{\max,i}$ basically depends on the monthly temperature T_i, the previous relationships are rewritten in the form:

$$\sigma(L_i) = \sqrt{\left(\frac{\mathrm{d}ET_{\max}}{\mathrm{d}T}\right)^2\bigg|_{T=T_i} \sigma^2(T_i) + \sigma^2(P_i)}, \quad i \in \text{dry season}, \tag{27}$$

$$\sigma(\Sigma_{w,i}) = \sqrt{\left(\frac{\mathrm{d}ET_{\max}}{\mathrm{d}T}\right)^2\bigg|_{T=T_i} \sigma^2(T_i) + \sigma^2(P_i)}, \quad i \in \text{wet season}, \tag{28}$$

In dimensionless form we obtain:

$$\sigma(\lambda_i) = \frac{\sigma(L_i)}{U}, \tag{29}$$

$$\sigma(\sigma_{w,i}) = \frac{\sigma(\Sigma_{w,i})}{U}. \tag{30}$$

From Equations (27) and (28), or alternatively from Equations (29) and (30), the annual values of $\sigma(L)$ and $\sigma(\Sigma_w)$ might be calculated with the following form:

$$\sigma(L) = \sqrt{\Sigma_i \sigma^2(L_i)}, \quad i \in \text{dry season} \tag{31}$$

$$\sigma(\Sigma_w) = \sqrt{\Sigma_i \sigma^2(\Sigma_{w,i})}, \quad i \in \text{wet season} \tag{32}$$

As a second step, we define the procedure to determine the deficit uncertainty $\sigma(D)$ on the basis of the interannual variabilities $\sigma(L)$ and $\sigma(\Sigma_w)$. According to the original application of the method the local climate is such that at the end of the wet season, either the soil always reaches the field capacity or the available water is always smaller than the field capacity. In this application, we will not relax this hypothesis in order to determine the deficit uncertainty. Therefore, two cases should be considered, either that at the beginning of the dry season, it is reached or not the field capacity of the soil.

For each jth year, Equation (26) holds in the form:

$$\delta(t_{w,j}) = -\lambda_{\min,j} + \alpha_{\min,j} - \alpha_{\max,j} \tag{33}$$

In the case $A_{\max} < U$ (and $\alpha_{\max} < 1$), by means of Equations (22), Equation (33) takes the form:

$$\delta(t_{w,j}) = -\lambda_{\min,j} + \alpha_{\min,j} - \alpha_{\min,j-1} - \sigma_{w,\max,j-1} \tag{34}$$

By taking the expectation of Equation (34) and assuming that $E[\alpha_{\min,j}] = E[\alpha_{\min,j-1}] = \alpha_{\min}$, we obtain:

$$\delta(t_w) = -\lambda_{\min} - \sigma_{w,\max} \tag{35}$$

In this case, an estimate of the uncertainty $\sigma(\delta(t_w))$ is given by:

$$\sigma(\delta(t_w)) = \sqrt{\sigma^2(\lambda_{\min}) + \sigma^2(\sigma_{w,\max})} \tag{36}$$

and the uncertainty $\sigma(D(t_w))$ is given by:

$$\sigma(D(t_w)) = U\,\sigma(\delta(t_w)) \tag{37}$$

In the case $A_{\max} = U$ ($\alpha_{\max} = 1$), Equations (26) and (33) are rewritten as:

$$\delta(t_w) = -\lambda_{\min} + \alpha_{\min} - 1 \tag{38}$$

$$\delta(t_{w,j}) = -\lambda_{\min,j} + \alpha_{\min,j} - 1 \tag{39}$$

By means of a first-order approximation around $\lambda_{\min}$ and by means of Equation (17), the yearly anomaly $\delta(t_{w,j}) - \delta(t_w)$ is:

$$\delta(t_{w,j}) - \delta(t_w) = -\lambda_{\min,j} + \alpha_{\min,j} - (-\lambda_{\min} + \alpha_{\min}) \tag{40}$$

$$\approx \left(-1 + \left.\frac{\mathrm{d}\alpha}{\mathrm{d}\lambda}\right|_{\lambda_{\min}}\right)(\lambda_{\min,j} - \lambda_{\min}) \tag{41}$$

$$= (-1 + \alpha^{m}_{\min})(\lambda_{\min,j} - \lambda_{\min}) \tag{42}$$

The uncertainty $\sigma(\delta(t_w))$ is, therefore, given by:

$$\sigma(\delta(t_w)) = \sqrt{(-1 + \alpha^{m}_{\min})^2 \sigma^2(\lambda_{\min})} \tag{43}$$

$$\sigma(\delta(t_w)) = |-1 + \alpha^{m}_{\min}|\sigma(\lambda_{\min}) \tag{44}$$

from which $\sigma(D(t_w))$ is obtained by means of Equation (37).

2.3. GIS application of the method

Traditional applications of Benfratello's method were performed at the local scale, even in the perspective of producing climatic maps. Yet GIS framework allows us to systematically apply it at large scales, accounting for the features of the local climate, soil and cultures. In order to do so, the following (minimal) set of maps is required: soil field capacity, land-use, monthly air temperature and precipitation values with interannual standard deviation as a measure of the interannual variability. The method is applied according to the numbered list presented in paragraph *Method application*.

Finally, the uncertainty of the yearly irrigation deficit $\sigma(D(t_w))$ is locally provided by Equations (36), (44) and by Equation (37).

We now want to define a spatially equivalent estimate of the expectation $D(t_w)$ and of its uncertainty $\sigma(D(t_w))$ over multiple cells of the investigated area. Let $D_{j,k}(t_w)$ be the annual deficit for the jth year and the kth cell and assume that it can be represented by:

$$D_{j,k}(t_w) = D_k(t_w) + z_{j,k}\sigma(D_k(t_w)) \tag{45}$$

where $D_k(t_w)$ and $\sigma(D_k(t_w))$ are the local expectation and uncertainty, respectively, and $z_{j,k}$ is the standard reduced variable corresponding to $D_{j,k}(t_w)$.

If we assume that each realization of temperature and precipitation has the same probability over the whole investigated district, also each $D_{j,k}(t_w)$ has the same probability for each kth cell and the value of $z_{j,k}$ is therefore independent from the cell. Let z_j be this value, so that Equation (45) is written as:

$$D_{j,k}(t_w) = D_k(t_w) + z_j\sigma(D_k(t_w)) \tag{46}$$

and the spatial average $\langle D_{j,k}(t_w)\rangle$ is provided by:

$$\langle D_{j,k}(t_w)\rangle = \langle D_k(t_w)\rangle + z_j\langle\sigma(D_k(t_w))\rangle \tag{47}$$

The equivalent values of the expected deficit and of its uncertainty over the whole area are therefore given by the spatial averages $\langle D_k(t_w)\rangle$ and $\langle\sigma(D_k(t_w))\rangle$, respectively, as they emerge from Equation (47).

2.4. Budyko's framework to highlight the relationship between *ET* and ET_{max}

In the Discussion section, in order to highlight the relationship between ET and ET_{max}, the results will be interpreted in the framework of the classical Budyko curve and of an irrigation-deficit curve obtained by transforming the Budyko curve, for the case of full irrigation. It is therefore worth briefly recalling the shape and the meaning of the two curves.

The Budyko curve (Budyko 1974) puts into relationship the actual annual evapotranspiration, in the form ET/P, with the corresponding maximum annual evapotranspiration ET_{max}/P, the latter being sometimes referred to as Budyko's aridity index. The Budyko curve has the following form:

$$\frac{ET}{P} = \sqrt{\frac{ET_{max}}{P} \tanh\left(\frac{P}{ET_{max}}\right)\left(1 - \exp\left(-\frac{ET_{max}}{P}\right)\right)} \tag{48}$$

and its upper bound is the identity line of the first quadrant (i.e. the slope of ET/P for $ET_{max}/P \to 0$) and the asymptote $ET/P = 1$. These two limiting curves represent the cases when ET is maximum, i.e. the case in which ET is energy-limited ($ET/P = ET_{max}/P$) and the case in which ET is water-limited ($ET/P = 1$).

The (climatic) water-deficit curve, in the form D/P, can be straightforwardly obtained by the Budyko curve as:

$$\frac{D}{P} = \frac{ET_{max} - ET}{P}. \tag{49}$$

The irrigation deficit curve is inferiorly bounded by the maximum between $D/P = 0$ and $D/P = (ET_{max}/P) - 1$, which represent the minimum irrigation deficit required, in case the actual evapotranspiration is optimal.

3. THE CAPITANATA CASE STUDY

The investigated area is the Bonifica della Capitanata (the Capitanata reclamation and irrigation district), an area of about 4,410 km^2 in Southern Italy (Apulia region, province of Foggia) which is one of the most important agricultural districts in Italy (Figure 4). The district is Northernly bounded by the Saccione river and Southernly by the Ofanto river. It includes the alluvial plain of the Tavoliere delle Puglie, the second largest plain in Italy with a surface of about 3,000 km^2, mainly cultivated with herbaceous crops, olives, fruit and grapevine trees (Ventrella *et al.* 2012). Part of the Bonifica della Capitanata is irrigated by the following three water-distribution schemes: the major one is the Fortore system in the Northern and Central part, and the others are the Sinistra Ofanto and the Carapelle systems in the Southern part. Particularly, on the Fortore river there is the Occhito reservoir with an active capacity of 250 hm^3, corresponding to about a depth of 210 mm for the total irrigated area. They are on-demand pressurized systems, equipped with water-meters and prepaid card devices for every single user (Lamaddalena *et al.* 2004). According to Lamaddalena *et al.* (2004), the covered area by the consortium distribution network is about 1,404 km^2 and about 90% of it (1,260 km^2) is effectively under-functioning, making the Capitanata one of the greatest and most important irrigation consortia in the Mediterranean region. However the irrigated land is slightly smaller and it is estimated to be about 1,210 km^2 (27.5% of the Bonifica della Capitanata consortium total area), and only 45% of it is irrigated through the consortium water distribution network. In fact, part of the land is directly irrigated by the farmers using private wells and basins as a supplementary or substitutive source.

For the study, monthly maximum and minimum temperature and monthly precipitation of 109 stations located in the Apulia Region, 19 of which fall within the area of interest for the Capitanata consortium, were collected for the period 1950–2007 from the hydrological annals[1].

The Köppen and Geiger climate type of the area, determined according to the procedure suggested by Peel *et al.* (2007), is quite variegated, being in most cases Cfa (Temperate without a dry season and with hot summer), with some Csa (Temperate with dry and hot summer) and a few BSk (Cold arid steppe). In the nearby Promontorio del Gargano (Gargano promontory) there are also some Cfb cases (Temperate without dry season and with warm summer), due to the higher altitude and lower

[1] https://protezionecivile.puglia.it/centro-funzionale-decentrato/rete-di-monitoraggio/annali-e-dati-idrologici-elaborati/annali-idrologici-parte-i/, checked 10 March 2021.

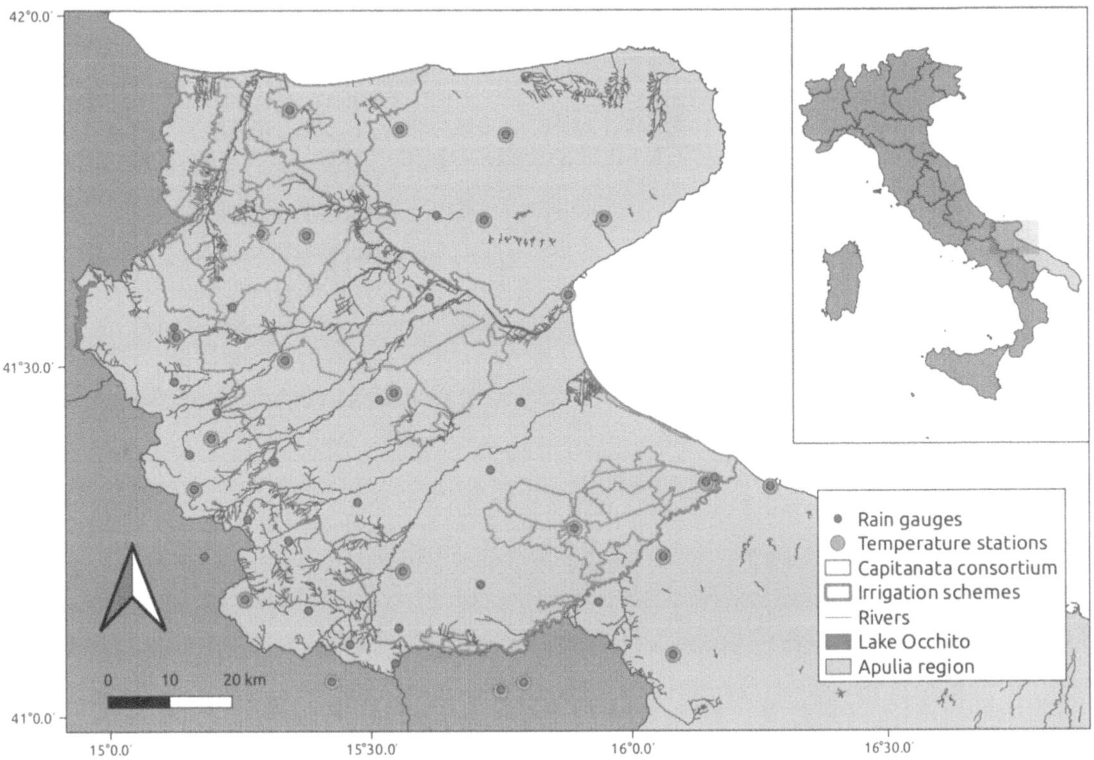

Figure 4 | The Bonifica della Capitanata district in the Northern part of Apulia region, Southern Italy.

temperatures. The precipitation regime is Mediterranean with mostly winter precipitation, which corresponds to the maritimal one among the classification of the Italian pluviometric types by Bandini (1931).

The average annual precipitation in the Capitanata ranges from 400 to 650 mm, and in the surrounding hilly areas it is greater and it reaches 1,160 mm. According to Johansson's continentality index and Kerner's oceanity index, the local climate is maritime but weakly – and mainly in the coastal areas (Lamaddalena *et al.* 2008) – it benefits from the presence of the Adriatic sea. According to both De Martonne's and Pinna-De Martonne's aridity indices the local climate is semiarid (see Baltas 2007, for details on the indices calculation). Finally, the typical summer proneness to water shortages and to droughts in the circum-Mediterranean areas might increase in the near future as stated by Vitale *et al.* (2010) who in recent years observed an increase in the average maximum monthly temperatures (0.35 °C per decade), of the number of tropical nights (4–5 days per decade), of the number of very hot summer days (6 days per decade), and of other indicators that seem to evidence anticipation of the summer dry season. Despite the fact that – as stated by the Authors – these results should be taken with caution because of the relatively short length of the analyzed data series (1979–2008), they are valuable because data were collected in an experimental station located in the Apulian Tavoliere at the centre of the Capitanata.

Also, the effects on groundwater salinity due to the reduced natural discharge from the aquifers are expected to be joined by the impact of the ongoing sea level rise in the Capitanata (Guyennon *et al.* 2017).

4. PRELIMINARY ACTIVITIES

4.1. Input maps preparation

In Table 3, we list the input maps used in order to run Benfratello's method within a GIS framework, which are monthly precipitation, monthly average, maximum and minimum temperature, field capacity and land-cover, needed to modulate the maximum evapotranspiration demand. In order to obtain these maps, we proceeded as follows. Field capacity maps, for different land-cover scenarios, were obtained by multiplying the map of water content at field capacity θ_c by the map of the hydrologically active soil layer. The latter was obtained by assigning a rooted depth to each land-cover class, according to Foxx *et al.* (1984) and Walker (1989). In the nearby Gargano area, despite the deep rooted depth obtained by the referred

Table 3 | Summary of the input layers

Content	Type	Resolution (m)	Ellipsoid and projection	Source
DEM	raster	250	WGS84–UTM32N	INGV
θ_c	raster	250	WGS84–UTM32N	ESDAC
CLC 1990	vector	–	ETRS89–LAEA Europe	GRP
CLC 2000	vector	–	ETRS89–LAEA Europe	ISPRA
CLC 2006	vector	–	ETRS89–LAEA Europe	ISPRA
CLC 2011	vector	–	ETRS89–LAEA Europe	GRP
MS	vector	–	Roma40–Gauss Boaga East	–
P	raster	250	WGS84–UTM32N	–
T, $T_{\max}$, $T_{\min}$	raster	250	WGS 84 UTM 32N	–
U	raster	250	WGS84–UTM32N	–

Note: DEM, Digital Elevation Model; INGV, Istituto Nazionale di Geofisica e Vulcanologia (http://tinitaly.pi.ingv.it); θ_c(cm^{3}cm^{-3}), Water content at field capacity; ESDAC, European Soil Data Centre (https://esdac.jrc.ec.europa.eu/resource-type/datasets); CLC, Corine Land Cover; GRP, Geoportale Regione Puglia (http://sit.puglia.it); ISPRA, Istituto Superiore per la Protezione e la Ricerca Ambientale (https://www.isprambiente.gov.it/it/attivita/suolo-e-territorio/copertura-del-suolo/corine-land-cover); MS, Meteorological stations location; P(mm), average monthly precipitation; T, $T_{\max}$, $T_{\min}$(°C), average, maximum and minimum monthly temperature; U(mm), Field capacity.

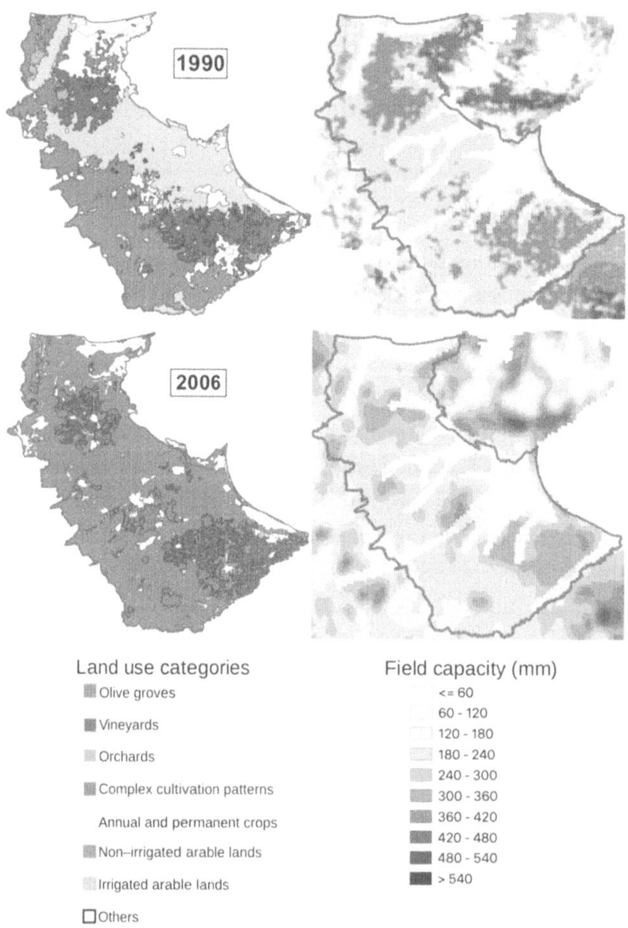

Figure 5 | Maps of the land-cover classification and of the resulting field capacity U of the Bonifica della Capitanata district, according to two different databases made available by the Apulia region administrative authority.

literature, we preferred to assign a shallow hydrologically active soil layer according to Ancona *et al.* (2010) (Figure 5). Monthly precipitation maps were generated from measured precipitation data by means of Thiessen–Voronoi tessellation (see, e.g. Figure 6, top-right map). Monthly temperatures available at the micrometeorological stations were interpolated by a linear function of the elevation in order to provide the temperature maps (see, e.g. Figure 6, top-left map).

4.2. Estimate of the maximum evapotranspiration

The monthly maximum required evapotranspiration was calculated as the crop evapotranspiration in standard conditions according to FAO procedure (Allen *et al.* 1998). In order to do so, the monthly reference evapotranspiration $ET_{\max,o}$ was first calculated by means of Hargreaves' formula:

$$ET_{\max,o} = 0.0023R_a(T + 17.8)\sqrt{T_{\max} - T_{\min}} \tag{50}$$

where $R_a\,(\text{mm}\,\text{d}^{-1})$ is the potentially evaporated water depth corresponding to solar radiation, and it is given as a geographical characteristic, and T, $T_{\max}$, $T_{\min}$ (°C) are the average, maximum and minimum daily temperatures, respectively. The crop evapotranspiration in standard conditions $ET_{\max}$ was obtained by correcting the former with the cultural coefficients k_c which account for the crop characteristics and for its phenological condition, on a monthly basis. Cultural coefficients were applied according to the tables reported by Allen *et al.* (1998) and updated by Nistor *et al.* (2018) in order to fit the Corine Land Cover classes. As an example, in Figure 6 we present the obtained map of the maximum required evapotranspiration for the month

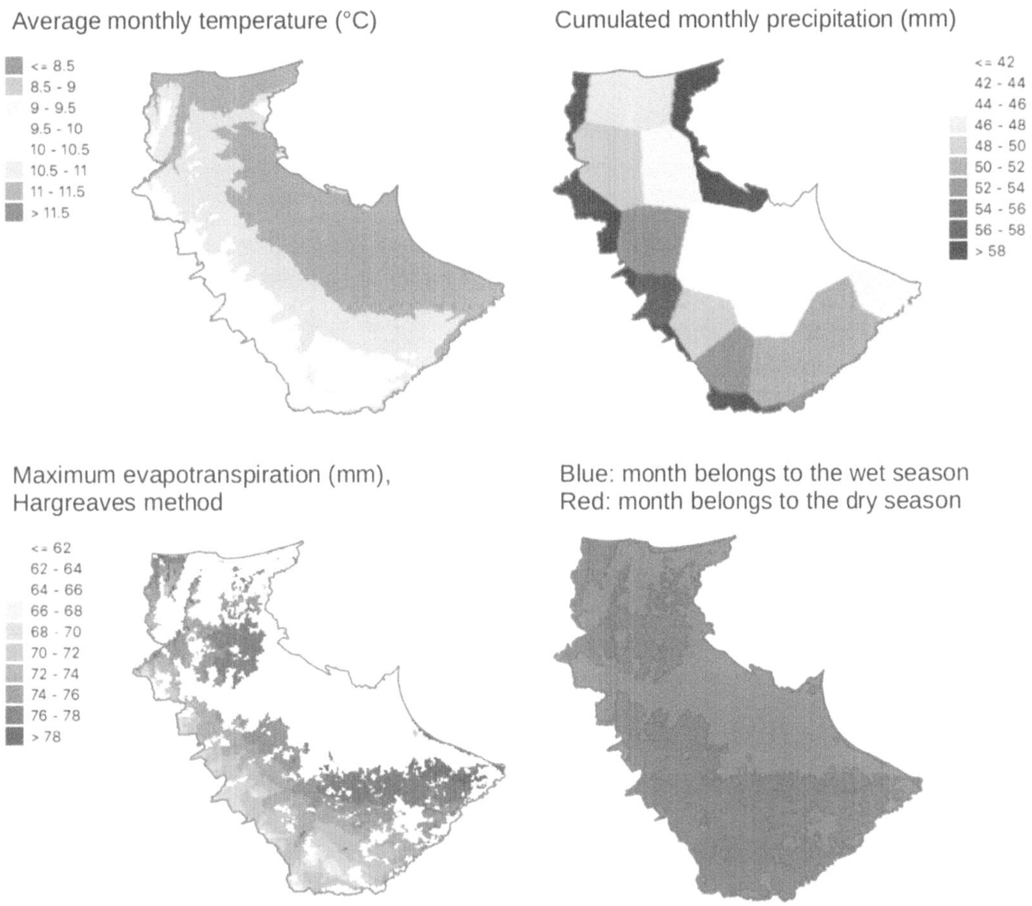

Figure 6 | Maps of the input layers required to run the simulations. As an example, the month of March, at the very beginning of the dry season in some areas of the district, is reported. From top left to bottom right: average monthly temperature T, cumulated monthly precipitation P and maximum evapotranspiration $ET_{\max}$, flag to identify the dry areas, i.e. the areas where $L = P - ET_{\max} < 0$. Please refer to the online version of this paper to see this figure in colour: http://dx.doi.org/10.2166/nh.2023.081.

of March with the 1990 version of the land-cover map (bottom-left map). The application of Benfratello's method largely depends on the beginning of the dry season, therefore it is necessary to assess for each point whether the month is wet or dry depending on the value taken by $L \gtrless 0$. In Figure 6 we reported as an example the precipitation map of the month of March (top-right map) and a map reporting the sign of L in March (bottom-right map). The latter shows the areas for which the dry season has already started in March ($L < 0$, red areas) and those which are still in the wet season ($L \geq 0$, blue areas).

Particularly referring to the arable lands, which are the main part of the district, according to the collected information we simulated the cultivation of spring varieties in the non-irrigated land and the cultivation of summer ones in the irrigated land. Spring varieties are therefore more active in March than summer ones and this results in greater maximum evapotranspiration in the South-Western part of the district, even if the temperature is a little smaller.

As a term of comparison, the monthly maximum required evapotranspiration was calculated also as potential evapotranspiration with Thornthwaite's method, as proposed by the original Benfratello's paper. Data are not provided here for the sake of brevity but the effect of this scenario on the deficit estimate will be reported in the Discussion section.

5. RESULTS AND DISCUSSION

In Figure 7, the main components of the dimensionless water balance (α_{max}, α_{min}, δ) and the dimensionless uncertainty of the irrigation deficit $\sigma(\delta)$ are presented for the 1990 version of the land-cover map.

Despite α_{max} values being quite great, many areas of the district do not reach the field capacity ($\alpha_{max} = 1$) at the end of the wet season and most of the district would spontaneously reach very small water content values ($\alpha_{min} < 0.1$) at the end of the dry season, if not irrigated. Comparing Figure 7 with the field capacity reported in Figure 5 we recognize that the area most prone to reach the smallest α_{min} values are those characterized by the smallest U. This behavior is reflected in the assessment

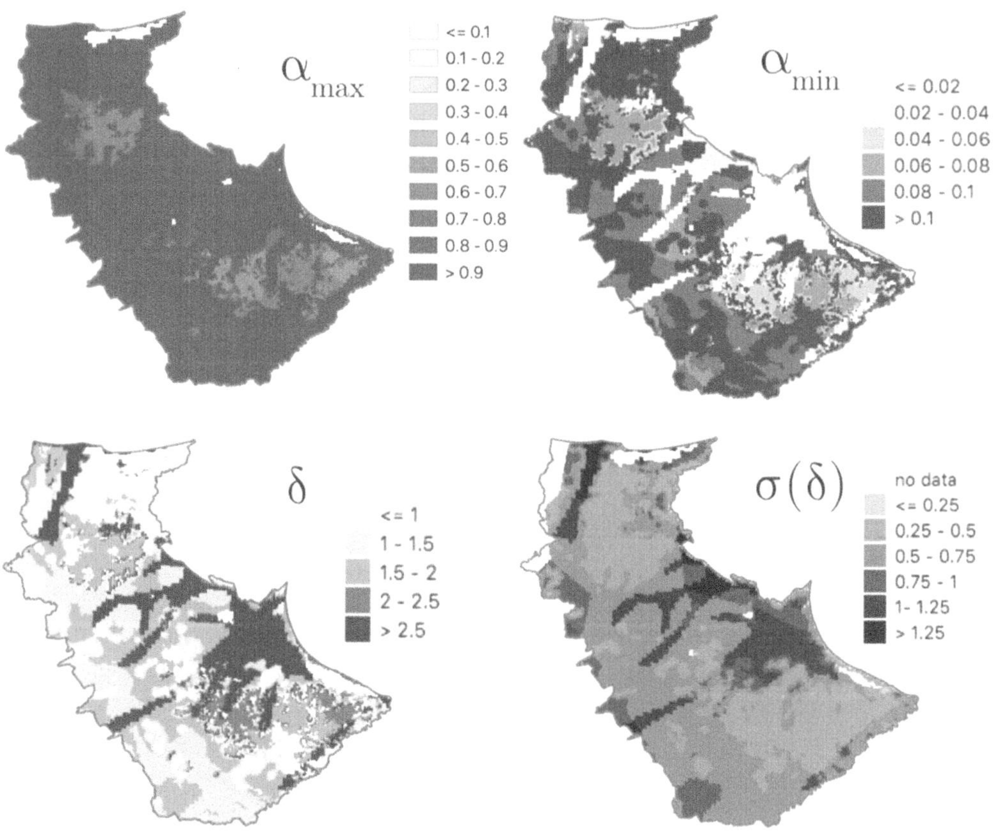

Figure 7 | Dimensionless components of Benfratello's water balance calculated for the land-cover classes of 1990. From top left to bottom right: maximum available soil water at the beginning of the dry season α_{max}, minimum available soil water at the end of the dry season α_{min}, expected annual deficit δ, uncertainty of the annual deficit $\sigma(\delta)$.

of the dimensionless irrigation deficit δ and of its uncertainty $\sigma(\delta)$, which results greatest where the field capacity U and α_{min} are smallest.

Figure 8 presents the expected annual irrigation deficit D and its uncertainty $\sigma(D)$ for the considered land-cover classes of 1990, 2000, 2006 and 2011. Comparing the four obtained spatial distributions of D, one sees how the areas characterized by higher values (i.e. $D > 600$ mm) have been decreasing since 1990, when they occupied the Northern and the South-Eastern parts of the district, to the point where they essentially disappeared in 2011. On the other hand, the central areas of the district experienced a first increase of D (from 200–300 to 300–400 mm and from 300–400 to 400–500 mm) from 1990 to 2006, followed by a decrease, as in 2011 there are just a few small areas with $D > 400$ mm. Recalling Figure 5 it is possible to relate the decrease in the width of the areas characterized by higher D values with a reduction in field capacity U in the Northern and South-Eastern parts of the district between 1990 and 2006, due mainly to an extensive change in cultivations typology, while there are no significant changes in U in the central part of the district despite some minor land-cover changes. If we consider the expected annual irrigation deficit uncertainty $\sigma(D)$ instead, Figure 8 depicts a situation which is essentially stationary between 1990 and 2011 and which appears to be greatly influenced by the precipitation dynamics. Indeed it shows spatial patterns that are very similar to those of the Thiessen–Voronoi tessellation applied for interpolating the precipitation data as reported in Figure 6 (top-right map).

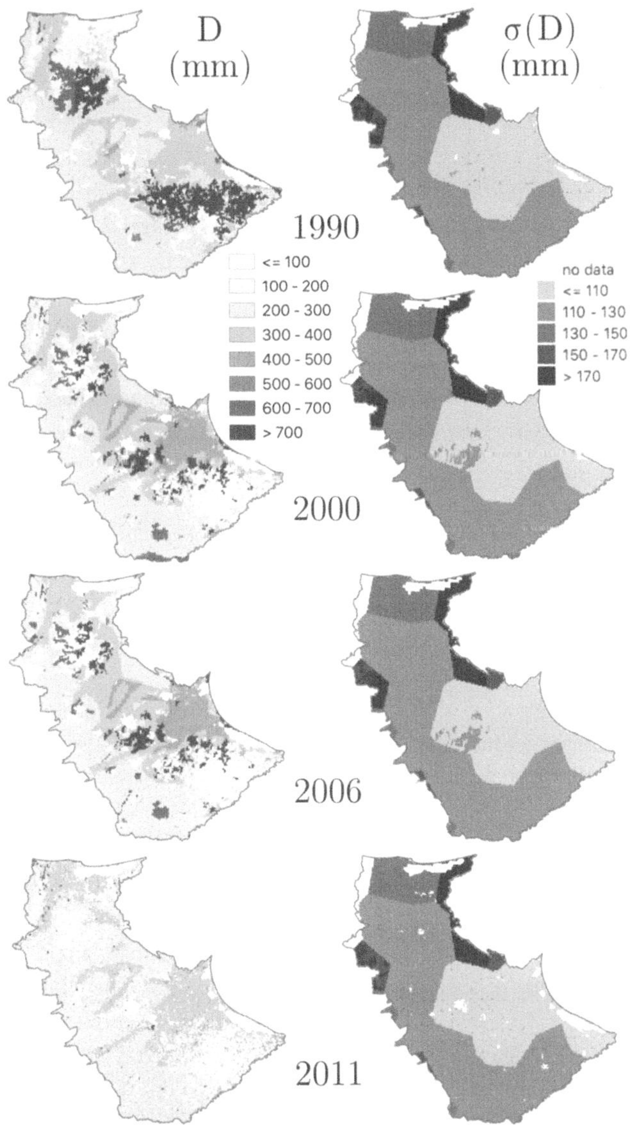

Figure 8 | Expected annual irrigation deficit D and its uncertainty $\sigma(D)$ for the land-cover classes of 1990, 2000, 2006 and 2011.

In Figure 9, we plot the distributions of crop-specific annual deficit values obtained by applying Benfratello's method in six different scenarios deriving from the combination of three different methods to assess the maximum required evapotranspiration (i.e. Hargreaves' method plus cultural coefficient, Hargreaves' method with no cultural coefficient and Thornwhaite's method) and two land-cover conditions (1990 and 2006).

The violin plots visualize the distribution densities while the white boxplots inside provide summary statistics (i.e. 25%, median and 75% quartiles). For the same maximum required evapotranspiration method, land-cover changes impact on the distribution densities but often do not affect the statistical summary in a meaningful way. Regarding the maximum required evapotranspiration methods, the cultural coefficients k_c bring the deficit values from Hargreaves' method closer to the ones from Thornwhaite's method for all the land-use classes except complex cultivation patterns.

In the figure, these results are also compared with both the average irrigation demands reported by Lamaddalena *et al.* (2004) and the erogated water volume ranges publicly stated by the authority managing the Capitanata reclamation and irrigation district, plotted with red dashed-dotted lines and black dashed lines, respectively (for the orchards land-use class the consortium provides only the maximum value of the erogated water). As it was not possible to provide a direct validation of the method in terms of comparison between a simulated and a measured sample of actual evapotranspiration data all over the Capitanata region, we propose a comparison between the average irrigation deficit values derived by the presented method (equal to 347 and 323 mm in 1990 and 2006, respectively) and the value estimated by Lamaddalena *et al.* (2004) for almost the same land-use classes (equal to 326 mm). By this comparison, we obtained a fair agreement, with an error of 6.4 and −1.1% if 1990 and 2006 land-use are considered, respectively. The results are also summarized in Table 4, where the ratio between the calculated deficit and the mean water supply provided by the consortium is reported. Apart from the cases of complex cultivation patterns and orchards, which are in many cases complex mosaic crops and would therefore require deeper insight into the actual cultivations, the estimated deficit is always smaller than the erogated water supply. Moreover, by comparing the average irrigation deficit with the recommended water volumes given by the Consortium, values of global irrigation efficiency of 82 and 92% are obtained for 1990 and 2006 land-use, respectively. These values are consistent with the drip- and micro-irrigation practices that are commonly and widely adopted in the Capitanata irrigation district.

In order to highlight the influence of the precipitation dynamics on the expected annual irrigation deficit uncertainty $\sigma(D)$, Figure 10 presents a plot of the ($\sigma(P)$, $\sigma(D)$) pairs of every cell within the Capitanata territory for the considered land-cover classes of 1990, 2000, 2006 and 2011, for a total of 266,359 records. Despite some values laying outside from the scatter line, the sample – due to the great number of values – is well fitted by a linear model with equation:

$$\sigma(D) = 3.627 + 0.986\sigma(P), \quad R^2 = 0.9807, \tag{51}$$

with $\sigma(D)$ and $\sigma(P)$ in (mm). This means that despite the different land-cover classes (and the consequently different evapotranspirative demand and hydrological behavior) the irrigation deficit uncertainty is practically given by the interannual variability of the annual precipitation.

In Figure 11, we present the classical Budyko curve and the irrigation-deficit curve derived by it with the average annual water balance, for each cell and for each land-cover scenario.

The calculated water balances of the four scenarios are quite different from each other according to the scenarios of land-cover depicted by the Regional Authority. In fact, the 1990 and 2011 land-cover maps split many arable land areas between irrigated and non-irrigated, whereas the 2000 and 2006 maps classify all the arable lands as just non-irrigated. According to the collected information, the non-irrigated land is mainly cultivated with spring varieties whereas the irrigated land is mainly cultivated with summer varieties, and we maintained this set up in all the scenarios in order to appreciate how the land-cover change could have affected the general behavior of the district water balance. As the precipitation regime is typically Mediterranean, with a winter rainy season (Figure 3), the spring varieties can benefit from the water stored in the soil at the end of the rainy season. On the contrary, the summer varieties benefit less from the water stored in the soil during the rainy season. As a consequence, the clouds of the water balances for the 2000 and 2006 scenarios are more compact and nearer to the limit curves than those of the 1990 and 2011 scenarios, that are more spread and far from the limit curves. This results in greater estimates of the irrigation deficit for the 1990 and 2011 scenarios with respect to the 2000 and 2006 ones.

These findings are clearly represented in Figure 12, which reports a detail of the Budyko curve and of the irrigation-deficit curve only for the 1990 land-cover scenario, with evidence of the different land-cover classes. Almost all the classes, with the

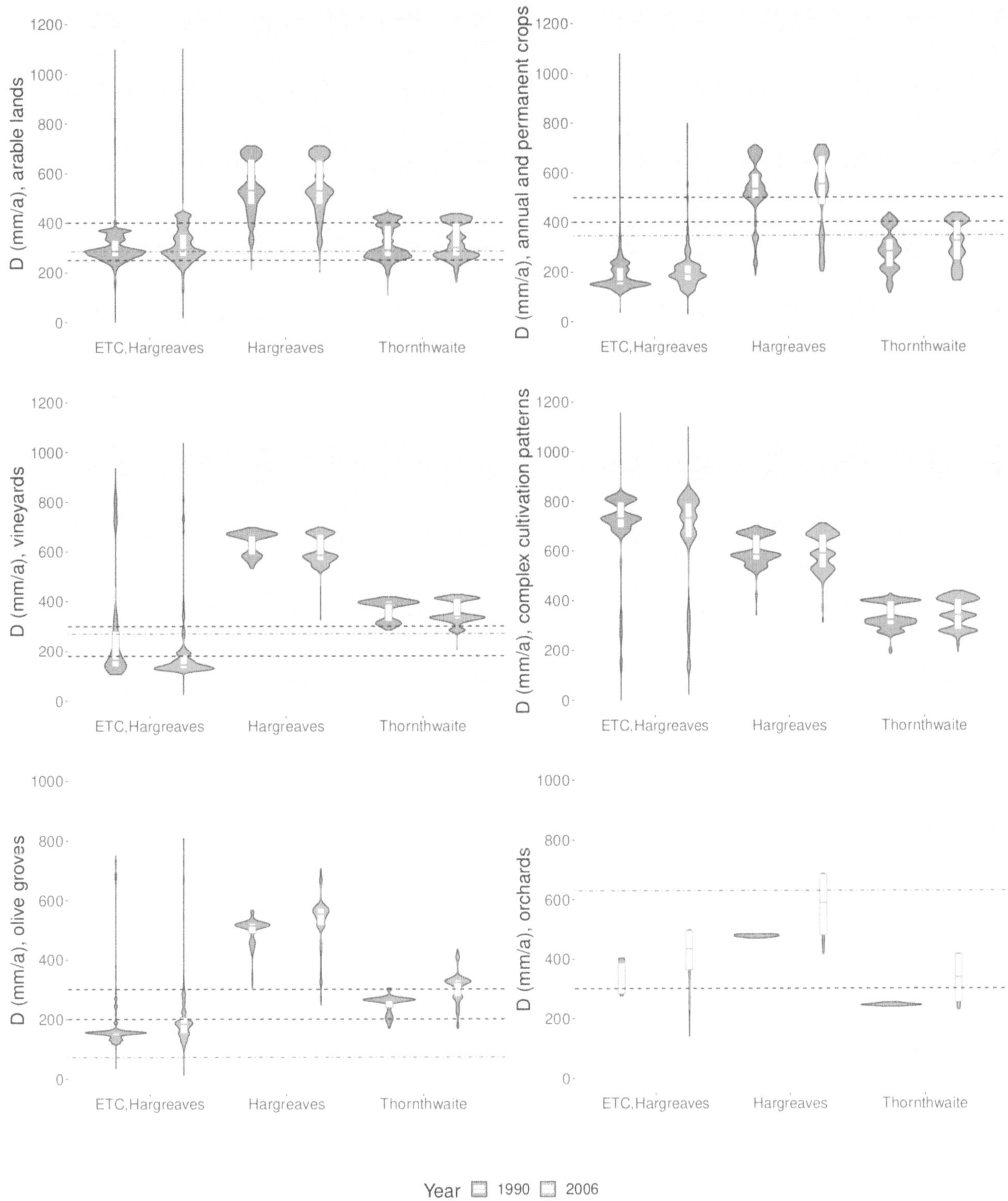

Figure 9 | Crop-specific annual deficit values obtained applying Benfratello's method in three different maximum evapotranspiration scenarios: violin plots visualize the distributions densities while the white boxplots provide summary statistics (25% median and 75% quartiles). Red dashed-dotted lines represent the average irrigation demands reported by Lamaddalena *et al.* (2004). Black dashed lines represent the ranges of crop-water requirements provided by the Capitanata reclamation and irrigation district authority (for the orchards land-use class the consortium provides only the maximum value of the erogated water). Please refer to the online version of this paper to see this figure in colour: http://dx.doi.org/10.2166/nh.2023.081.

Table 4 | Summary data of the calculated irrigation deficits for 1990 and 2006 land-use scenarios

	WS			1990 land-use					2006 land-use				
Land-use class	**Max (mm)**	**Min (mm)**	**Mean (mm)**	**Area %**	**Med(*D*) (mm)**	**⟨*D*⟩ (mm)**	**⟨σ(*D*)⟩ (mm)**	**⟨*D*⟩/⟨*WS*⟩ (mm/mm)**	**Area %**	**Med(*D*) (mm)**	**⟨*D*⟩ (mm)**	**⟨σ(*D*)⟩ (mm)**	**⟨*D*⟩/⟨*WS*⟩ (mm/mm)**
Arable lands	400	250	325	64.88	286	299		0.92	68.26	293	315		0.97
Annual cr.	500	400	450	12.63	162	192		0.43	3.31	190	206		0.46
Vineyards	300	180	240	1.26	164	269		1.12	9.32	144	184		0.77
Complex patt.				15.26	731	700			8.01	732	665		
Olive groves	300	200	250	1.25	154	183		0.73	4.53	184	198		0.79
Orchards			300	0.03	388	353		1.18	0.21	434	409		1.36
Other				4.71					6.35				
Capitanata				100	287	347	122	0.82	100	287	323	122	0.92

Note: WS, water supply provided by the Capitanata consortium; Area, percentual coverage of each land-use class; Med(*D*), (Spatial) median value of the calculated deficit *D*; ⟨*D*⟩, Spatial average of *D*; ⟨σ(D)⟩, spatial average of the uncertainty σ(*D*). Abbreviated land-use classes are as follows: Annual cr., Annual and permanent crops; Complex patt., Complex cultivation patterns.

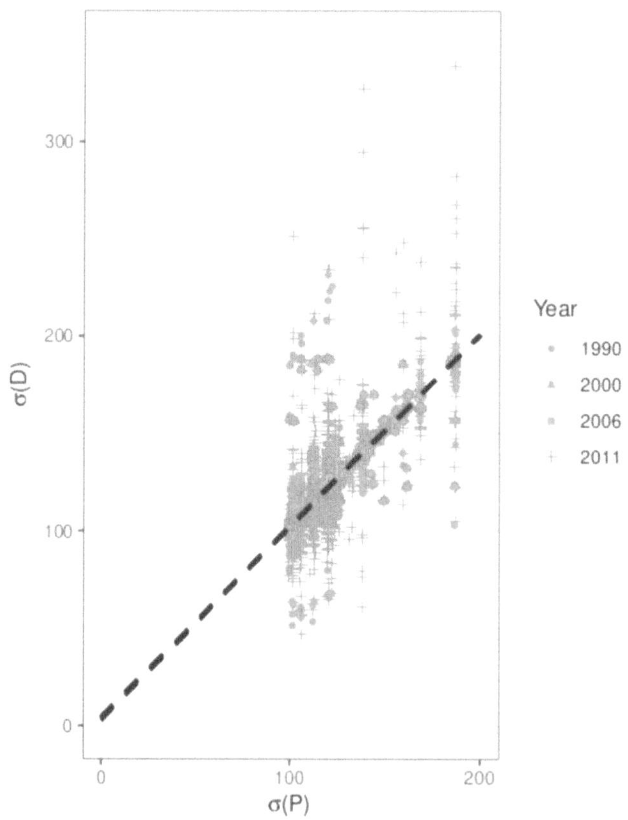

Figure 10 | Influence of the precipitation interannual variability $\sigma(P)$ on the expected annual irrigation deficit uncertainty $\sigma(D)$: $\sigma(D) = 3.627 + 0.986\sigma(P)$, $R^2 = 0.9807$.

exception of the irrigated arable land, lay either along the Budyko curve or between the Budyko curve and the limit curve. The irrigated arable land, instead, is spontaneously characterized by a small evapotranspiration and lays under the Budyko curve, consequently needing a greater irrigation demand, with respect to that compatible with the Budyko curve.

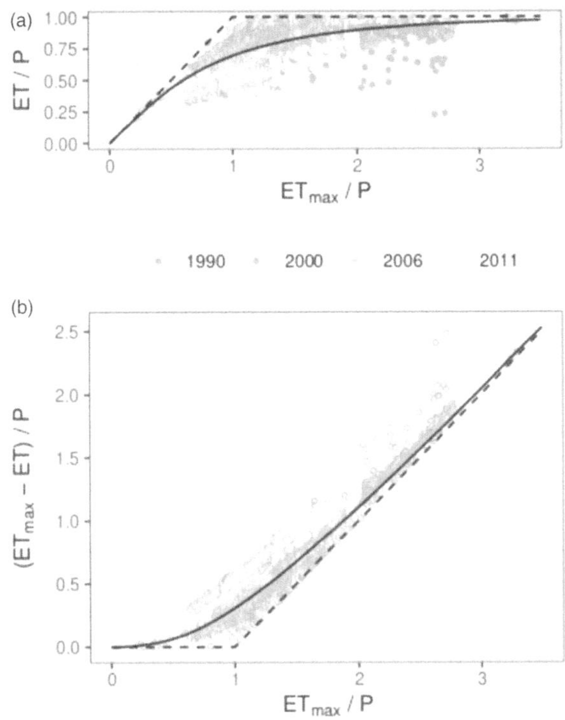

Figure 11 | (a) Budyko curve and calculated water balance components for each cell of the Capitanata district in the 1990, 2000, 2006, 2011 scenarios of land-cover classification. (b) Irrigation-deficit curves based on the Budyko curve and on its limit curve, and calculated irrigation deficits for each cell of the Capitanata district in the aforementioned land-cover classification scenarios.

6. CONCLUSIONS

Our aim was to provide a robust and easy-to-use tool that allows us to make distributed estimates of irrigation deficits at the district level, in order to estimate the expected variations of the deficit in both climate and anthropogenic change scenarios. For this purpose we have proposed a GIS-based implementation of Bentratello's method, which allows us to estimate the deficit as a climatic characteristic on the basis of the knowledge of the precipitation regimes, the maximum required evapotranspiration and of the soil field capacity. The presented model couples the solidity of the results with a low computational burden, and could work as a webGIS tool for water management and irrigation advisory. The method was originally designed for Mediterranean semiarid climates and it is suitable for all semiarid climates. In order to extend its potential, we have also defined relationships that allow estimation of the deficit uncertainty. The model was applied to the case study of Capitanata agricultural district, for which the reference climate was defined on the basis of data collected at 19 stations with data series lengths ranging from 15 to 62 years. In compliance with the original simplicity of the method, the maps of the precipitation and temperature regimes were obtained with the Thiessen–Voronoi tessellation and with an altitudinal gradient, respectively, and the maximum required evapotranspiration was determined both with Thornthwaite's method and Hargreaves' method, modulated with the FAO procedure proposed by Allen *et al.* (1998). However, the method is flexible and its application is independent of how the input maps are generated. Simulations were carried out in four different land-use scenarios available for the case study area, with reference to the years 1990, 2000, 2006 and 2011. The estimated irrigation deficit, summarized in Table 4 for different land-use classes, was compared with both the values reported by Lamaddalena *et al.* (2004) and the irrigation supply declared by the Consortium, obtaining an overall good agreement: the first comparison produced an error of 6.4 and −1.1% if 1990 and 2006 land-use were considered, respectively, while the second one obtained values of global irrigation efficiency of 82 and 92% (for 1990 and 2006 land-use, respectively), consistent with the drip- and micro-irrigation practices that are commonly and widely adopted in the Capitanata irrigation district. Uncertainty maps were produced and it was observed how the deficit uncertainty is practically attributable to the precipitation interannual variability, while the effects of the interannual variability of the evapotranspiration demand – resulting from the temperature variability according to the applied model – are minimal. The results were then represented on a Budyko curve and on a deficit curve (as

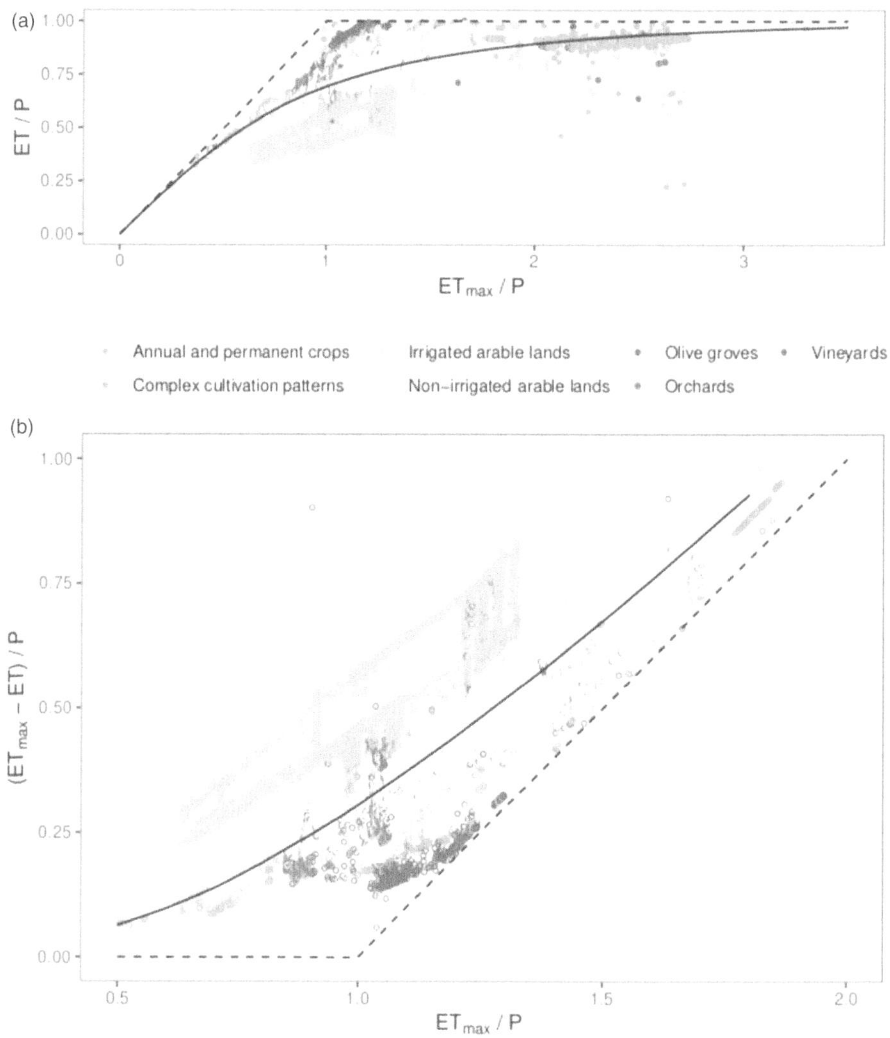

Figure 12 | (a) Budyko curve and calculated water balance components for each cell of the Capitanata district in the 1990 land-cover scenario, clustered for the land-cover classes. (b) Irrigation-deficit curves based on the Budyko curve and on its limit curve, and calculated irrigation deficits for the 1990 land-cover scenario, clustered for the land-cover classes (zoom on the range $0.5 < ET_{max}/P < 2$).

a function of the aridity index) derived from the Budyko curve. It was thus observed that most of the crops (apart from the crops grown on irrigated arable land) are well identified by the area between the Budyko curve and the limit curves, the latter of which corresponds to the minimization of the irrigation deficit. The case is instead different for the irrigated arable land which, being able to benefit from irrigation, deviates from the natural hydrological behavior of the soil, thus showing a greater irrigation demand. It is our main conclusion that, due to its simplicity and the small number of needed parameters, Benfratello's method (although developed in the 1960s) might still be regarded as an effective tool to assess the effects of climatic, land-use and anthropogenic change scenarios on the soil water balance and on the irrigation deficit.

ACKNOWLEDGEMENTS

This study was supported by the project *SWaRM-Net Smart Water Resource Management Networks*[2], by the Water Research Institute (IRSA) of the Italian National Research Council (CNR) and by the University of Brescia. We greatfully thank the Centro Funzionale Decentrato of the Apulia Region for providing the data, and Dr Emanuele Romano (CNR-IRSA). This

[2] https://www.cnr.it/it/progetti-di-ricerca/progetto/39395/swarm-net-dta-ad002-554

paper is dedicated to the memory of our friend and colleague Prof. Giuseppe Provenzano, who passed away in December 2022. He was part of the School of Agricultural Hydrology in Palermo, founded by Guglielmo Benfratello, and when we started to deal with Benfratello's water balance method he gave us inspiring and encouraging suggestions to develop our research.

DATA AVAILABILITY STATEMENT

Data cannot be made publicly available; readers should contact the corresponding author for details.

CONFLICT OF INTEREST

The authors declare there is no conflict.

REFERENCES

Allen, R. G., Pereira, L. S., Raes, D. & Smith, M. 1998 Crop evapotranspiration guidelines for computing crop water requirements-FAO irrigation and drainage paper 56. *FAO, Rome* **300** (9), D05109.

Ancona, V., Bruno, D. E., Lopez, N., Pappagallo, G. & Uricchio, V. F. 2010 A Modified Soil Quality Index to Assess the Influence of Soil Degradation Processes on Desertification Risk: The Apulia Case. *Italian Journal of Agronomy* **5** (53), 45–56.

Arias, P., Bellouin, N., Coppola, E., Jones, R., Krinner, G., Marotzke, J., Naik, V., Palmer, M., Plattner, G.-K., Rogelj, J., Rojas, M., Sillmann, J., Storelvmo, T., Thorne, P., Trewin, B., Rao, K. A., Adhikary, B., Allan, R., Armour, K., Bala, G., Barimalala, R., Berger, S., Canadell, J., Cassou, C., Cherchi, A., Collins, W., Collins, W., Connors, S., Corti, S., Cruz, F., Dentener, F., Dereczynski, C., Luca, A. D., Niang, A. D., Doblas-Reyes, F., Dosio, A., Douville, H., Engelbrecht, F., Eyring, V., Fischer, E., Forster, P., Fox-Kemper, B., Fuglestvedt, J., Fyfe, J., Gillett, N., Goldfarb, L., Gorodetskaya, I., Gutierrez, J., Hamdi, R., Hawkins, E., Hewitt, H., Hope, P., Islam, A., Jones, C., Kaufman, D., Kopp, R., Kosaka, Y., Kossin, J., Krakovska, S., Lee, J.-Y., Li, J., Mauritsen, T., Maycock, T., Meinshausen, M., Min, S.-K., Monteiro, P., Ngo-Duc, T., Otto, F., Pinto, I., Pirani, A., Raghavan, K., Ranasinghe, R., Ruane, A., Ruiz, L., Sallée, J.-B., Samset, B., Sathyendranath, S., Seneviratne, S., Sörensson, A., Szopa, S., Takayabu, I., Tréguier, A.-M., van den Hurk, B., Vautard, R., von Schuckmann, K., Zaehle, S., Zhang, X. & Zickfeld, K. 2021 Technical summary. In: Masson-Delmotte V., Zhai P. , Pirani A., Connors S. L., Péan C., Berger S., Caud N., Chen Y., Goldfarb L., Gomis M. I., Huang M., Leitzell K., Lonnoy E., Matthews J. B. R., Maycock T. K., Waterfield T., Yelekçi O., Yu R. & Zhou B. (eds). *Climate Change 2021: The Physical Science Basis. Contribution of Working Group I to the Sixth Assessment Report of the Intergovernmental Panel on Climate Change.* Cambridge University Press, Cambridge, United Kingdom; New York, NY, USA, pp. 33–144.

Baltas, E. 2007 Spatial distribution of climatic indices in northern Greece. *Meteorological Applications* **14** (1), 69–78. Available from: https://rmets.onlinelibrary.wiley.com/doi/abs/10.1002/met.7

Bandini, A. 1931 Tipi pluviometrici dominanti sulle regioni italiane. In: *Il Servizio Idrografico Italiano. XV Congresso Internazionale di Navigazione.* Min LL. PP., Regno d'Italia, Tipografia del Senato, Roma, pp. 111–118.

Bartolini, E., Allamano, P., Laio, F. & Claps, P. 2011 Runoff regime estimation at high-elevation sites: a parsimonious water balance approach. *Hydrology and Earth System Sciences* **15** (5), 1661–1673. Available from: https://hess.copernicus.org/articles/15/1661/2011/

Benfratello, G. 1961 Contributo allo studio del bilancio idrologico del terreno agrario. *L'Acqua* (2), 34–53.

Benfratello, G. 1964 Programmazione di un computo di idrologia su calcolatrice elettronica numerica. In: *Atti della Accademia di Scienze, Lettere e Arti di Palermo, Serie IV, Vol. XXIII, Parte I, Anno Accademico 1962–1963.* Accademia di Scienze, Lettere e Arti di Palermo, Palermo, Italy, pp. 197–235.

Benfratello, G., Melisenda Giambertoni, I. & Santoro, M. 1979 Hydrogeological research aimed at the exploitation of aquifers in some areas in sicily. In: *Proceedings of the 18th IAHR World Congress 'Hydraulic Engineering in Water Resources Development and Management', vol. 5: Subject D – Physical and Mathematical Models for Watersheds.* IAHR, Cagliari, pp. 135–143.

Braca, G., Bussettini, M., Ducci, D., Lastoria, B. & Mariani, S. 2019 Evaluation of national and regional groundwater resources under climate change scenarios using a Gis-based water budget procedure. *Rendiconti Lincei. Scienze Fisiche E Naturali* **30** (1), 109–123. https://doi.org/10.1007/s12210-018-00757-6.

Braca, G., Bussettini, M., Lastoria, B., Mariani, S. & Piva, F. 2021 *Il Bilancio Idrologico Gis Based a scala Nazionale su Griglia regolare – BIGBANG: metodologia e stime. Rapporto sulla disponibilità naturale della risorsa idrica.* Rapporto 339/21, Istituto Superiore per la Protezione e la Ricerca Ambientale, Roma.

Brussolo, E., Palazzi, E., von Hardenberg, J., Masetti, G., Vivaldo, G., Previati, M., Canone, D., Gisolo, D., Bevilacqua, I., Provenzale, A. & Ferraris, S. 2022 Aquifer recharge in the Piedmont Alpine zone: historical trends and future scenarios. *Hydrology and Earth System Sciences* **26** (2), 407–427. Available from: https://hess.copernicus.org/articles/26/407/2022/

Budyko, M. I. 1974 *Climate and Life. No. Vol. 18 in International Geophysics Series.* Academic Press. Available from: https://proxy.unibs.it/login?url=https://search.ebscohost.com/login.aspx?direct=true&db=nlebk&AN=297011&lang=it&site=eds-live&scope=site

Casadei, S., Mannocchi, F., Mecarelli, P., 1993 Operational hydrology and reservoir management in the upper tiber river basin. In: *Extreme Hydrological Events: Precipitation, Floods and Droughts, no. 213 in IAHS Publications* (Kundzewicz, Z., Rosbjerg, D., Simonovic, S. & Takeuchi, K., eds). Int Assoc Hydrol Sci; Int Assoc Meteorol & Atmospher Phys; UNESCO, Int Assoc Hydrological Sciences. International Symposium on Extreme Hydrological Events: Precipitation, Floods and Droughts, Yokohama, Japan, Jul 20–23, 1993, pp. 393–402.

Chen, D., Shams, S., Carmona-Moreno, C. & Leone, A. 2010 Assessment of open source GIS software for water resources management in developing countries. *Journal of Hydro-Environment Research* **4** (3), 253–264. Available from: https://www.sciencedirect.com/science/article/pii/S1570644310000511

da Silva Tavares, P., Giarolla, A., Chou, S. C., de Paula Silva, A. J. & de Arruda Lyra, A. 2018 Climate change impact on the potential yield of Arabica coffee in southeast Brazil. *Regional Environmental Change* **18** (3), 873–883.

De Varennes e Mendonça, P. 1958 Sobre o novo método de balanço hidrológico do solo de Thornthwaite-Mather. *Anais do Instituto Superior de Agronomia* **22**, 271.

de Vito, R., Pagano, A., Portoghese, I., Giordano, R., Vurro, M. & Fratino, U. 2019 Integrated Approach for Supporting Sustainable Water Resources Management of Irrigation Based on the WEFN Framework. *Water Resources Management* **33** (4), 1281–1295.

Dourado-Neto, D., van Lier, J., d, Q., Metselaar, K., Reichardt, K. & Nielsen, D. R. 2010 General procedure to initialize the cyclic soil water balance by the Thornthwaite and Mather method. *Scientia Agricola* **67**, 87–95.

Dragonetti, G., Sengouga, A., Comegna, A., Lamaddalena, N., Basile, A., Coppola, A. 2020 SIRR-MOD–A Decision Support System for Identifying Optimal Irrigation Water Needs at Field and District Scale. In: *MID-TERM AIIA 2019: Innovative Biosystems Engineering for Sustainable Agriculture, Forestry and Food Production, vol. 67 of Lecture Notes in Civil Engineering* (Coppola, A., Di Renzo, G. C., Altieri, G. & D'Antonio, P., eds). Springer International Publishing, Cham, pp. 117–124.

Dumitru, S., Chendes, V., Cojocaru, G. & Simota, C. 2009 Developing an integrated crop-meteorological model for hilly terrains using digital terrain models. *Scientific Papers. Series A* **LII**, 147–152.

D'Urso, G., Menenti, M. & Santini, A. 1999 Regional application of one-dimensional water flow models for irrigation management. *Agricultural Water Management* **40** (2-3), 291–302.

FAO 2021 *Tracking Progress on Food and Agriculture-Related SDG Indicators 2021: A Report on the Indicators Under FAO Custodianship*. FAO, Rome. https://doi.org/10.4060/cb6872en.

Farthing, M. W. & Ogden, F. L. 2017 Numerical solution of Richards' equation: a review of advances and challenges. *Soil Science Society of America Journal* **81** (6), 1257–1269. Available from: https://acsess.onlinelibrary.wiley.com/doi/abs/10.2136/sssaj2017.02.0058

Foxx, T. S., Tierney, G. D. & Williams, J. M. 1984 *Rooting Depths of Plants Relative to Biological and Environmental Factors*. Tech. rep., Los Alamos National Lab., NM, USA.

Guyennon, N., Romano, E. & Portoghese, I. 2016 Long-term climate sensitivity of an integrated water supply system: the role of irrigation. *Science of The Total Environment* **565**, 68–81. Available from: https://www.sciencedirect.com/science/article/pii/S0048969716308518

Guyennon, N., Salerno, F., Portoghese, I. & Romano, E. 2017 Climate change adaptation in a Mediterranean semi-arid catchment: testing managed aquifer recharge and increased surface reservoir capacity. *Water* **9** (9). Available from: https://www.mdpi.com/2073-4441/9/9/689

Lamaddalena, N., Hamdy, A., Todorovic, M. & Quarto, A. 2004 Participatory water management in Italy: Case study of the Consortium Bonifica della Capitanata. *Options Mediterranennes Series B* **48**, 159–169.

Lamaddalena, N., Caliandro, A., Daccache, A., D'Agostino, D., Scardigno, A. & De Santis, S. 2008 Riorientamenti produttivi del territorio agricolo pugliese per uno sviluppo rurale sostenibile. *Regione Puglia and CIHEAM-IAMB*, 93pp.

Licciardello, F., Rossi, C. G., Srinivasan, R., Zimbone, S. M. & Barbagallo, S. 2011 Hydrologic evaluation of a Mediterranean watershed using the SWAT model with multiple PET estimation methods. *Transactions of the ASABE* **54** (5), 1615–1625. https://doi.org/10.13031/2013.39840.

Mantica, I. 1975 Primi risultati di un'Indagine circa la correlazione tra afflussi e deflussi nei bacini abruzzesi con corso d'acqua con foce sul litorale adriatico. In: *Notizie dell'economia teramana: pubblicazione mensile della Camera di commercio, industria e agricoltura di Teramo*. C.E.T.I., Teramo, pp. 20–47.

Melisenda, I. 1964 Sui calcoli idrologici per il terreno agrario: influenza del 'clima'. *L'Acqua* (4), 3–20.

Melisenda, I. 1965 Sul calcolo dell'evapotraspirazione nell'ambiente siciliano. In: *Atti della Accademia di Scienze, Lettere e Arti di Palermo, Serie IV, Vol. XXIV, Parte I, Anno Accademico 1963–1964*. Accademia di Scienze, Lettere e Arti di Palermo, Palermo, Italy, pp. 291–328.

Melisenda, I. 1967a Possibilité de déduction des besoins en eau d'irrigation à partir de la loi de dessèchement du sol. *La Houille Blanche* **53** (8), 829–839.

Melisenda, I. 1967b Su una classificazione del clima basata sui parametri dell'idrologia agraria. In: *Atti della Accademia di Scienze, Lettere e Arti di Palermo, Serie IV, Vol. XXVI, Parte I, Anno Accademico 1965–1966*. Accademia di Scienze, Lettere e Arti di Palermo, Palermo, Italy, pp. 381–417.

Nistor, M.-M., Man, T. C., Benzaghta, M. A., Nedumpallile Vasu, N., Dezsi, Ş. & Kizza, R. 2018 Land cover and temperature implications for the seasonal evapotranspiration in Europe. *Geographia Technica* **13** (1), 85–108.

Peel, M. C., Finlayson, B. L. & McMahon, T. A. 2007 Updated world map of the Köppen-Geiger climate classification. *Hydrology and Earth System Sciences* **11** (5), 1633–1644. Available from: https://hess.copernicus.org/articles/11/1633/2007/

Pereira, L., Paredes, P. & Jovanovic, N. 2020 Soil water balance models for determining crop water and irrigation requirements and irrigation scheduling focusing on the FAO56 method and the dual kc approach. *Agricultural Water Management* **241**, 106357. Available from: https://www.sciencedirect.com/science/article/pii/S0378377420303930

Portoghese, I., Uricchio, V. & Vurro, M. 2005 A GIS tool for hydrogeological water balance evaluation on a regional scale in semi-arid environments. *Computers & Geosciences* **31** (1), 15–27. Available from: https://www.sciencedirect.com/science/article/pii/S009830040400144X

Quignones, R. 1974 Correlazione tra afflussi e deflussi dei principali bacini siciliani. *L'ingegnere* **49**, 243–265.

Romano, N. 2014 Soil moisture at local scale: measurements and simulations. *Journal of Hydrology* **516**, 6–20. Determination of soil moisture: Measurements and theoretical approaches. Available from: https://www.sciencedirect.com/science/article/pii/S0022169414000328

Santini, A. 1979 Water balance of the soil-plant system: the use of a mathematical simulation model. *Rivista di ingegneria agraria* **10** (2), 63–76.

Santini, A. 1992 Modelling water dynamics in the soil-plant-atmosphere system for irrigation problems. *Excerpta of the Italian Contributions to the Field of Hydraulic Engineering* **6**, 133–166.

Santoro, M. 1970 Sul calcolo numerico del bilancio idrologico del terreno agrario. In: *Atti della Accademia di Scienze, Lettere e Arti di Palermo, Serie IV, Vol. XXIX, Parte I, Anno Accademico 1968–1969*. Accademia di Scienze, Lettere e Arti di Palermo, Palermo, Italy, pp. 359–394.

Santoro, M. 1991 Il bilancio idrico del terreno agrario. In: Corso di Aggiornamento: Modelli Idrologici Superficiali nella Pianificazione di Bacino, 10-14 giugno 1991. Programma di Istruzione Permanente del Politecnico di Milano, CittàStudi, Milano, pp. 443–487.

Scalenghe, R. & Ferraris, S. 2009 The First Forty Years of a Technosol. *Pedosphere* **19** (1), 40–52. Available from: https://www.sciencedirect.com/science/article/pii/S100201600860082X

Thornthwaite, C. W. 1948 An approach toward a rational classification of climate. *Geographical Review* **38** (1), 55–94. Available from: http://www.jstor.org/stable/210739

Thornthwaite, C. & Mather, J. 1955a The Water Budget and Its Use in Irrigation. In: *The Yearbook of Agriculture 1955: Water*. U.S. Government Printing Office, Washington, DC, pp. 346–358.

Thornthwaite, C. W. & Mather, J. R 1955b The water balance. In: *Publications in Climatology*, **VIII** (1), Drexel Institute of Technology, Laboratory of Climatology, Centerton, New Jersey.

Thornthwaite, C. W. & Mather, J. R. 1957 Instructions and tables for computing potential evapotranspiration and the water balance. In: *Publications in Climatology*, **X** (3), Drexel Institute of Technology, Laboratory of Climatology, Centerton, New Jersey.

UN–Water 2022 *The United Nations World Water Development Report 2022: Groundwater: Making the Invisible Visible. The United Nations World Water Development Report*. United Nations Educational, Scientific and Cultural Organization, Paris, France.

Ventrella, D., Charfeddine, M., Moriondo, M., Rinaldi, M. & Bindi, M. 2012 Agronomic adaptation strategies under climate change for winter durum wheat and tomato in southern Italy: irrigation and nitrogen fertilization. *Regional Environmental Change* **12**, 407–419. https://doi.org/10.1007/s10113-011-0256-3.

Verma, S., Verma, R. K., Singh, A., Naik, N. S., 2012 Web-Based GIS and Desktop Open Source GIS Software: An Emerging Innovative Approach for Water Resources Management. In: *Advances in Computer Science, Engineering & Applications* (Wyld, D. C., Zizka, J. & Nagamalai, D., eds). Springer Berlin Heidelberg, Berlin, Heidelberg, pp. 1061–1074.

Viparelli, C. 1956a Bilanci idrologici di alcuni gruppi di bacini impermeabili dell'Italia meridionale e insulare. *L'Acqua*.

Viparelli, C. 1956b Possibilità di regolazione dei bacini impermeabili dell'Italia meridionale. *L'Acqua*.

Viparelli, R. & Versace, P. 1979 *Distribuzione di probabilità dei deflussi D. Tech. Rep. 421*. Università di Napoli, Istituti Idraulici.

Vitale, D., Rana, G. & Soldo, P. 2010 Trends and Extremes Analysis of Daily Weather Data from a Site in the Capitanata Plain (Southern Italy). *Italian Journal of Agronomy* **5** (2), 133–144.

Walker, W. 1989 *Guidelines for Designing and Evaluating Surface Irrigation Systems*. FAO, Rome.

Xu, C.-Y. & Singh, V. P. 1998 A Review on Monthly Water Balance Models for Water Resources Investigations. *Water Resources Management* **12** (1), 20–50. https://doi.org/10.1023/A:1007916816469.

First received 30 July 2022; accepted in revised form 21 February 2023. Available online 10 March 2023

doi: 10.2166/nh.2022.100

Modelling annual maximum daily rainfall with the STORAGE (STOchastic RAinfall GEnerator) model

Andrea Petroselli [a], Davide Luciano De Luca [b], Dariusz Młyński [c,*] and Andrzej Wałęga [c]

[a] Department of Economy, Engineering, Society and Business Organization (DEIM), Tuscia University, Viterbo, Italy
[b] Department of Informatics, Modelling, Electronics and System Engineering, University of Calabria, Arcavacata di Rende, Italy
[c] Department of Engineering Sanitary and Water Management, University of Agriculture in Krakow, Krakow, Poland
*Corresponding author. E-mail: dariusz.mlynski@urk.edu.pl

AP, 0000-0003-4943-0928; DLD, 0000-0001-5255-7343; DM, 0000-0002-8925-2591; AW, 0000-0001-6839-4745

ABSTRACT

In this work, the capability of STORAGE (STOchastic RAinfall GEnerator) model for generating long and continuous rainfall series for the upper Vistula basin (southern Poland) is tested. Specifically, in the selected area, only parameters of depth–duration–frequency curves are known for sub-daily rainfall heights (which are usually estimated in an indirect way by using Lambor's equations from daily data), while continuous daily series with a sufficient sample size are available. Attention is focused on modelling the sample frequency distributions of daily annual maximum rainfall. The obtained results are promising for further elaborations, concerning the use of STORAGE synthetic continuous rainfall data as input for a continuous rainfall-runoff approach, to be preferred with respect to classical event-based modelling.

Key words: annual maximum daily rainfall, continuous approach, STORAGE model

HIGHLIGHTS

- STORAGE model as an effective tool in rainfall modelling.
- Possibility of modelling rainfall with different characteristics.
- Promising results in the upper Vistula basin, paving the way for a continuous approach.

GRAPHICAL ABSTRACT

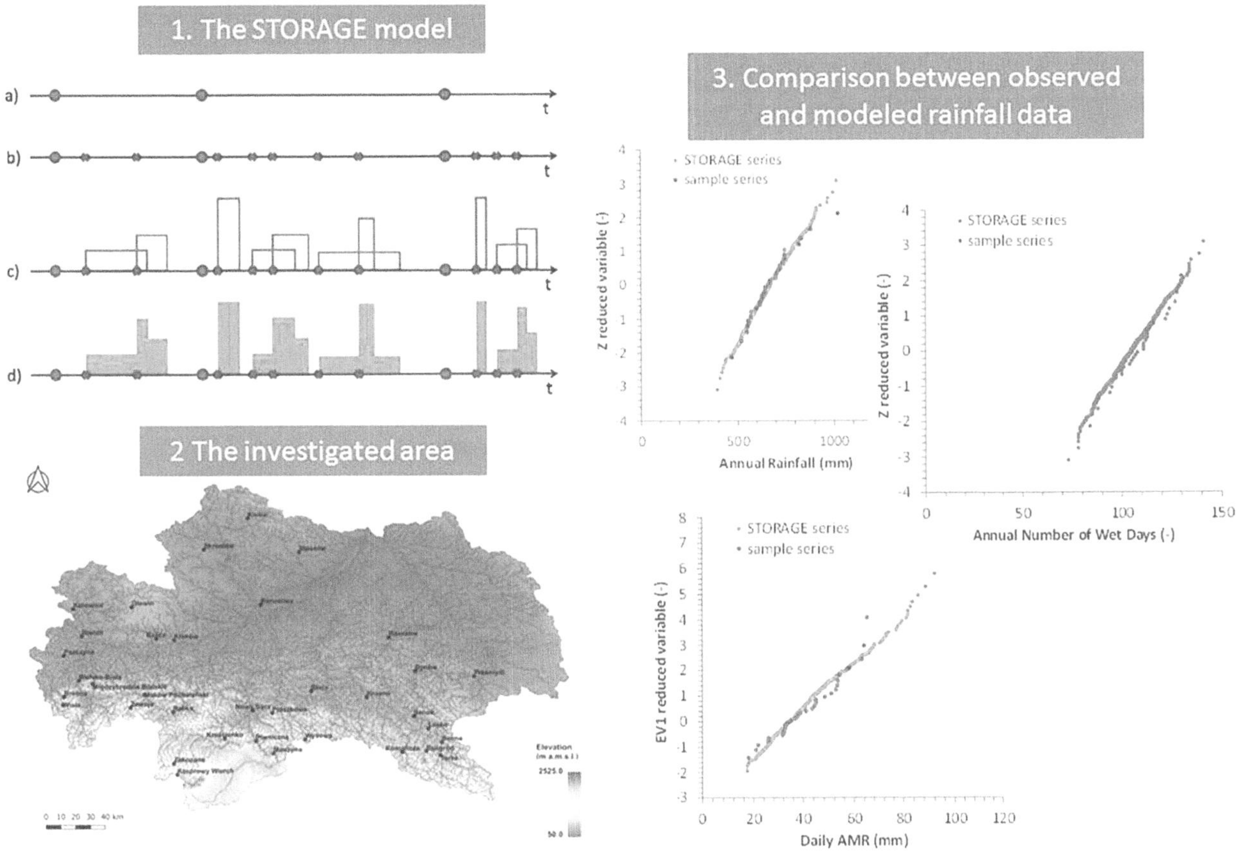

1. INTRODUCTION

Hydro-meteorological data availability plays a crucial role in planning and managing the optimum use of water resources. In recent years, high maintenance costs and the need of trained staff have led to a consistent decrease in the number of observations in hydro-meteorological networks (Turnipseed & Sauer 2010). Hence, the modelling of hydrological or meteorological factors assumes an increasingly important role, for instance in the calculation of design discharges and flood prone areas in scarcely gauged catchments, for which the knowledge of design rainfall is mandatory (Sabzevari 2017). Rainfall temporal distribution, frequency, and duration are the basic information necessary to correctly estimate the runoff amount in such catchments (Wałęga *et al.* 2016; Petroselli *et al.* 2020). This is particularly important when using hydrological and hydrodynamic models, which are sensitive to changes in the input parameter values (Mannina *et al.* 2006; Sabzevari *et al.* 2009).

When sufficiently long observed rainfall information is available, statistical methods based on theoretical distribution functions are commonly used to assess the design rainfall. In particular, concerning annual maximum time series, commonly adopted functions are generalized extreme value (GEV), Log-Normal (LN), Weibull, Gumbel, generalized exponential distribution, generalized logistic distribution (Fitzgerald 2005; Boni *et al.* 2006; Villarini *et al.* 2011; Kaźmierczak & Kotowski 2012, 2015; Młyński *et al.* 2019a, 2019b).

Another solution is taking advantage of radar observations that also consider the spatial variability of the rainfall field, in order to reconstruct the sequences of rainfall characteristics, and then using them to determine the depth–duration–frequency (DDF) curves (Wright *et al.* 2013). Since there is a lack of radar meteorological data availability in some areas of the world, there is often the need to use the more widely available satellite-based products (Siddique-E-Akbor *et al.* 2014). Generally, the satellite and radar products underestimate extreme rainfall in comparison to gauge-based observations (Bharti *et al.* 2016; Chen *et al.* 2020). This is mainly due to the different spatial resolutions, i.e. satellite or radar outputs are average estimates on areal pixels, while raingauge data refer to measures in specific points. Spatial resolution refers to the size of one pixel on the land. Landsat data, for example, has a 30 m resolution which is a medium-resolution image. It could cover bigger area alone, but this resolution level is not enough to distinguish individual objects like houses or cars. Currently, the following resolution levels are provided by commercial satellites: low (over 60 m per pixel), medium (from 30 to 10 m per pixel) and high to very high (from 5 m to 30 cm per pixel). For instance, Derin *et al.* (2019) investigated the ability of six satellite-based products to estimate extreme rainfall in nine mountainous locations with a dense raingauge network, determining a constant underestimation of extreme rainfall values. Consequently, satellite or radar estimates should require downscaling or bias correction methods (Willems *et al.* 2012; Maraun 2016) for hydrological analyses. Other methods using satellite products to indirectly estimate rainfall from soil moisture can be used like SM2RAIN (Brocca *et al.* 2018; Dari *et al.* 2020).

However, the evaluation of the design rainfall as input for the calculation of the design discharge (also named as event-based approach) can be affected by several relevant assumptions: the hyetograph shape, the critical rainfall duration, the pre-event soil moisture condition and the implicit hypothesis, assuming that the design storm and the design hydrograph have the same return period. So, in order to address the major drawbacks of event-based approaches, continuous rainfall-runoff models have been recently developed (Grimaldi *et al.* 2012, 2021; Biondi & De Luca 2015), but they require long and continuous rainfall series at high resolutions (which are not so long or available in many areas of the world) as input.

In this context, synthetic rainfall generators (SRGs) are able to produce long and continuous rainfall series at the desired time resolution. Their calibration is carried out using the usually scarce information derived from data at sub-daily resolution (which should correspond to long high-resolution continuous data, but, in many cases, it only refers to the knowledge of annual maxima and DDF curves) and the more available and longer continuous daily time series.

For instance, Iwański & Kuchar (2003) described a SRG using daily series of meteorological data for a single geographical site with a known climatic profile. Pierzga & Jarosińska (2014) presented a SRG that used annual maximum daily rainfall data that later were disaggregated into hourly values, achieving to generate many synthetic hyetographs. Agilan & Umamahesh (2019) proposed a modified version of a K-nearest neighbour weather generator that incorporates nonstationarity in the extreme rainfall series. Zhao *et al.* (2019) described a daily, spatially distributed, SRG based on a first-order Markov chain model to simulate three types of rainfall, i.e. convective, frontal and tropical storms.

Among the SRGs, the Neyman–Scott Rectangular Pulse (NSRP) (Rodriguez-Iturbe *et al.* 1987; Cowpertwait 1991; Burton *et al.* 2010) is a commonly used model, which is suitable for generating continuous rainfall series at several time scales, from fine (e.g. 1 or 5 min) to coarse ones (e.g. average annual), and capable of preserving the statistical properties of the observed rainfall data (Fowler *et al.* 2000). In a stationary context, De Luca & Galasso (2019) calibrated a NSRP model by using information from extreme value distributions. Their results were promising, because they found acceptable relationships among extreme value distributions and rainfall statistical properties of continuous series, and this constitutes an interesting alternative to calibration when only annual maximum data are available. In a climate change context, De Luca *et al.* (2020a, 2020b) investigated the modelling of climate changes with a NSRP model, while Wasko & Sharma (2017) proposed a novel method for simulating continuous rainfall sequences for future warming climate using one-dimensional NSRP model.

Recently, De Luca & Petroselli (2021) developed a modified NSRP model named as STORAGE (STOchastic RAinfall GEnerator). Its main advantages with respect to a classic NSRP model are:

1. The model calibration is carried out using summary statistics from annual maxima series (from sub-hourly/hourly to daily scales), annual and seasonal cumulative rainfall heights, and annual number of wet days, which are usually longer than continuous observed high-resolution series (mainly adopted for SRGs calibration but typically very short or absent in many locations) and
2. The seasonality is modelled by using series of goniometric functions. This approach makes STORAGE more parsimonious with respect to the use of monthly or seasonal sets for parameters.

In the present work, we investigated the performances of the STORAGE model for the raingauge network of the upper Vistula basin, which is located in Poland. In the selected study area, only parameters of DDF curves are known for sub-daily rainfall heights, which are usually estimated in an indirect way by using Lambor's equations from daily data (Młyński *et al.* 2020), while continuous daily series with a sufficient sample size are available. In particular, the goal of the present work is to assess the capability of STORAGE in reconstructing the annual maxima of daily rainfall, which can constitute a starting point for further elaborations, related to the use of STORAGE synthetic high-resolution rainfall data as input for a continuous approach aiming in evaluating the design discharge in scarcely gauged basins.

2. STUDY AREA

The study was conducted for 36 rainfall stations that are located in the upper Vistula basin, Poland (Figure 1). This area covers about 25% of the entire Vistula basin and about 15% of the whole Poland. The region is divided into three main physiographic units: Carpathian's, uplands and flats. The water resources of the upper Vistula basin are the highest in the entire Poland. The water cycle in the mountain part is very dynamic, due to steep relief and low permeability of the ground, what causes that floods events develop rapidly here (Pociask-Karteczka 2016). The elevation of rainfall stations ranges from 171 to 1,991 m a.s.l. The used data were continuous daily rainfall series, which were recorded in the period 1989–2018, and the parameters of the DDF curves, which were estimated in an indirect way by using Lambor's equations from daily data (Młyński *et al.* 2020). The data were obtained from the public database of the Institute of Meteorology and Water Management, National Research Institute in Warsaw, Poland. Table A1 in Supplementary Material, Appendix 1 presents names, coordinates (ETRS89/Poland CS92, authority ID: EPSG:2180), elevation of analysed stations, and the mean annual precipitation (MAP), the mean annual number of wet days (MANWD) and the mean values for the seasons DJF (December–January–February), MAM (March–April–May), JJA (June–July–August) and SON (September–October–November). In this work, a wet day is defined when the daily rainfall height is at least equal to 1 mm.

3. METHODOLOGY

For each investigated raingauge, the study involved the following steps: (1) extracting the correspondent observed series of daily annual maximum rainfall (AMR, also indicated with P_{max} in the following) and determining its statistical properties; (2) analysing the trends in the observed P_{max} time series, in order to justify the use of STORAGE model in a stationary context and (3) generating the P_{max} using the STORAGE model and analysing the uncertainty due to the sample size.

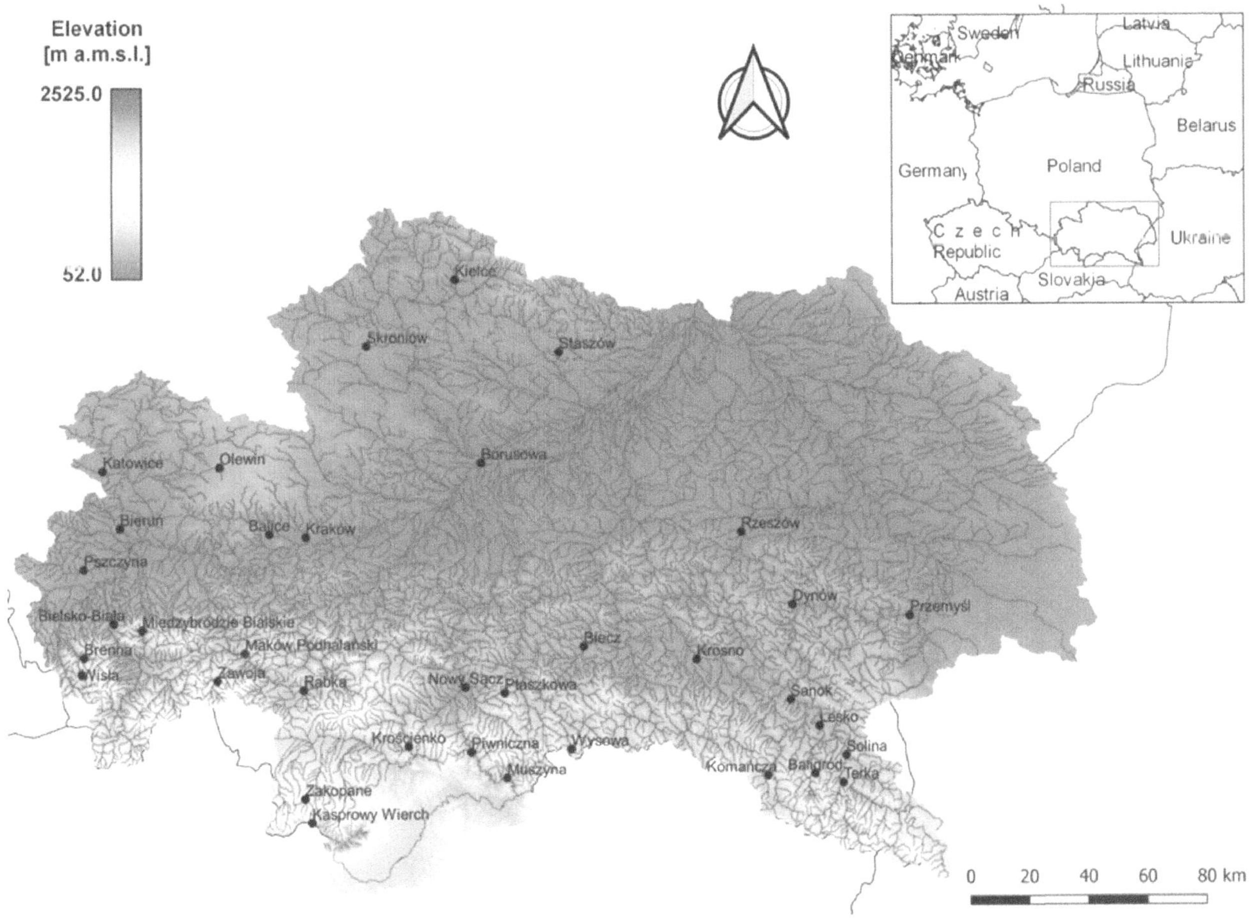

Figure 1 | Location of analysed rainfall stations in the upper Vistula basin.

Then, focusing on some specific raingauges as examples, STORAGE performances (in terms of comparison among observed and generated data for annual maxima, cumulative annual and seasonal rainfall data, plus annual number of wet days) are shown and discussed.

3.1. Statistical analysis

The statistical analysis of P_{max} observed time series regarded the estimation of minimum (*min*), mean, maximum (*max*), standard deviation (*s*) and coefficient of variation (C_v) of the sample, together with the modelling of these series with the LN distribution (Kottegoda & Rosso 2008), which is commonly adopted for annual maxima data in Poland. However, Extreme Value Type 1 (EV1) distribution (Kottegoda & Rosso 2008) is also suitable for P_{max} time series in Poland, as shown in Section 4 for some raingauge data series.

3.2. Trend analysis

Trend analysis for P_{max} observed time series was conducted using the modified Mann–Kendall (MK) test. The test considers the relative magnitudes of sample data rather than the data values themselves, and it does not require the data to follow any specific probability distribution. The null hypothesis H_0 of test indicates a lack of monotonic trend for the time series. The alternative hypothesis H_1 claims its existence. The *S* statistic of the modified Mann–Kendall test is expressed as (Wałęga *et al.* 2016; Ali *et al.* 2019; Agbo *et al.* 2021) follows:

$$S = \sum_{k=1}^{n-1} \sum_{j=k+1}^{n} sgn(x_j - x_k) \tag{1}$$

$$sgn(x_j - x_k) = \begin{cases} 1 \text{ for } (x_j - x_k) > 0 \\ 0 \text{ for } (x_j - x_k) = 0 \\ -1 \text{ for } (x_j - x_k) < 0 \end{cases} \tag{2}$$

where n is the number of elements of the time series, x_j is the observation at time j and x_k is the observation at time k.

The *S* is approximately normally distributed with the mean $E[S]$ equal to 0 and the variance $Var(S)$ is computed as follows:

$$Var(S) = \frac{1}{18} \cdot (n \cdot (n-1) \cdot (2 \cdot n + 5)) \tag{3}$$

The standardized test statistic for the modified Mann–Kendall test $Z(S)$ was calculated based on the following equation:

$$Z(S) = \frac{S - sgn(S)}{Var(S)^{1/2}} \tag{4}$$

Positive values of Z statistic indicate increasing trends, while negative values of Z show decreasing trends. Moreover, once chosen a specific significance level and calculated the theoretical quantile $Z_{1-(\alpha/2)}$, the null hypothesis of no trend is rejected when $|Z_S| > Z_{1-(\alpha/2)}$. For the frequently adopted value of $\alpha = 0.05$, the null hypothesis of no trend is rejected when $|Z_S| > 1.96$.

The main assumption of the modified Mann–Kendall test is the lack of autocorrelation in analysed time series. Because the autocorrelation leads to underestimation of the variance *S*, the following adjustment for variance corrections was included only for data where the significant autocorrelation was detected (Sridhar & Raviraj 2017; Młyński *et al.* 2019a, 2019b):

$$Var^*(S) = Var(S) \cdot \frac{n}{n_s^*} \tag{5}$$

where n/n_s^* is a correction factor, which is calculated as follows:

$$\frac{n}{n_s^*} = 1 + \frac{2}{n(n-1)(n-2)} \cdot \sum_{k=1}^{n-1} (n-k)(n-k-1)(n-k-2)\rho_k \tag{6}$$

where n is the number of elements of the time series, n_s^* is the effective number of observations to account for the autocorrelation in the data and ρ_k is the autocorrelation function for the ranks of the observations.

It should be remarked that a more comprehensive trend investigation should regard fine and coarse time scales and not only the extremes, but if high-resolution data are scarce or absent, then trend/no trend hypothesis for specific fine time scales can only be assumed on the basis of the obtained MK test results for daily data. Moreover, since the present study aims at evaluating the design discharge, the extreme analysis is sufficient for investigation purposes.

3.3. Generation of daily AMR using the NSRP model and analysis of uncertainty due to the sample size

Authors adopted a modified version of the NSRP (Rodriguez-Iturbe *et al.* 1987; Cowpertwait 1991; Burton *et al.* 2010) model, named STORAGE – STOchastic RAinfall GEnerator (De Luca *et al.* 2020a, 2020b; De Luca & Petroselli 2021). The NSRP model usually presents a set of parameters, which represent mean values for (1) inter-arrival time between the origins of two consecutive storms; (2) the number of rain cells, also indicated as bursts or pulses, inside a specific storm; (3) waiting time between a specific burst origin and the origin of the associated storm; (4) intensity and (5) duration of the bursts. Such parameters are estimated by minimizing an objective function (OF), which is defined as the sum of residuals between the considered summary statistics of the observed data at chosen time resolutions and their theoretical expressions. The summary statistics are typically referred to high-resolution continuous time series, which unfortunately are very short or absent in many locations. Moreover, the main limitation of the basic version of NSRP is that the model fitting to a particular rainfall time series involves highly nonlinear equations, regarding analytical and sample moments, and then the parameters estimation needs nontrivial solutions. Even when the nonlinear sophisticated methods are used, the fitting process is a trial-and-error process consisting in trying different initial parameter values. The process is finished when physically reasonable parameter final values are found that provide a good fit (Morrissey 2009). For this reason, STORAGE was developed, being characterized by two main innovations as respect to the basic versions of the NSRP model. First, the parametric calibration is carried out using data series (e.g. annual maxima rainfall, annual and monthly cumulative rainfall and annual number of wet days), which are usually longer than observed high-resolution series. Second, the seasonality is modelled using series of goniometric functions. This approach makes STORAGE strongly parsimonious as respect to basic NSRP models.

In the present work, STORAGE was applied, for all the considered raingauge stations, by employing as input data, the continuous rainfall daily data, recorded in the period 1989–2018, and the parameters of the DDF curves. MAP, MANWD and mean values of seasonal rainfall (DJF, MAM, JJA and SON) were calculated from daily data for each raingauge (see Table A1 in Supplementary Material, Appendix): they constitute (together with parameters of DDF curves) the input information for STORAGE calibration (De Luca & Petroselli 2021).

Then, a 500-year continuous rainfall series with a resolution of 60 min was generated for each raingauge. It must be highlighted that both stationary and non-stationary time series can be generated with STORAGE, depending on the observed/

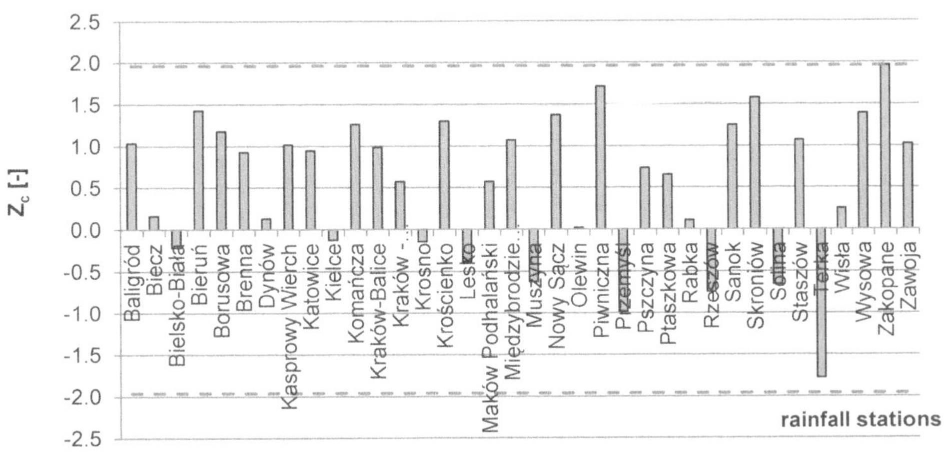

Figure 2 | Results of the *Z* statistics for the Mann–Kendall test.

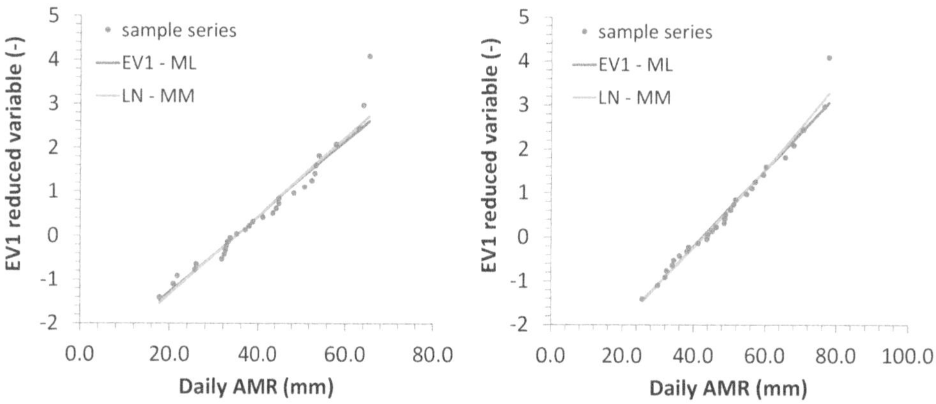

Figure 3 | EV1-ML and LN-MM fitting of sample daily AMR distribution for Kraków-Balice (left) and Solina (right).

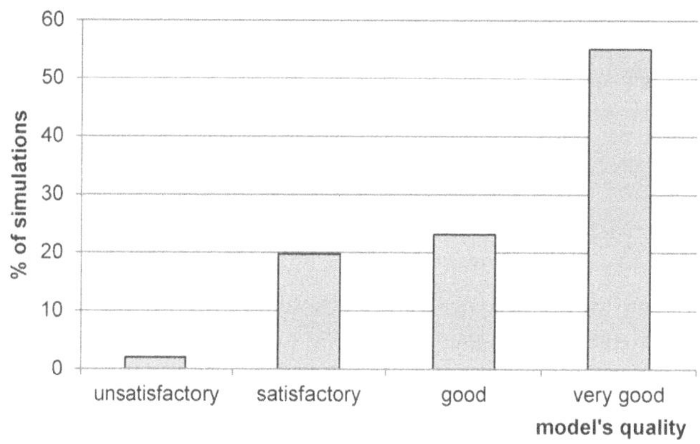

Figure 4 | Summary of the quality of the STORAGE model for the entire upper Vistula basin.

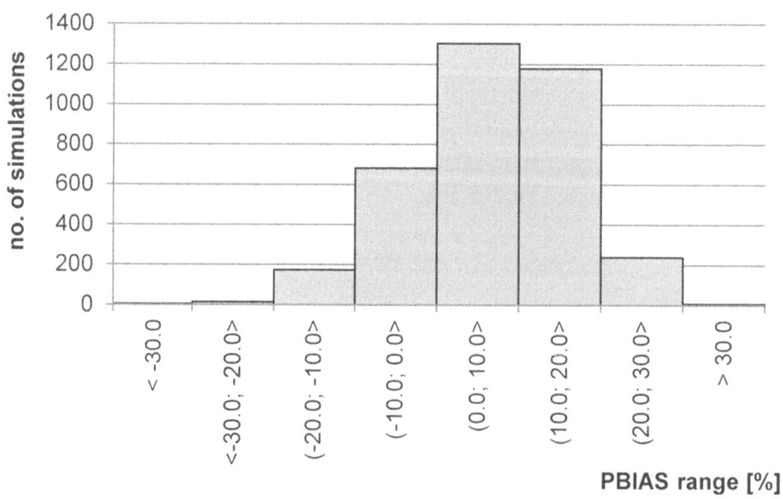

Figure 5 | The values of PBIAS for the upper Vistula basin.

predicted trend. In this work, the stationary version was adopted based on results from the modified Mann-Kendall test (MMK) test. For further details related to the STORAGE application, see De Luca & Petroselli (2021).

Next, for each raingauge, STORAGE daily AMR series were extracted and statistically examined, also considering uncertainty analysis due to the sample size. In detail, from the 500-year series of annual maxima, 100 samples, each one with 30

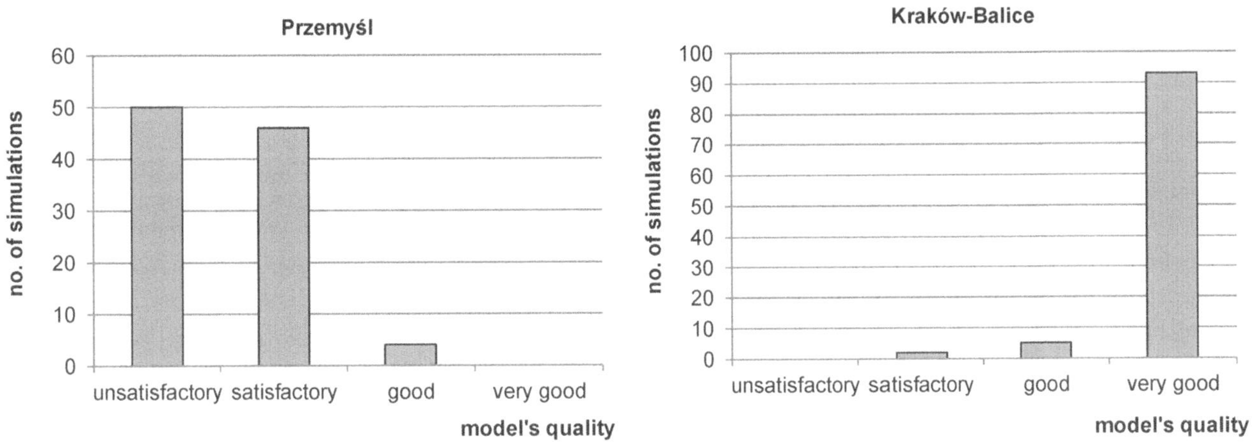

Figure 6 | Summary of the quality of the STORAGE model for the station with the worst (Przemyśl) and best (Kraków-Balice) results.

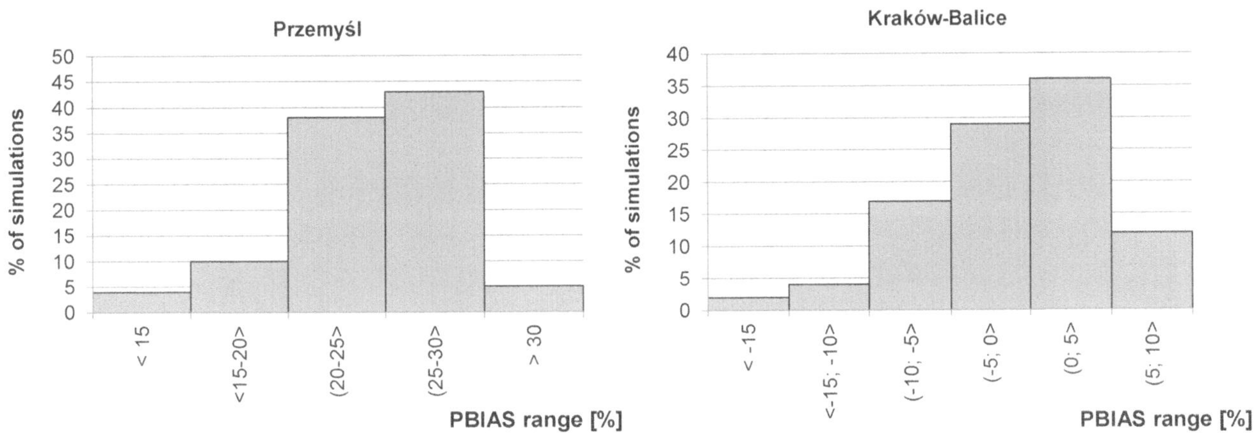

Figure 7 | The values of PBIAS for the station with the worst (Przemyśl) and best (Kraków-Balice) results.

elements, were randomly sampled. After that, the comparison between the observed 30-year time series and each 30-year series generated using the STORAGE model was made by calculating the mean value of *PBIAS* (Kim & Kim 2016; Młyński 2020):

$$PBIAS = \frac{P_T - P_{T_{STORAGE}}}{P_T} \cdot 100[\%] \tag{7}$$

where P_T is the daily design rainfall, for the assumed return period, which is calculated from observed time series, and $P_{T_{STORAGE}}$ is the daily design rainfall, for the assumed return period, which is calculated from the STORAGE model.

Both P_T and $P_{T_{STORAGE}}$ are evaluated by using LN distribution. The mean of *PBIAS* (indicated as m*PBIAS* in the following) was estimated by considering the obtained values from specific return periods (2, 5, 10, 20, 50 and 100 years).

Table 1 | Annual and seasonal summary statistics for Kraków-Balice and Solina raingauges

Raingauge	MAP (mm)	MANWD (–)	DJF (mm)	MAM (mm)	JJA (mm)	SON (mm)
Kraków-Balice	673.2	108.0	100.3	163.9	251.9	157.1
Solina	940.3	129.6	140.1	235.4	343.9	220.9

The performance rating for mean values m*PBIAS* can be expressed as in the following: |m*PBIAS*| < 10 very good; 10 ≤ |m*PBIAS*| < 15 good; 15 ≤ |m*PBIAS*| < 25 satisfactory; |m*PBIAS*| ≥ 25 unsatisfactory (Ajmal *et al.* 2020).

4. RESULTS

4.1. Initial analysis of the observed daily AMR

The preliminary analysis included the calculation of summary statistics of the observed time series: minimum (*min*), mean, maximum (*max*), standard deviation (*s*) and coefficient of variation (C_v). The results are presented in Table A2 in Supplementary Material, Appendix. Analysing the coefficient of variation, it was concluded that 78% of all stations were characterized by medium dynamics of changes in AMR, which was described by C_v values in range from 0.2 to 0.4. Only for the eight stations: Bielsko-Biała, Brenna, Kielce Krościenko, Maków Podhalański, Międzybrodzie Bielskie, Skroniów, Wisła and Zawoja, the values of C_v are greater than 0.4, which means high dynamics of changes in their AMR.

4.2. Analysis of trend in the observed daily AMR

As previously mentioned, trend investigation was carried out for the observed daily AMR series using the modified Mann–Kendall test.

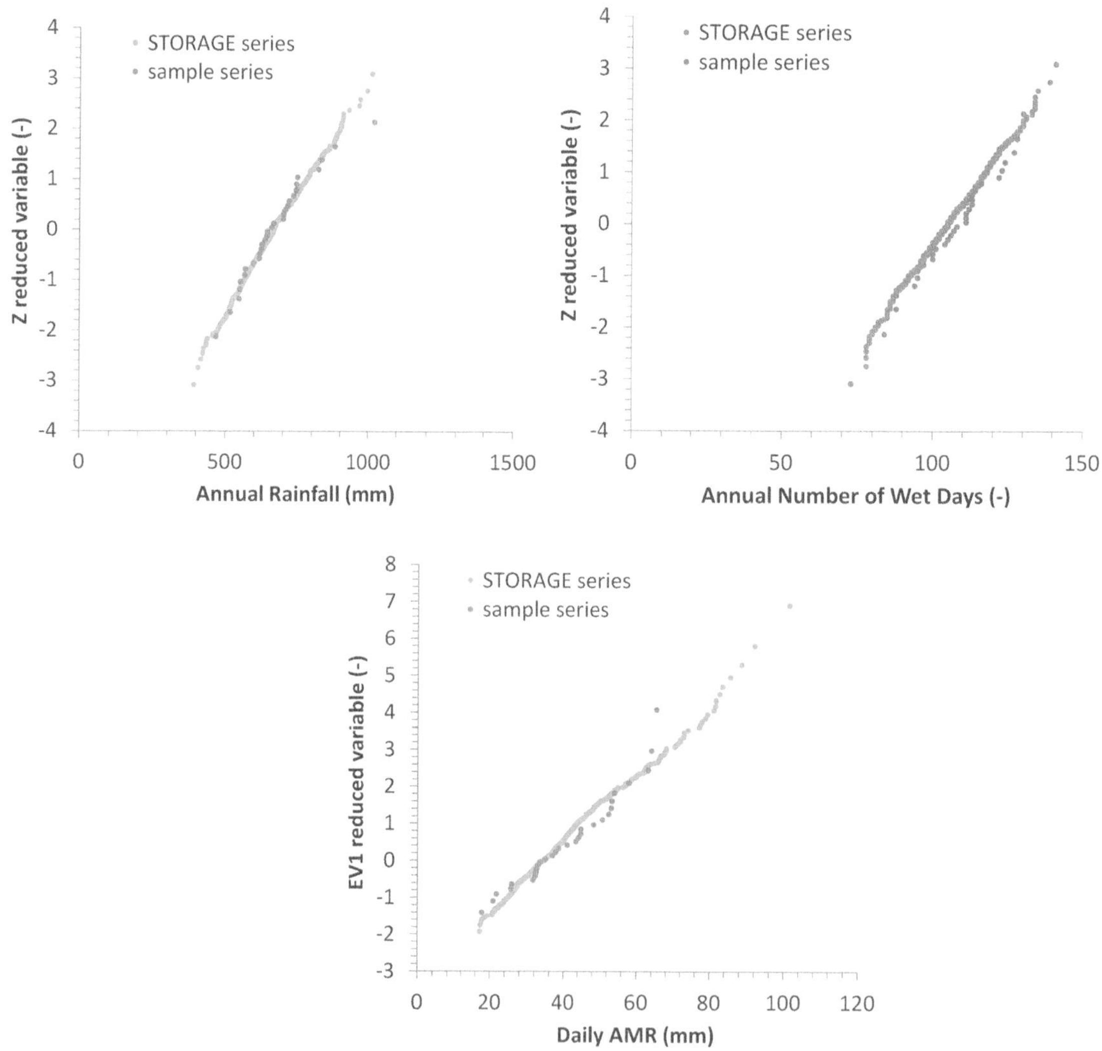

Figure 8 | Kraków-Balice raingauge: STORAGE performances in terms of reproduction of sample frequency distribution for annual rainfall, annual number of wet days and daily AMR.

From the obtained results (Figure 2), it should be emphasized that there was no case where the $|Z|$ statistic is higher than 1.96, and then the hypothesis of stationary processes cannot be rejected. Similar results were obtained by Młyński *et al.* (2018), where significant trends in the annual daily maximum rainfall for the upper Vistula basin have not been found. The study conducted by Kubiak-Wójcicka (2020) has shown a slight increase of rainfall for the Vistula basin, mostly in particular seasons. Seasonal trends were analysed also by Pińskwar *et al.* (2019). The analysis of rainfall extremes showed that maximum daily rainfall mostly increased for the summer half-year and that the increases during summer half-year are more numerous than in winter half-year. However, many changes were weak and statistically insignificant. Finally, Niedźwiedź *et al.* (2014) published similar results for southern Poland, which indicated growing but not significant trends in daily rainfall.

Concerning P_{max} modelling, under the hypothesis of stationarity, both EV1 and LN distributions (Kottegoda & Rosso 2008) usually provide comparable performances. As examples, the EV1 and LN fittings are illustrated in Figure 3 for Kraków-Balice and Solina raingauges using EV1 probabilistic plots. In detail, EV1 distribution is estimated using the maximum-likelihood (ML) method, while the method of moments (MM) is adopted for LN function. The observed time series for Kraków-Balice and Solina raingauges are in-depth analysed in Section 4.4, where examples of the STORAGE application are discussed.

4.3. STORAGE uncertainty analysis due to the sample size

As previously mentioned in Section 3.3, the uncertainty analysis of the STORAGE model consisted in selecting randomly 30-element series from the 500-year synthetic AMR series, for each raingauge station, and comparing with the observed annual

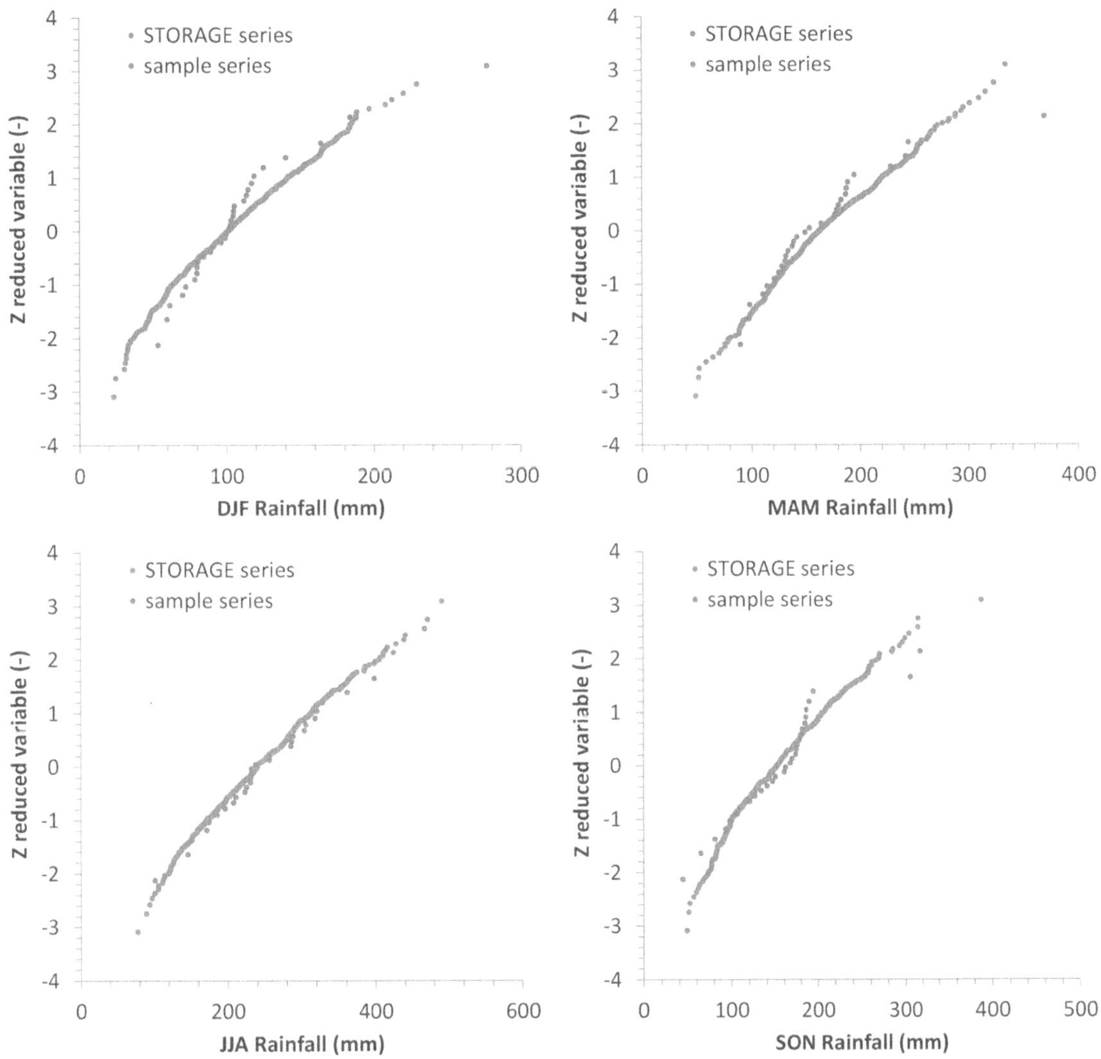

Figure 9 | Kraków-Balice raingauge: STORAGE performances in terms of reproduction of sample frequency distribution for seasonal rainfall.

maxima. Next, the values of *PBIAS* were determined. The operation was repeated 100 times. The results are presented in Figures 4 and 5 and in Table A3 in Supplementary Material, Appendix.

Based on the obtained results, it was concluded that the STORAGE model generated the daily AMR series usually with good or very good quality. The best results were obtained for Bieruń, Borusowa, Brenna, Kraków-Balice and Międzybrodzie Bialskie rainfall stations, where the PBIAS quality was very good in more than 90 out of 100 simulations. It should be emphasized that in 75% of all analysed rainfall stations, the unsatisfactory quality of model was not observed. Only for the Przemyśl station, there was 50 out of 100 unsatisfactory cases. Generally, analysing the model's quality for the entire upper Vistula basin (Figure 4), it was concluded that only 2% of all simulations were unsatisfactory, 20% was satisfactory, 23% was good and 55% was very good, i.e. 98 and 78% of simulations were at least satisfactory and good, respectively. Figure 6 presents the summary of quality for the station with best (Kraków-Balice) and worst results (Przemyśl). Figure 7 presents the values of PBIAS for the station with best and worst results. Figure 6 shows that that the very good quality of the model was not obtained for the Przemyśl station. The most common PBIAS values were in range from 25 to 30%. Moreover, for the Przemyśl station, the values of PBIAS were only positive. For the Kraków-Balice station, the unsatisfactory quality of model was not observed and the most common PBIAS values were in the range from 0 to 5%. In this case, negative values of PBIAS were also observed.

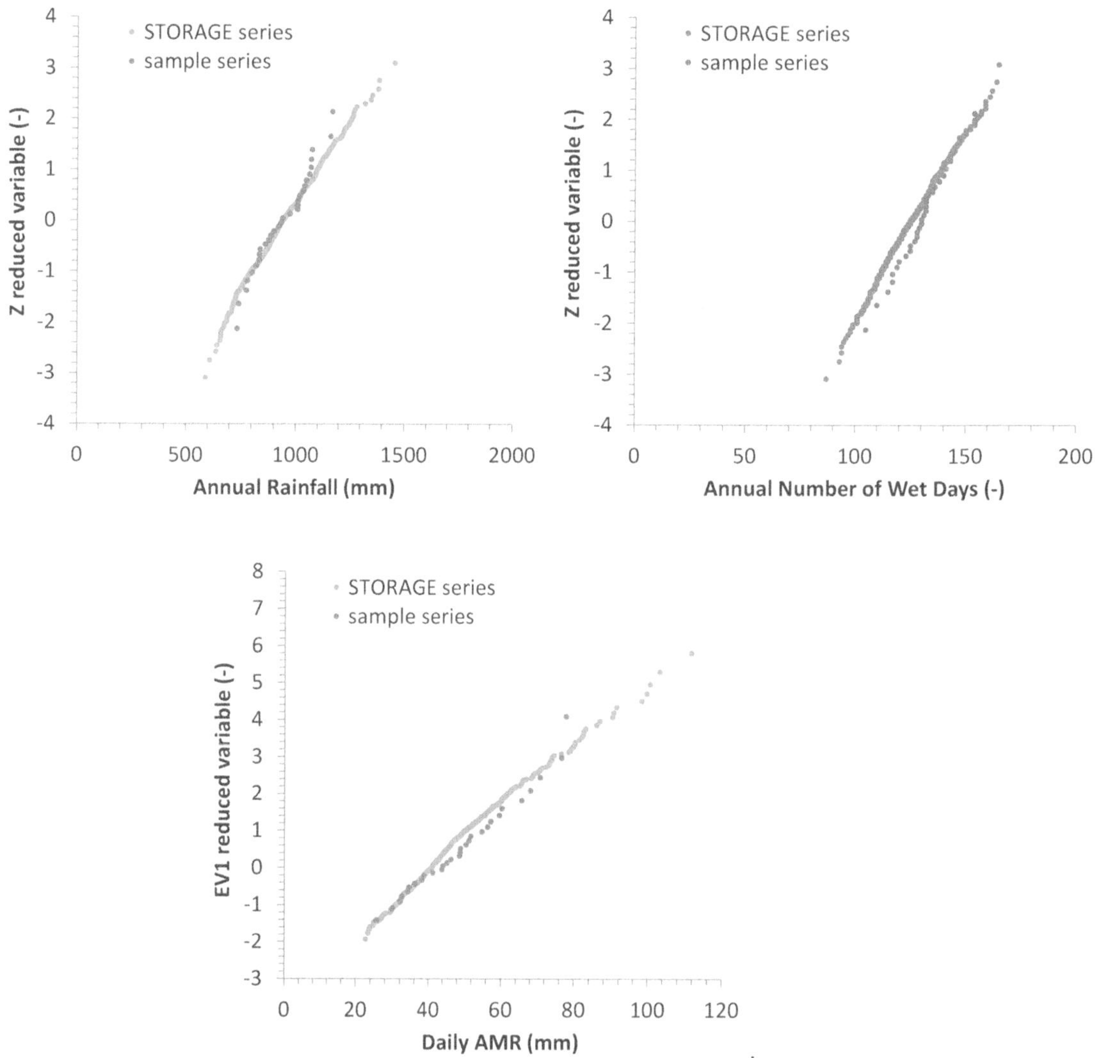

Figure 10 | Solina raingauge: STORAGE performances in terms of reproduction of sample frequency distribution for annual rainfall, annual number of wet days and daily AMR.

4.4. Examples of STORAGE generation for two observed time series

As examples of in-depth analyses of STORAGE application, we focused on continuous daily series for Kraków-Balice and Solina raingauges, for which the summary statistics concerning annual and seasonal rainfall are reported in Table 1. These selected time series present good performances in terms of mPBIAS and are located in different areas of the investigated Vistula basin (Figure 1).

In detail, from the generated 500-year synthetic series with a resolution of 60 min, we carried out model validation by analysing the reproduction of sample frequency distributions of daily AMR, annual and seasonal rainfall, and annual number of wet days (Figures 8–11).

Concerning annual and seasonal frequency distributions, STORAGE provided for both raingauges (on Gaussian probabilistic plots), a very good fitting for annual rainfall and a slight underestimation for annual number of wet days, while a good/acceptable fitting is obtained for seasonal rainfall heights. Sample daily AMR distributions are fitted in good/acceptable way as well (on EV1 probabilistic plots).

These specific results for Kraków-Balice and Solina raingauges clearly constitute a first promising step for further investigations of rainfall fields in Poland employing the STORAGE model, which will specifically focus on the high resolutions and

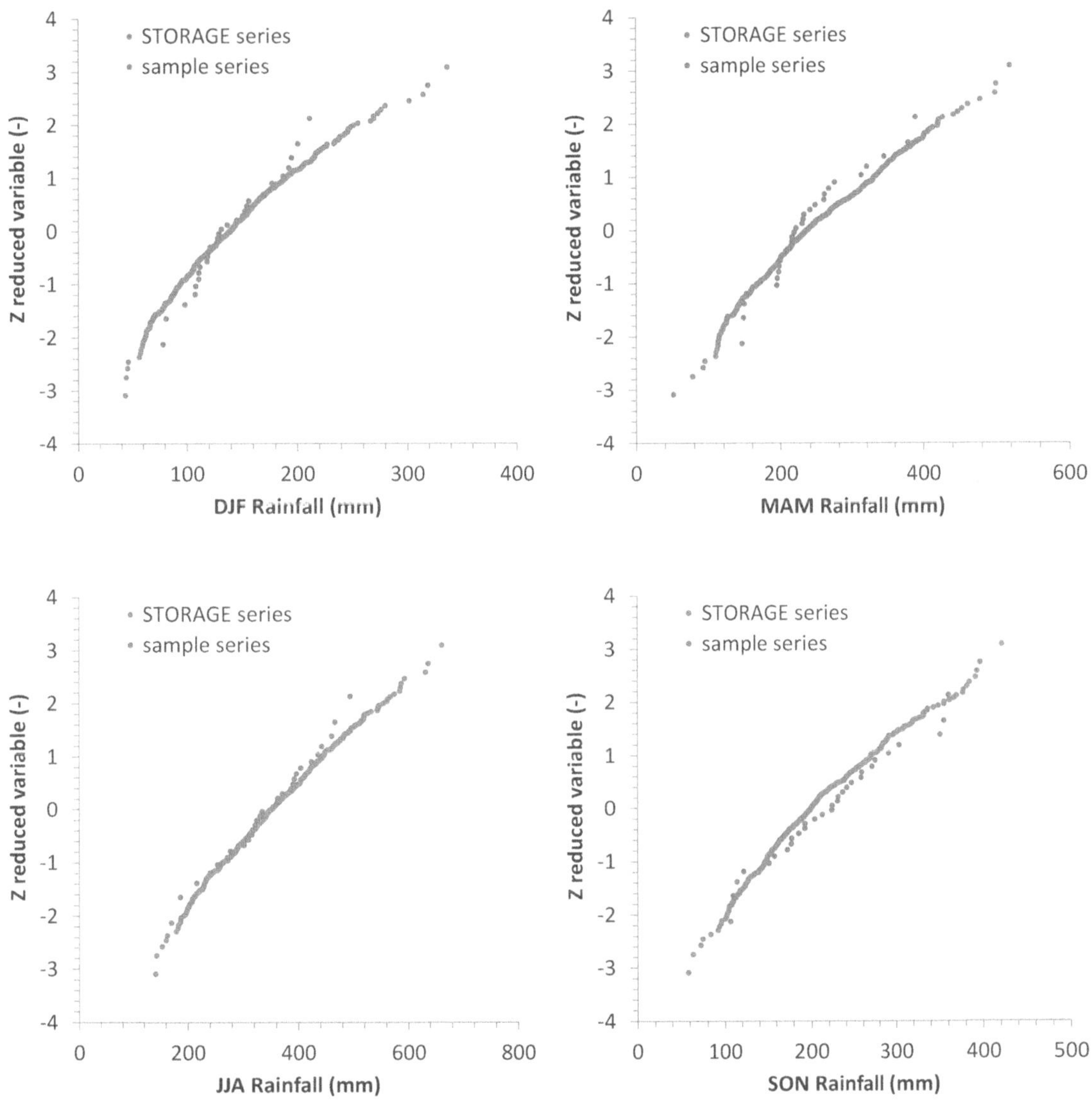

Figure 11 | Solina raingauge: STORAGE performances in terms of reproduction of sample frequency distribution for seasonal rainfall.

on the use of STORAGE synthetic rainfall data for a continuous approach aiming at evaluating the design discharge in ungauged or scarcely gauged basins.

5. CONCLUSION

This work aimed at modelling, for 36 raingauge stations located in the upper Vistula basin (Poland), the daily annual maximum rainfall (AMR) time series, extracted from synthetic continuous rainfall data which were generated using the STORAGE model. The present analysis constitutes a preliminary step of a more comprehensive study, focused on the use of the synthetic high-resolution rainfall series as input for a continuous rainfall-runoff modelling, aiming at evaluating the design discharge in ungauged or scarcely gauged basins.

For all the investigated raingauge stations, a lack of significant trends for the observed data series was assessed; hence, a stationary version of STORAGE was adopted for the successive steps. Moreover, the uncertainty analysis, due to sample size, for STORAGE output highlighted that the generated values of daily AMR are usually characterized by good or very good quality (in terms of mPBIAS), with respect to the observed data. Good or very good qualities have been confirmed from the in-depth analysis of STORAGE performances for two raingauges, chosen as examples: Kraków-Balice and Solina stations. The obtained reconstruction of sample frequency distributions, in terms of daily AMR, annual and seasonal rainfall heights and annual number of wet days, allows to consider STORAGE as a promising tool for modelling Polish rainfall fields. Further steps of the more comprehensive study of interest (i.e. design discharge evaluation for ungauged or scarcely gauged basins in Poland) are encouraged.

AUTHORS CONTRIBUTIONS

A.P., D.L.D.L., D.M. and A.W. conceptualized and investigated the study; A.P. and D.L.D.L. did the data curation, formulated the study methodology and provided software and validation; D.M. provided resources; D.L.D.L. and A.W. provided supervision; D.M. did the visualization; writing of the original draft was done by D.M. and A.W., and writing of review & editing was done by A.P. and D.L.D.L.

CONFLICTS OF INTEREST

The authors declare no conflicts of interest.

DATA AVAILABILITY STATEMENT

Data cannot be made publicly available; readers should contact the corresponding author for details.

REFERENCES

Agbo, E. P., Ekpo, C. M. & Edet, C. O. 2021 Analysis of the effects of meteorological parameters on radio refractivity, equivalent potential temperature and field strength via Mann-Kendall test. *Theoretical and Applied Climatology* **143**, 1437–1456.

Agilan, V. & Umamahesh, N. V. 2019 Rainfall generator for nonstationary extreme rainfall condition. *Journal of Hydrologic Engineering* **24** (9), 1–13.

Ajmal, M., Waseem, M., Kim, D. & Kim, T.-W. 2020 A pragmatic slope-adjusted curve number model to reduce uncertainty in predicting flood runoff from steep watersheds. *Water* **12**, 1–15.

Ali, R., Kuriqi, A., Abubaker, S. & Kisi, O. 2019 Long-term trends and seasonality detection of the observed flow in Yangtze River using Mann-Kendall and Sen's innovative trend method. *Water* **11**, 1–17.

Bharti, V., Singh, C., Ettema, J. & Turkington, T. A. R. 2016 Spatiotemporal characteristics of extreme rainfall events over the Northwest Himalaya using satellite data: spatiotemporal characteristics of extreme rainfall events. *International Journal of Climatology* **36**, 3949–3962.

Biondi, D. & De Luca, D. L. 2015 Process-based design flood estimation in ungauged basins by conditioning model parameters on regional hydrologic signatures. *Natural Hazards* **79** (2), 1015–1038.

Boni, G., Parodi, A. & Rudari, R. 2006 Extreme rainfall events: learning from raingauge time series. *Journal of Hydrology* **327**, 304–314.

Brocca, L., Tarpanelli, A., Filippucci, P., Dorigo, W., Zaussinger, F., Gruber, A. & Fernández-Prieto, D. 2018 How much water is used for irrigation? A new approach exploiting coarse resolution satellite soil moisture products. *International Journal of Applied Earth Observation and Geoinformation* **73**, 752–766.

Burton, A., Fowler, H. J., Blenkinsop, S. & Kilsby, C. G. 2010 Downscaling transient climate change using a Neyman-Scott rectangular pulses stochastic rainfall model. *Journal of Hydrology* **381** (1–2), 18–32.

Chen, S., Liu, B., Tan, X. & Wu, Y. 2020 Inter-comparison of spatiotemporal features of precipitation extremes within six daily precipitation products. *Climate Dynamics* **54**, 1057–1076.

Cowpertwait, P. S. P. 1991 Further developments of the Neyman-Scott clustered point process for modeling rainfall. *Water Resources Research* **27**, 1431–1438.

Dari, J., Brocca, L., Quintana-Seguí, P., Escorihuela, M. J., Stefan, V. & Morbidelli, R. 2020 Exploiting high-resolution remote sensing soil moisture to estimate irrigation water amounts over a Mediterranean region. *Remote Sensing* **12** (16), 1–22.

De Luca, D. L. & Galasso, L. 2019 Calibration of NSRP models from extreme value distributions. *Hydrology* **6**, 1–20.

De Luca, D. L. & Petroselli, A. 2021 STORAGE (STOchastic RAinfall GEnerator): a user-friendly software for generating long and high-resolution rainfall time series. *Hydrology* **8**, 1–32.

De Luca, D. L., Petroselli, A. & Galasso, L. 2020a Modelling climate changes with stationary models: is it possible or is it a paradox? In: *Numerical Computations: Theory and Algorithms* (Sergeyev, Y. & Kvasov D, D., eds). Springer, Cham.

De Luca, D. L., Petroselli, A. & Galasso, L. 2020b A transient stochastic rainfall generator for climate changes analysis at hydrological scales in Central Italy. *Atmosphere* **11** (12), 1–21.

Derin, Y., Nikolopoulos, E. & Anagnostou, E. N. 2019 Chapter seven – estimating extreme precipitation using multiple satellite-based precipitation products. In: *Extreme Hydroclimatic Events and Multivariate Hazards in a Changing Environment* (Maggioni, V. & Massari, C., eds). Elsevier, Cambridge, pp. 163–190.

Fitzgerald, D. L. 2005 Analysis of extreme rainfall using the log-logistic distribution. *Stochastic Environmental Research and Risk Assessment* **19**, 249–257.

Fowler, H. J., Kilsby, C. G. & O'Connell, P. E. 2000 A stochastic rainfall model for the assessment of regional water resource systems under changed climatic conditions. *Hydrology and Earth System Sciences* **4**, 261–280.

Grimaldi, S., Petroselli, A. & Serinaldi, F. 2012 Design hydrograph estimation in small and ungauged watersheds: continuous simulation method versus event-based approach. *Hydrological Process* **26**, 3124–3134.

Grimaldi, S., Nardi, F., Piscopia, R., Petroselli, A. & Apollonio, C. 2021 Continuous hydrologic modelling for design simulation in small and ungauged basins: a step forward and some tests for its practical use. *Journal of Hydrology* **595**. http://doi.org/10.1016/j.jhydrol.2020.125664.

Iwański, S. & Kuchar, L. 2003 Spatial generation of daily meteorological data. *Acta Scientarum Polonorum Formatio Circumiectus* **2** (1), 113–121.

Kaźmierczak, B. & Kotowski, A. 2012 A depth–duration–frequency rainfall model for dimensioning and modelling of Wrocław drainage systems. *Environment Protection Engineering* **38**, 127–138.

Kaźmierczak, B. & Kotowski, A. 2015 The suitability assessment of a generalized exponential distribution for the description of maximum precipitation amounts. *Journal of Hydrology* **525**, 345–351.

Kim, S. & Kim, H. 2016 A new metric of absolute percentage error for intermittent demand forecast. *International Journal of Forecasting* **32**, 669–679.

Kottegoda, N. T. & Rosso, R. 2008 *Applied Statistics for Civil and Environmental Engineers*. Wiley-Blackwell, pp. 736.

Kubiak-Wójcicka, K. 2020 Variability of air temperature, precipitation and outflows in the Vistula Basin (Poland). *Resources* **9**, 1–26.

Mannina, G., Freni, G., Viviani, G., Saegrov, S. & Hafskjold, L. S. 2006 Integrated urban water modeling with uncertainty analysis. *Water Science and Technology* **54** (6–7), 379–386.

Maraun, D. 2016 Bias correcting climate change simulations – a critical review. *Current Climate Change Reports* **2** (4), 211–220.

Młyński, D. 2020 Analysis of problems related to the calculation of flood frequency using rainfall-runoff models: a case study in Poland. *Sustainability* **12** (17), 1–18.

Młyński, D., Cebulska, M. & Wałęga, A. 2018 Trends, variability, and seasonality of maximum annual daily precipitation in the Upper Vistula Basin, Poland. *Atmosphere* **2018** (9), 1–14.

Młyński, D., Wałęga, A., Bugajski, P., Operacz, A. & Kurek, K. 2019a Verification of empirical formulas for calculating mean low flow in reflect to affecting on disposable water resources. *Acta Scientarum Polonorum Formatio Circumiectus* **18** (2), 83–92.

Młyński, D., Wałęga, A., Petroselli, A., Tauro, F. & Cebulska, M. 2019b Estimating maximum daily precipitation in the Upper Vistula Basin, Poland. *Atmosphere* **10**, 1–17.

Młyński, D., Wałęga, A., Ozga-Zielinski, B., Ciupak, M. & Petroselli, A. 2020 New approach for determining the quantiles of maximum annual flows in ungauged catchments using the EBA4SUB model. *Journal of Hydrology* **589**, 125198.

Morrissey, M. L. 2009 Superposition of the Neyman-Scot rectangular pulses model and the Poisson white noise model for the representation of tropical rain rates. *Journal of Hydrometeorology* **10** (2), 395–412.

Niedźwiedź, T., Łupikasza, E., Pińskwar, I., Kundzewicz, Z. W., Stoffel, M. & Małarzewski, Ł. 2014 Climatological background of floods at the northern foothills of the Tatra Mountains. *Theoretical and Applied Climatology* **119**, 273–284.

Petroselli, A., Asgharinia, S., Sabzevari, T. & Saghafian, B. 2020 Comparison of design peak flow estimation methods for ungauged basins in Iran. *Hydrological Sciences Journal* **65** (1), 127–137.

Pierzga, K. & Jarosińska, E. 2014 Annual maximum daily rainfall generator. In: *Hydrology in Environment Protection and Development* (Magnuszewski, A., ed.). PAN, Warsaw, pp. 57–66.

Pińskwar, I., Choryński, A., Graczyk, D. & Kundzewicz, Z. W. 2019 Observed changes in extreme precipitation in Poland: 1991–2015 versus 1961–1990. *Theoretical and Applied Climatology* **135**, 773–787.

Pociask-Karteczka, J. 2016 The Upper Vistula Basin – a geographical overview. In: *Flood Risk in the Upper Vistula Basin* (Kundzewicz, Z., Stoffel, M., Niedźwiedź, T. & Wyżga, B., eds). Springer, Cham.

Rodriguez-Iturbe, I., Cox, D. R. & Isham, V. 1987 Some models for rainfall based on stochastic point processes. *Proceedings of the Royal Society A: Mathematical, Physical* **410**, 269–288.

Sabzevari, T. 2017 Runoff prediction in ungauged catchments using the gamma dimensionless time-area method. *Arabian Journal of Geosciences* **10** (6), 1–11.

Sabzevari, T., Ardakanian, R., Shamsaee, A. & Talebi, A. 2009 Estimation of flood hydrograph in no statistical watersheds using HEC-HMS model and GIS (Case study: Kasilian watershed). *Journal of Water Process Engineering* **4**, 1–11.

Siddique-E-Akbor, A. H. M., Hossain, F., Sikder, S., Shum, C. K., Tseng, S., Yi, Y., Turk, F. J. & Limaye, A. 2014 Satellite precipitation data-driven hydrological modeling for water resources management in the Ganges, Brahmaputra, and Meghna Basins. *Earth Interactions* **18**, 1–25.

Sridhar, S. I. & Raviraj, A. 2017 Statistical trend analysis of rainfall in Amaravathi river basin using Mann-Kendall test. *Current World Environment* **12** (1), 89–96.

Turnipseed, D. P. & Sauer, V. B. 2010 *U.S. Geological Survey Techniques and Methods*, Vol. A8. USGS, Ch. Discharge Measurements at Gaging Stations, 87.

Villarini, G., Smith, J. A., Baeck, M. L., Vitolo, R., Stephenson, D. B. & Krajewski, W. F. 2011 On the frequency of heavy rainfall for the Midwest of the United States. *Journal of Hydrology* **400**, 103–120.

Wałęga, A., Młyński, D., Bogdał, A. & Kowalik, T. 2016 Analysis of the course and frequency of high water stages in selected catchments of the Upper Vistula Basin in the South of Poland. *Water* **8**, 1–15.

Wasko, C. & Sharma, A. 2017 Continuous rainfall generation for a warmer climate using observed temperature sensitivities. *Journal of Hydrology* **544**, 575–590.

Willems, P., Arnbjerg-Nielsen, K., Olsson, J. & Nguyen, V. T. V. 2012 Climate change impact assessment on urban rainfall extremes and urban drainage: methods and shortcomings. *Atmospheric Research* **103**, 106–118.

Wright, D. B., Smith, J. A., Villarini, G. & Baeck, M. L. 2013 Estimating the frequency of extreme rainfall using weather radar and stochastic storm transposition. *Journal of Hydrology* **488**, 150–165.

Zhao, Y., Nearing, M. A. & Guertin, D. P. 2019 A daily spatially explicit stochastic rainfall generator for a semi-arid climate. *Journal of Hydrology* **574**, 181–192.

First received 20 September 2021; accepted in revised form 21 March 2022. Available online 6 April 2022

IWA Publishing's authorised EU representative for General Product Safety Regulations is Diane D'Arras, 15 rue Duret, 75116 Paris, France, e-mail: safety@iwap.co.uk.

Printed and bound by CPI Group (UK) Ltd, Croydon, CR0 4YY

12/05/2026

02108868-0005